VOLUME EIGHTY ONE

ADVANCES IN BOTANICAL RESEARCH

Insect-Plant Interactions in a Crop Protection Perspective

ADVANCES IN BOTANICAL RESEARCH

Series Editors

VOLUME EIGHTY ONE

ADVANCES IN BOTANICAL RESEARCH

Insect-Plant Interactions in a Crop Protection Perspective

Volume Editor

NICOLAS SAUVION

INRA,UMR BGPI 0385 (INRA-CIRAD-SupAgro), Montpellier, France

DENIS THIÉRY

INRA, UMR SAVE 1065, Bordeaux Sciences Agro, Centre INRA de recherches de Bordeaux- Aquitaine, Institut des Sciences de la Vigne et du Vin, Villenave d'Ornon, France

PAUL-ANDRÉ CALATAYUD

IRD UMR EGCE (Evolution, Génome, Comportement, Ecologie), CNRS-IRD-Univ. Paris-Sud, IDEEV, Université Paris-Saclay, Gif-sur-Yvette, France;
IRD c/o ICIPE, Nairobi, Kenya

ELSEVIER

ACADEMIC PRESS

An imprint of Elsevier
elsevier.com

Academic Press is an imprint of Elsevier
125 London Wall, London EC2Y 5AS, United Kingdom
The Boulevard, Langford Lane, Kidlington, Oxford OX5 1GB, United Kingdom
50 Hampshire Street, 5th Floor, Cambridge, MA 02139, United States
525 B Street, Suite 1800, San Diego, CA 92101-4495, United States

First edition 2017

Notices
Knowledge and best practice in this field are constantly changing. As new research and experience broaden our understanding, changes in research methods, professional practices, or medical treatment may become necessary.

Practitioners and researchers must always rely on their own experience and knowledge in evaluating and using any information, methods, compounds, or experiments described herein. In using such information or methods they should be mindful of their own safety and the safety of others, including parties for whom they have a professional responsibility.

To the fullest extent of the law, neither the Publisher nor the authors, contributors, or editors, assume any liability for any injury and/or damage to persons or property as a matter of products liability, negligence or otherwise, or from any use or operation of any methods, products, instructions, or ideas contained in the material herein.

ISBN: 978-0-12-803318-0
ISSN: 0065-2296

Publisher: Zoe Kruze
Acquisition Editor: Kirsten Shankland
Editorial Project Manager: Hannah Colford
Production Project Manager: Magesh Mahalingam
Designer: Christian Bilbow

Typeset by TNQ Books and Journals

CONTENTS

CONTRIBUTORS

K. Bermudez-Torres
National Polytechnical Institute, Yautepec, Mexico

R. Bommarco
Swedish University of Agricultural Sciences, Uppsala, Sweden

P.-A. Calatayud
IRD UMR EGCE (Evolution, Génome, Comportement, Ecologie), CNRS-IRD-Univ. Paris-Sud, IDEEV, Université Paris-Saclay, Gif-sur-Yvette Cedex, France; IRD c/o ICIPE, Nairobi, Kenya

S. Colazza
University of Palermo, Palermo, Italy

J. Cook
Western Sydney University, NSW, Australia

E. Corcket
Université de Bordeaux, INRA, UMR1202 BioGeCo, Pessac, France

A.-M. Cortesero
UMR 1349 IGEPP (Institute for Genetics, Environment and Plant Protection), Université de Rennes 1, Rennes, France

F. Dedeine
Université François-Rabelais de Tours, Centre National de la Recherche Scientifique, Tours, France

E. Desouhant
University of Lyon, Claude Bernard University, CNRS, Laboratory of Biometry and Evolutionary Biology UMR 5558, Villeurbanne, France

G. Dubreuil
Université François-Rabelais de Tours, Centre National de la Recherche Scientifique, Tours, France

B. Ekbom
Swedish University of Agricultural Sciences, Uppsala, Sweden

B. Frérot
INRA, UMR 1392, Institut d'Ecologie et des Sciences de l'Environnement de Paris, Versailles, France

B. Giffard
Bordeaux Sciences Agro, Université de Bordeaux, Gradignan, France

D. Giron
IRBI (Insect Biology Research Institute) – UMR 7261 CNRS/Université François-Rabelais de Tours, Tours, France; Université François-Rabelais de Tours, Centre National de la Recherche Scientifique, Tours, France

M. Goubault
IRBI (Insect Biology Research Institute) – UMR 7261 CNRS/Université François-Rabelais de Tours, Tours, France

A.T. Groot
University of Amsterdam, Amsterdam, Netherlands

M.O. Harris
North Dakota State University, Fargo, ND, United States

J.K. Holopainen
University of Eastern Finland, Kuopio, Finland

E. Huguet
Université François-Rabelais de Tours, Centre National de la Recherche Scientifique, Tours, France

A. Jarrige
IRBI (Insect Biology Research Institute) – UMR 7261 CNRS/Université François-Rabelais de Tours, Tours, France

E. Jousselin
INRA – CBGP (INRA, IRD, CIRAD, Montpellier SupAgro), Montferrier-sur-Lez, France

L. Kaiser
IRD UMR EGCE (Evolution, Génome, Comportement, Ecologie), CNRS-IRD-Université Paris-Saclay, Gif-sur-Yvette, France

G.J. Kergoat
INRA – CBGP (INRA, IRD, CIRAD, Montpellier SupAgro), Montferrier-sur-Lez, France

F. Kjellberg
CEFE (Centre d'Ecologie Fonctionnelle et Evolutive), CNRS, Université de Montpellier – Université Paul-Valéry Montpellier – EPHE, Montpellier, France

L. Legal
ECOLAB, UMR 5245 (INP, UPS, CNRS), Ecologie Fonctionnelle et Environnement, Toulouse, France

E. Leppik
INRA, UMR 1392, Institut d'Ecologie et des Sciences de l'Environnement de Paris, Versailles, France

F. Lieutier
INRA, Laboratoire de Biologie des Ligneux et des Grandes Cultures, Université d'Orléans, Orléans, France

P. Louâpre
University of Bourgogne Franche-Comté, Evolutionary Ecology Team, UMR 6282 Biogéosciences, 6 Bd Gabriel, Dijon, France

B. Martinet
INRA, UR0406 Abeilles et Environnement, Domaine St Paul, Avignon, France

F. Menu
University of Lyon, Claude Bernard University, CNRS, Laboratory of Biometry and Evolutionary Biology UMR 5558, Villeurbanne, France

A.S. Meseguer
INRA – CBGP (INRA, IRD, CIRAD, Montpellier SupAgro), Montferrier-sur-Lez, France

J. Moreau
University of Bourgogne Franche-Comté, Evolutionary Ecology Team, UMR 6282 Biogéosciences, 6 Bd Gabriel, Dijon, France

L. Mouton
Université de Lyon 1, Centre National de la Recherche Scientifique, Villeurbanne, France

A. Nel
Institut de Systématique, Évolution, Biodiversité, ISYEB – UMR 7205 – CNRS, MNHN, UPMC, EPHE, Muséum national d'Histoire naturelle, Sorbonne Universités, Paris, France

P. Ode
Colorado State University, Fort Collins, CO, United States

Y. Outreman
Agrocampus Ouest, Institut National de la Recherche Scientifique, Université de Rennes 1, Université Bretagne-Loire, Rennes, France

S. Pincebourde
IRBI (Insect Biology Research Institute) – UMR 7261 CNRS/Université François-Rabelais de Tours, Tours, France

E. Rajon
University of Lyon, Claude Bernard University, CNRS, Laboratory of Biometry and Evolutionary Biology UMR 5558, Villeurbanne, France

S. Rasmann
University of Neuchâtel, Neuchâtel, Switzerland

P. Rasmont
INRA, UR0406 Abeilles et Environnement, Domaine St Paul, Avignon, France

G. Rodet
AE, INRA, Avignon, France

A. Rusch
INRA, UMR SAVE 1065, Bordeaux Sciences Agro, Centre INRA de recherches de Bordeaux- Aquitaine, Institut des Sciences de la Vigne et du Vin, Villenave d'Ornon Cedex, France

A. Sallé
INRA, Laboratoire de Biologie des Ligneux et des Grandes Cultures, Université d'Orléans, Orléans, France

N. Sauvion
INRA, UMR BGPI 0385 (INRA–CIRAD–SupAgro), Campus International de Baillarguet, Montpellier, France

B. Schatz
CEFE (Centre d'Ecologie Fonctionnelle et Evolutive), UMR 5175 CNRS, Université de Montpellier – Université Paul-Valéry Montpellier – EPHE, Montpellier, France

R.F.H. Sforza
USDA-ARS-European Biological Control Laboratory, Montferrier-sur-Lez, France

J.-C. Simon
INRA, UMR 1349 IGEPP INRA-Agrocampus Ouest Rennes-Université Rennes 1, Le Rheu, France

A. Thiel
University of Bremen, FB 02, Institute of Ecology, Bremen, Germany

D. Thiéry
INRA, UMR SAVE 1065, Bordeaux Sciences Agro, Centre INRA de recherches de Bordeaux- Aquitaine, Institut des Sciences de la Vigne et du Vin, Villenave d'Ornon Cedex, France

M. Unbehend
Max Planck Institute for Chemical Ecology, Jena, Germany

J. van Baaren
UMR-CNRS 6553 Ecobio, Université de Rennes 1, Université Bretagne-Loire, Rennes, France

S. van Nouhuys
University of Helsinki, Helsinki, Finland

F. Vavre
Université de Lyon 1, Centre National de la Recherche Scientifique, Villeurbanne, France

PREFACE

> *Man has always had his troubles with insects. When he first emerged as man he already had fleas and lice and was fed on by mosquitoes and pestered by flies... Gradually, as the more stark enemies of primeval life, such as the leopard and tiger, ceased to be a great threat to primitive man, insects became increasingly important as a challenge to his success.*[1]

Insects always were an enormous factor human activity had to deal with, and especially with food and health. The Bible explicitly referred to several scourges such as locust invasions, destroying crops or complete destruction of grapes by grape worms.

Current estimates of crop losses or post harvest due to insects' pest are estimated around 50% according to crops.[2] Thus, biggest challenges of the 21st-century agriculture will obviously concern climate changes, sustainable crop protection and reduction in the massive use of pesticides used during the last 50 years.

But the picture is far from so simple. The arthropods among which insects impair agriculture by different ways: as blight directly consuming crops or store products, or transmitting plant pathogens inducing severe economic losses, particularly in developing countries; conversely, others are beneficial to crops as biocontrol agents or pollinators, that contribute to the survival of vegetable and fruit crops.

Scientists are dealing with the diversity and the complexity of nature, trying to find over time the different pieces of the puzzle that human often perceive as wonderful, and that our knowledge of the insect world is a permanent discovery, and most likely a never-ending story. A glance in the rear-view mirror of science reminds several 19th-century scientists such as

1 Ross, H. H. (1956). *A textbook of entomology* (2nd ed.). New York: J. Wiley & Sons, London: Chapman et Hall.

2 Oerke, E. C., & Dehne, H. W. (2004). Safeguarding production—losses in major crops and the role of crop protection. *Crop Protection, 23*, 275–285.
Aggarwalim, P. K., Kalra, N., Chander, S., & Pathak, H. (2006). InfoCrop: a dynamic simulation model for the assessment of crop yields, losses due to pests, and environmental impact of agro-ecosystems in tropical environments. I. Model description. *Agricultural Systems, 89*, 1–25.

Louis Pasteur (1822–95),[3] Charles Valentine Riley (1843–95)[4] and more recently Karl Von Frisch (1886–1982).[5] These pioneers were mostly multidisciplinary scientists.

The present book humbly attempts to put some light on multidisciplinarity by associating renowned scientists. We have thus endeavoured to link different scientific disciplines (palaeontology, evolutionary biology, chemical ecology, ethology, agronomy, animal and plant physiology, etc.) to emphasize on current scientific approaches to tackle down crop pests, while limiting the use of pesticides. In Chapter 1, Bertrand Schatz, Nicolas Sauvion, Finn Kjellberg and André Nel retrace the main steps of the past history of diversification of pollination interactions with the emergence of some major pollinator groups of insects and propose some insights into the selective processes acting on the evolution of the entomophilous pollination. To complete this first overview, Gaël Kergoat, Andrea Meseguer and Emmanuelle Jousselin (Chapter 2) review recent works in the phylogenetic investigations of plant–insect interactions focussing on the evolutionary history of the associations with plants, the diversification dynamics of herbivorous insects and the biogeographic history of herbivorous insects. Then, in Chapter 3, François Lieutier, Kalina Bermudez-Torres, James Cook, Marion O. Harris, Luc Legal, Aurélien Sallé, Bertrand Schatz and David Giron describe how the phytophagous insects have developed mechanisms of various complexity levels (from plant exploitation to mutualism) to utilize plants in spite of the barriers that plants have built to resist aggressions.

Trophic relationships between plants and insects are not confined to biological interactions such as herbivory (i.e., direct consumption of one primary producer by a predator). In an ecological approach, many other interactions, trophic or even nontrophic, may influence plant herbivory by insects. Emmanuel Corcket, Brice Giffard and René Sforza (Chapter 4) describe these biological interactions within food webs, and some of the drivers that may influence plant and insects.

3 The microbiologist Louis Pasteur is often referred for his important work in human health. He also achieved significant research on the entomopathogenic microorganisms, and especially he is recognized to have identified a variety of pathogens (e.g., *Beauveria bassiana*, *Bacillus bombycis* and *Nosema bombycis*) infecting the European silk moth production.

4 Charles V. Riley was a USDA-applied entomologist, well known for impressive work in agricultural entomology from which having resolved the *Phylloxera* crisis in Europe.

5 Karl Von Frisch was an Austrian zoologist best known for his studies of bee behaviour, perception and communication, discovering the 'Dance of the Bees'. In 1973, he was awarded the Nobel Prize in Physiology or Medicine with Konrad Lorenz and Nikolaas Tinbergen.

Chemical signals are important cues throughout the life of an insect especially for mate location and for prey and host finding. In Chapter 5, Brigitte Frérot, Ene Leppik, Astrid Groot, Melanie Unbehend and Jarmo Holopainen focus on the plasticity of pheromone emission, which is a recent progress in the field of pheromones, the specific odorscape produced by the cultivated plants which is now obviously related to the plant physiological state and health and factors of variability of the chemical signals released by plants.

The biological control of insect pests has proved its usefulness and efficiency for over one century, its origin being probably the biological programmes achieved in California by Albert Koebele (1898) to control citrus scales by *Metaphycus albertii*. Many entomophagous species are now used for this approach worldwide, hence significantly reducing insecticides use. Laure Kaiser, Paul Ode, Saskya van Nouhuys, Paul-André Calatayud, Stefano Colazza, Anne-Marie Cortesero, Andra Thiel and Joan van Baaren (Chapter 6) provide an updated review of mechanisms underlying the interactions between plants and entomophagous insects, including those governing life history traits at the individual level, as well as those acting on population and community structure and dynamics.

Symbiosis is an obvious example of intimate plant–insect interactions. Indeed, there is growing evidence that microorganisms are important 'hidden players' in insect–plant interactions. For example, microorganisms as symbionts can directly interfere with the plant physiology and also insects' life history traits and host selection process. In Chapter 7, David Giron, Franck Dedeine, Géraldine Dubreuil, Elisabeth Huguet, Laurence Mouton, Yannick Outreman, Fabrice Vavre and Jean-Christophe Simon highlight the excitement that surrounds the investigations on the plant–insect–microbe interactions and the promise they hold for a better understanding of the functional, ecological and evolutionary impacts of symbionts on plant–insect interactions.

Host plants possibly represent a main selection pressure for the evolution of reproductive traits in phytophagous insects. In Chapter 8, Jérôme Moreau, Emmanuel Desouhant, Philippe Louâpre, Marlène Goubault, Etienne Rajon, Alicia Jarrige, Frédéric Menu and Denis Thiéry review how plant quality affects both female and male reproductive strategies, and how insects adapt to unpredictable environments (bet-hedging strategies).

Global change is resetting the spatial and ecological equilibrium of complex coevolutionary relationships between plants and their insect herbivores. In Chapter 9, Sylvain Pincebourde, Joan van Baaren, Sergio Rasmann,

Pierre Rasmont, Guy Rodet, Baptiste Martinet and Paul-André Calatayud review the mechanisms at play in the responses of plant–insect interactions to global changes, including increased temperature and atmospheric CO_2 concentrations, modification of land use and pollution.

Integrating, supporting and regulating ecosystem functions provided by several components of biodiversity into cropping systems has been proposed as a promising way to decrease agrochemical inputs and negative environmental impacts, while maximizing crop productivity. In Chapter 10, Adrien Rusch, Riccardo Bommarco and Barbara Ekbom provide an integrated landscape view illustrating how agroecological global knowledge can be used to revisit crop protection and insect pest management based on conservation biological control.

In the future, the challenge that agriculture has to face is feeding soon *ca* 9 billions humans with probably reduced cultivated areas due to climate changes and access to water. New pests regularly emerge as the results of biological invasions which are often related to human trades or become resistant to control methods. All the 53 coauthors of this volume wish that their contribution can be useful to some extend for future modern and healthy crop protection.

Nicolas Sauvion, Denis Thiéry and Paul-André Calatayud
19 October 2016

CHAPTER ONE

Plant–Insect Interactions: A Palaeontological and an Evolutionary Perspective

B. Schatz*[,1], N. Sauvion[§], F. Kjellberg*, A. Nel[¶]
*CEFE (Centre d'Ecologie Fonctionnelle et Evolutive), CNRS - Université de Montpellier - Université Paul-Valéry Montpellier – EPHE, Montpellier, France
[§]INRA, UMR BGPI 0385 (INRA–CIRAD–SupAgro), Campus International de Baillarguet, Montpellier, France
[¶]Institut de Systématique, Évolution, Biodiversité, ISYEB – UMR 7205 – CNRS, MNHN, UPMC, EPHE, Muséum national d'Histoire naturelle, Sorbonne Universités, Paris, France
[1]Corresponding author: E-mail: Bertrand.SCHATZ@cefe.cnrs.fr

Contents

1. PALAEO-ENTOMOLOGY OR HOW WE CAN RECONSTRUCT THE EVOLUTIONARY HISTORY OF PLANT–INSECT INTERACTIONS

The first clue used to infer past plant–insect interactions are fossilized insects or parts of insects such as appendices, buccal pieces, ovipositors, wings, etc (Fig. 1). These remains allow us to determine taxonomic position and, with that information, infer mode of life and feeding habit, by analogy with modern species. For instance a reduviid hemipteran will be considered a predator, while a pentatomid hemipteran will be assumed to be phytophagous, but with a weaker probability, because all modern reduviids are predators while some pentatomids are not phytophagous.

Advances in Botanical Research, Volume 81
ISSN 0065-2296
http://dx.doi.org/10.1016/bs.abr.2016.11.002

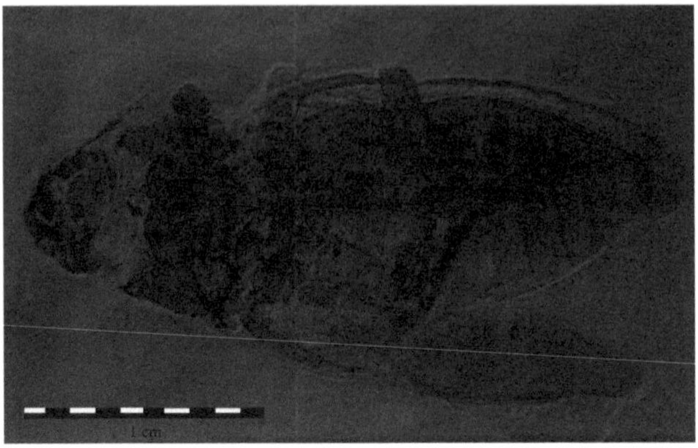

Figure 1 *Chalcophora oligocenica* Theobald, 1937 (Coleoptera, Buprestidae), Oligocene (~33.9–23.03 million years), Aix-en-Provence (France), André Nel©.

This first-line inference can be improved (Nel, 1997). The morphological analysis of buccal pieces provides better supported and direct evidence on feeding habits because morphology is directly affected by feeding habits of the insect; for example, the mandibles of predatory insects are very different from those of phytophagous ones. But it is necessary to remain very prudent in the inferences because sometimes only weak differences are sufficient for an adaptation to a completely different feeding habits, viz., the case of a modern Orthoptera Gryllacrididae adapted to nectar feeding but having mouthparts very similar to those of the predator species in this family (Krenn, Fournel, Bauder, & Hugel, 2016). Furthermore, morphological analysis can provide more precise information due to functional constraints. For instance, the mandibles of a Poaceae consumer are quite different from those of dicotyledonous plant consumers. Sometimes, the content of the digestive tube is preserved, providing fossilized pollen or spores.

Traces of insect-feeding behaviour on fossilized plant organs are another type of remain of more complex interpretation. These traces can be missing parts of a leaf, traces of stinging but also galls that provide more or less precise signatures of the activity of the phytophagous insects on a plant that was subsequently fossilized (Labandeira, Kvacek, & Mostovski, 2007). The abundance of fossilized plants in the palaeontological freshwater record (Beerling, 2007, pp. 1–288) allows making systematic searches for such evidence of the activities of phytophagous organisms and of their impact on plants at different periods. For instance, the analysis of traces of consumption by phytophagous insects on fossil angiosperm leaves during the Palaeocene–Eocene transition some 55 million years ago shows a positive correlation between insect damages to plants and increasing mean temperatures (Currano et al., 2008). Nevertheless attributing damage to a particular zoological group can be tricky. Galls may be caused by insects or by mites, and traces of herbivory on leaves may be due to insects, mites or terrestrial molluscs. Even today, in warm, intertropical habitats, it is difficult to separate fresh traces of herbivory left by insects (caterpillars, orthopterans, coleopterans, etc.) from those left by the numerous species of molluscs living in the same habitats. Damages or traces of activity observed on fossilized leaves may also result from the activity of aquatic organisms, such as mites, that feed on dead leaves fallen into the water while traces of fungal attacks on leaves may resemble damages due to insect activities adding to potential sources of confusion.

Insects have left few fossil remains in comparison to their abundance and diversity in the modern terrestrial ecosystems, despite their excellent capacity of fossilization (the complete bodies of the insects are generally preserved, which is not the case for vertebrates, molluscs, etc.). The incompleteness of the insect fossil record is due to the fact that they generally fossilize in resins or in lacustrine sediments. Nevertheless with increasing numbers of investigated fossiliferous sites of continental origin and with new methods for investigating fossil remains, more and more material allows documenting the antiquity, the fossil record and the history of many taxa (e.g., Huang et al., 2016). These new data coupled with insights stemming from the rapid progress of molecular phylogenetic reconstructions allows proposing evolutionary scenarios for the origin and diversification of many taxa (Fig. 2). In some cases, confronting molecular and palaeontological data allow to date the origin of plant–insect interactions (see Fig. 3 for pollination, and Fig. 7 in Labandeira, 2006 for herbivory). For instance, combining information from the phylogenies of *Ficus* and of their pollinating agaonid wasps from the fossil record, and from palaeobiogeographic events, has allowed demonstrating that this extreme example of plant–insect codiversification originated c. 75 million years ago (Cruaud et al., 2012).

2. THE FIRST STEPS IN THE EVOLUTIONARY HISTORY OF PLANT–INSECT INTERACTIONS

The first steps of the evolution of hexapods in general and of insects in particular are poorly documented in the fossil record. Molecular studies suggest that hexapods originated during the Silurian period, 430 million years ago (Deuve, 2001; Fig. 2), a proposition supported by what has been

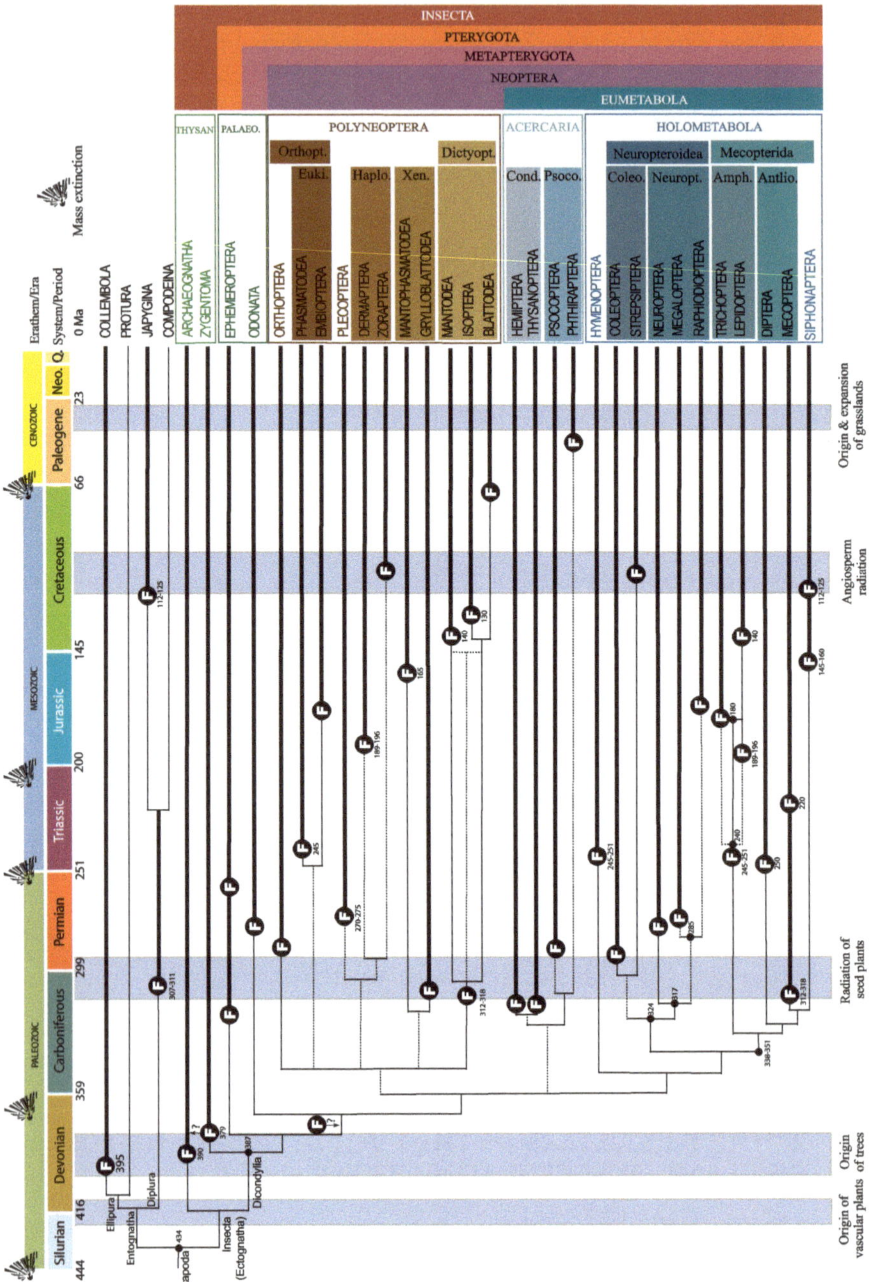
INSECTA
PTERYGOTA
METAPTERYGOTA
NEOPTERA
EUMETABOLA
THYSAN
PALAEO.
POLYNEOPTERA
ACERCARIA
HOLOMETABOLA
Orthopt.
Dictyopt.
Neuropteroidea
Mecopterida
Euki.
Haplo.
Xen.
Cond.
Psoco.
Coleo.
Neuropt.
Amph.
Antlio.
Mass extinction
Erathem/Era
System/Period
0 Ma
COLLEMBOLA
PROTURA
JAPYGINA
COMPODEINA
ARCHAEOGNATHA
ZYGENTOMA
EPHEMEROPTERA
ODONATA
ORTHOPTERA
PHASMATODEA
EMBIOPTERA
PLECOPTERA
DERMAPTERA
ZORAPTERA
MANTOPHASMATODEA
GRYLLOBLATTODEA
MANTODEA
ISOPTERA
BLATTODEA
HEMIPTERA
THYSANOPTERA
PSOCOPTERA
PHTHIRAPTERA
HYMENOPTERA
COLEOPTERA
STREPSIPTERA
NEUROPTERA
MEGALOPTERA
RAPHIDIOPTERA
TRICHOPTERA
LEPIDOPTERA
DIPTERA
MECOPTERA
SIPHONAPTERA
CENOZOIC
MESOZOIC
PALEOZOIC
Q.
Neo.
Paleogene
Cretaceous
Jurassic
Triassic
Permian
Carboniferous
Devonian
Silurian
23
66
145
200
251
299
359
416
444
Origin & expansion of grasslands
Angiosperm radiation
Radiation of seed plants
Origin of trees
Origin of vascular plants
Hexapoda
Entognatha
Ellipura
Diplura
Insecta (Ectognatha)
Dicondylia

Figure 2 ***Relationship between the modern orders of Hexapodes (Hexapoda).*** Consensus tree based on recent phylogenetic studies, and particularly on Misof et al. (2014) and Wheeler, Whiting, Wheeler, and Carpenter (2001) (Hexapoda), Trautwein, Wiegmann, Beutel, Kjer, and Yeates (2012) and Yeates, Cameroun, and Trautwein (2012) (Insecta), Terry and Whiting (2005) (Mantophasmatodea and Polyneoptera), Whiting (1998) (Strepsiptera), Whiting (2002) and Wiegmann et al. (2009) (Holometabola), Winterton, Hardy, and Wiegmann (2010) (Neuropterida). The *dotted lines* correspond to lack of consensus in branching order between different reconstructions or to ambiguous monophyly of terminal clades. Dates of divergence, and for each order, age of the oldest known fossil (letter F in *black circle*) follow the reviews by Regier, Shultz, and Kambic (2004) (Hexapoda, 480 million years), Gaunt and Miles (2002) (Insecta: ~434 million years; Pterygota: ~387 million years), Grimaldi and Engel (2005) and Mayhew (2012) (Insecta), Grimaldi (2001) and Wiegmann et al. (2009) (Holometabola), Winterton et al. (2010) (Neuropterida), Whalley (1986) (Lepidoptera–Trichoptera), Wiegmann et al. (2000) (Lepidoptera), Savard et al. (2006) and Ronquist et al. (2012) (Hymenoptera), Schlüter (1986) (Neuropterida) and Shcherbakov (2000) and Nel et al. (2013) (Hemiptera). The geological timescale follows the International Chronostratigraphic Chart (http://www.stratigraphy.org, v2016/04). *Amph*, Amphiesmenoptera; *Antlio*, Antliophora; *Coleo*, Coleopterida; *Cond*, Condylognatha; *Dictyopt*, Dictyoptera; *Euki*, Eukinolabia; *Haplo*, Haplocercata; *Neuropt*, Neuropterida; *Orthopt*, Orthopteroidea; *Palaeo*, Palaeoptera; *Psoco*, Psocodea; *Thysan*, Thysanoptera; *Xen*, Xenonomia.

interpreted as first traces of arthropod activity on plants. The earliest fossil hexapods (Entognatha *Rhyniognatha hirsti*, Collembola *Rhyniella praecursor*) date back to the mid-Devonian period (c. 400 million years; Engel & Grimaldi, 2004), while the first Pterygota and Holometabola would date back to the beginning of the Upper Carboniferous period (c. 350 million - years; Nel et al., 2013). These groups are probably much older as they were already quite diversified during the Upper Carboniferous period. Fig. 2 presents a synthetic reconstruction of the evolutionary history of hexapods in general and insects in particular. The most ancient documented fossils for each order have been placed on this phylogenetic tree. It is a consensus tree combining information from several phylogenetic studies (see figure legend for more details). Nevertheless the topology of the tree is still debated within some groups such as the Polyneoptera.

The transition from detritivorous to phytophagous feeding habit, a crucial step in the evolution of hexapods, is not directly documented in the fossil record. It can, however, be inferred from the morphological analysis of buccal pieces and from the interpretation of damages on fossil plant organs. As early as 1971, Carpenter proposed an ecological interpretation of observations on insect morphology (buccal pieces, ovipositor, wings and legs). His conclusion was that, during the Palaeozoic, hexapods had already diversified ecologically within terrestrial habitats, similarly to the modern hexapods. According to Labandeira (2006), phytophagy would have evolved at the end of the Silurian period and the beginning of the Devonian period (c. 420 million years ago), but the oldest records of phytophagous fossils correspond to the orthopteroids and blattoids at the end of the Mississippian subperiod and the middle of the Pennsylvanian subperiod (see Fig. 7 in Labandeira, 2006).

Modern orders of insects presenting phytophagous species are deeply rooted in geological times. The first documented Coleoptera appear at the beginning of the Permian period, Hemiptera and Thripida (sensus Nel et al., 2012; a super order including the Thysanoptera) during the Pennsylvanian, Heteroptera during the Triassic period and Lepidoptera at the beginning of the Jurassic period. Hence insects were certainly largely diversified before the appearance of the modern elements of our flora.

Labandeira (2006) separated the evolution of the interactions between arthropods and plants into four distinct phases:

- A first phase extends from the Silurian period to the Devonian period, covering 60 million years. It is marked by the consumption of vascular plants and primitive fungi by myriapods and apterous hexapods.

- A second phase extends from the late Mississippian subperiod to the end of the Permian period. During this phase, mites, Apterygota and numerous and well-diversified Pterygota were feeding on Pteridophytes and basal Gymnosperms.
- The third phase extends from the Middle Triassic period to present. During this phase mites, orthopteroids, hemipteroids and basal clades of Holometabola (all these clades originated during the Upper Carboniferous) colonized Pteridophytes and Gymnosperms.
- The fourth phase extends from the beginning of the Cretaceous period to present. It corresponds to the development of modern faunas on Angiosperms.

These scenarios sketching out the main lines of the history of the interactions between arthropods and their host plants cross several time periods of mass extinction in the history of Earth. These mass extinctions affected both plants and animals (Fig. 2). The response of insects to these crises differed among groups. The transition from the Permian period to the Triassic period is the first major episode of renewal of faunas and floras with the disappearance of some groups of hexapods. For instance Palaeodictyoptera disappeared from the fossil record near the end of the Permian period. The Triassic period is particularly important in the history of insects with the diversification of Diptera, Hymenoptera, Thysanoptera, Psocoptera (whose basal groups date back to the Upper Carboniferous period), Heteroptera, Trichoptera and Lepidoptera. Hence modern orders of insects have been present for over 150 million years and for some of them for over 200 million years. During the Mesozoic era, many groups remained stable. We observe the first representatives of several orders of insects (Mantodea, Zoraptera, Embioptera) and of modern eusocial insects (Isoptera, Apoidea, Vespoidea, Formicoidea) during the Cretaceous.

The transition from the Cretaceous period to the Cenozoic era was marked by a major extinction crisis that saw the disappearance of the large dinosaurs (still present as birds) as well as of numerous other terrestrial and aquatic lineages. Nevertheless insects seem to have been marginally affected by that crisis, at least at the family level (Labandeira & Sepkoski, 1993). Instead the major upheaval of insect faunas occurred at the middle of the Cretaceous period as quite a number of Jurassic lineages are no longer found in the Upper Cretaceous (Grimaldi & Engel, 2005; Nel, Petrulevičius, & Béthoux, 2005).

The first confirmed Angiosperm fossils date back to 135 million years ago in the Middle Jurassic or early Lower Cretaceous (Liu & Wang, 2016) and began their diversification during the Lower Cretaceous (Crane, Friis, & Pedersen, 1995; Crepet, 1995, 2000; Friis & Crepet, 1987; Judd, Campbell, Kellogg, Stevens, & Donoghue, 2002). The middle part of the Cretaceous is marked by a transition from Gymnosperm floras to Angiosperm floras. This transition is accompanied by the diversification of numerous modern insect families (Grimaldi & Engel, 2005). The impact of the establishment of Angiosperm-dominated floras is generalized, affecting purely phytophagous insects but also many other groups of insects such as the aquatic predators represented by Odonata (Nel et al., 2010), or ants as documented through dated phylogenetic reconstructions (Moreau, Bell, Vila, Archibald, & Pierce, 2006). The Cretaceous period is also of major importance in the biogeographic history of extant lineages as the current configuration of continents is beginning to appear with the fragmentation of the Gondwana (Sereno, Wilson, & Conrad, 2004).

In Africa, Gymnosperms disappeared almost totally during the Maastrichtian–Palaeocene transition (65 million years) (Maley, 1996). Only few but remarkable species of Gymnosperms have survived until today, including *Welwitschia mirabilis* Hook.f., 1862 (Welwitschiaceae) and African *Podocarpus* (Pseudocarpaceae). The modern flora and the associated groups of phytophagous insects become established during the Cenozoic (Boltenhagen, Dejax, & Salard-Cheboldaeff, 1985). Already at the beginning of the Palaeocene–Eocene epoch, fossil insects are very similar to the modern ones as indicated by the study of insects trapped in amber and lacustrine sediments which are particularly abundant and diversified in Western and Northern Europe (Nel, De Ploëg, Millet, Menier, & Waller, 2004). The main difference is in their biogeographic distributions as Europe and Northern America presented warm, intertropical climate floras and faunas.

3. THE APPEARANCE OF ENTOMOPHILOUS POLLINATION

Pollen and hence pollination is a major evolutionary innovation of seed plants. Seed plants originated some 350 million years ago during the Palaeozoic period. Pollination was anemophilous (i.e., pollen distributed by wind) in early Gymnosperms. Male gametophytes (pollen) were produced by male cones. This pollen was wind dispersed and was intercepted by female cones in which growth of the gametophytes allowed ovule fertilization. It is the typical pollination system of the final 150 million years of the Palaeozoic period (Fig. 3).

Entomophily (i.e., insect pollination) began to play an essential role in plant diversification during the Mesozoic period (~252–66 million years). There was, before the appearance of the Angiosperms, a first radiation of insects presenting mouthparts capable of reaching pollination drops in male and female cones and in parallel the evolution of pollen types consistent with entomophily in the major groups of seed plants, notably in the Lyginopteridales (improperly called seed ferns) and in the Pinales. While today, within Gymnosperms, all Pinales are wind pollinated, Cycadales are pollinated by Coleoptera and Thysanoptera and Gnetales are pollinated by flies and moths. Only 120 million years ago (Fig. 3) did entomophilous pollination appear in the Angiosperms. Out of these multiple evolutions of entomophilous pollination, only the Angiosperm one was followed by an impressive species diversification (235,000–250,000 extant species of plants) resulting in the replacement of the major part of the flora of the Palaeozoic (Rodet, 2013). A first diversification of insects occurred at the beginning of the Mesozoic (Labandeira & Sepkoski, 1993; see earlier). A second phase was initiated by the spread of Angiosperms and the diversification of anthophilous insects (i.e., living at the expense of flowers by feeding on nectar or pollen) (Pellmyr, 2002). These insects present convergent evolutionary innovations mainly within Hymenoptera, Lepidoptera and Diptera, such as the transformation of buccal pieces into tubular tongues. Anthophilous insects are rather large and good flyers. They bear hairs that can be modified so as to facilitate pollen collection, in particular in the Hymenoptera Apoidea (Rodet, 2013).

Within a species in entomophilous spermaphytes, flowers producing pollen and flowers at female receptivity are selected to attract the same insects thus allowing intraspecific pollen transfer and subsequent fertilization. The evolutionary advantage of entomophilous pollination is directional pollen transfer that does not depend on wind. Typically, comparatively to wind pollination, it facilitates pollination at low density or in habitats in which wind is absent such as the forest understory. There is, however, no free ride in nature. These advantages come at a cost in terms of signalling and of rewards offered to the pollinators, but also in terms of transmission of diseases and of dependence on the continued presence of sufficiently abundant pollinators (Fig. 4).

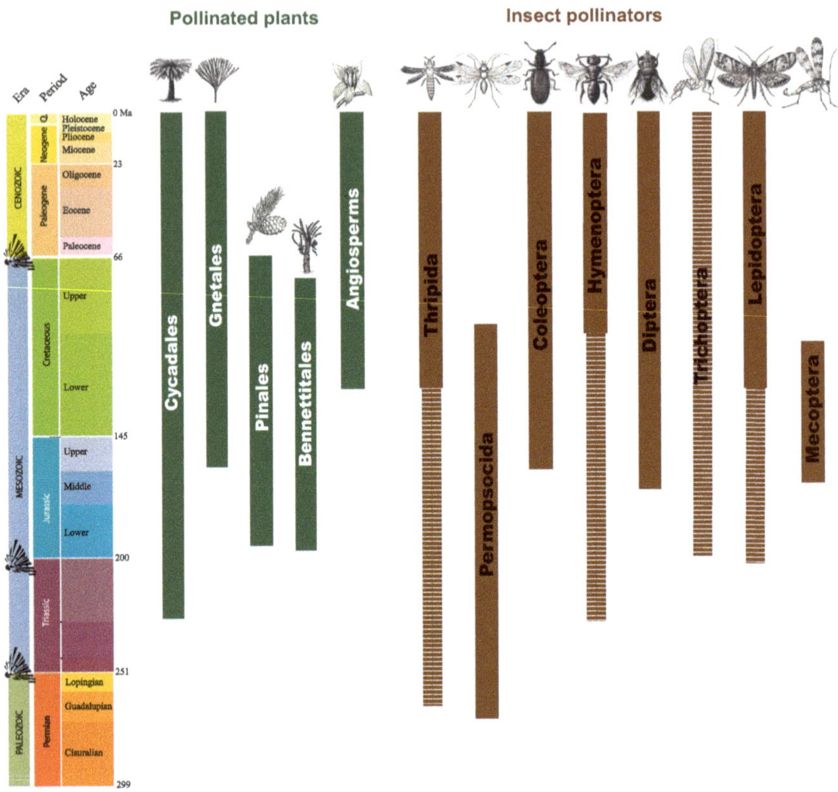

Figure 3 ***Plant-pollinating insect associations since the Mesozoic (~251 million years).*** Pollen, and hence pollination, appeared with the Spermatophytes about 350 million years ago. Male gametophytes (pollen) were wind dispersed and were intercepted by female cones (anemophilous pollination). This form of pollination is characteristic of the last 150 million years of the Palaeozoic (~444–251 million - years). After that insects played a major role in pollination. Pollination by insects played a major role in the now extinct Bennettilales and in Pinales up until the massive extinction at the end of the Cretaceous (~65 million years). Coleoptera and Diptera played an early role in pollination (as early as ~160 million years, during the mid-Jurassic). During 62 million years, some families of Mecoptera, today extinct, were also associated with the pollination of ancient Gymnosperms before the radiation of Angiosperms. Thripida (sensus Nel et al., 2012; a super order including the Thysanoptera), Permopsocida (new fossil order recently resolves as sister group of Thripida + Hemiptera), Hymenoptera, Trichoptera and Lepidoptera were certainly early contributors to pollination (Lower Jurassic-Late Triassic, maybe as early as the Permian for the Thripida) as they fed on pollen and nectar (*dotted lines*). For more information see Labandeira et al. (2007), Ren et al. (2009), Nel et al. (2012), and Huang et al. (2016). The relationships between the different insect orders are illustrated in Fig. 2.

Figure 4 Butterfly feeding on a fig, a behaviour that may lead to the transfer of yeasts and other potential pathogens into flower stigmas.

4. FUNCTIONAL GROUPS OF INSECT POLLINATORS AND POLLINATION SYNDROMES

Once initiated, entomophilous pollination led to huge diversification in insects and plants. Globally, modern insect pollinators can be classified into functional groups according to what resource they feed on and how they access these resources. Therefore, the shape of their buccal pieces plays a major role. For instance one may distinguish long-tongued bees, short-tongued bees, other Hymenoptera, Diptera, Coleoptera, Lepidoptera, Hemiptera, Neuroptera. Late Jurassic and Cretaceous long-tongued Diptera, Mecoptera, Thysanoptera and Permopsocida (sister groups of the Thripida and the Hemiptera) were certainly exploiting gymnosperm pollen and nectar (Huang et al., 2016). While some of these insects were pollinators (Peñalver et al., 2012), it is very difficult to establish which ones because some of these insects probably only visited male cones never transferring pollen to female cones and were hence parasites of the system. In many insect-pollinated angiosperms this problem is circumvented as many species present hermaphroditic flowers so that any insect visiting flowers in search of pollen is also visiting flowers presenting receptive stigmas. In that perspective, the angiosperm hermaphroditic flowers constitute a major evolutionary innovation. In parallel, angiosperm flowers have evolved traits that allow insect attraction such as visual cues (shape, size, colour), fragrance and the production of rewards (generally nectar and pollen or more rarely wax or oil). Flowering phenology, the presentation of flowers in inflorescences, period of flowering and the density of flowering individuals also play an important role in insect attraction (Schatz, Hossaert-McKey, & Kjellberg, 2013). Corolla shape is often determinant in insect choice, and two particularly important parameters are symmetry and depth (Neal, Dafni, & Giurfa, 1998). Ancestral symmetry was radial or actinomorphic (several planes of symmetry), while zygomorphic (mirror-symmetric) symmetry only evolved later. Selection for deeper corollas has led to the evolution of tubular corollas and spurs selecting for insects presenting more or less elongated tongues as found for instance in butterflies (Fig. 5).

Evolution of floral morphology may be canalized by preferential pollination by a functional group of insect pollinators, resulting in morphological and ecological convergence with other groups of plants pollinated by the same insects. This functional convergence results in a number of characteristics which define pollination syndromes (Faegri & van der Pijl, 1979).

Although often challenged (e.g., Ollerton et al., 2009), the notion of pollination syndrome is informative and recent studies support its validity (Danieli-Silva et al., 2012). It facilitates the analysis of communities and processes (Fenster, Armbruster, Wilson, Dudash, & Thomson, 2004) since it is used by field biologists as an easy proxy to determine the type of pollinators most likely to be associated with each plant species. The presence of pollination syndromes coupled with the possibility to define functional groups of pollinators is the result of a diffuse coevolution process between pollinating insect communities and plant communities. For any particular

Figure 5 ***Hovering* Macroglossum stellatarum *(Lepidoptera, Sphingidae) nectaring in a tubular flower of lavendel,* Lavandula angustifolia *(Lamiaceae).*** The short tubular corolla allows pollination by insect presenting much shorter mandibular pieces: the flowers are specialized, but to a large group of pollinators.

species of plant presenting generalist pollination, it is, however, difficult to demonstrate that its traits are currently evolving in response to a local community of pollinating insects. Indeed the relative abundance of the different species of pollinators visiting a plant is often highly variable among years and along the distribution area. Similarly, it is complicated to demonstrate that local processes of selection by local floras affect the long-term evolution of somewhat generalist pollinators. In a world marked by past climate shifts resulting in the displacement of species communities and with composition turnovers, it is almost impossible to define precisely under which interaction conditions current traits have evolved. One may hope that studies on the structure and evolution of communities will advance our knowledge on these topics.

5. THE MUTUALISM BETWEEN PLANTS AND POLLINATING INSECTS AND THE RADIATION OF ANGIOSPERMS

As noted by Darwin (1877), insects are capable of learning. If, during its visit to a flower, an insect benefits from a reward on leaving the flower, it will be interested in other flowers of the same species, and in the process it will affect within-species pollen transfer. There is hence a simple process by which flowers are selected to produce rewards for pollinating insects and to produce distinctive signals of their location, thus enabling the insect to find flowers of the same species providing the same reward. This signal should be distinctive and we can predict that in most cases it will constitute an honest signal. Hence, we expect, globally, convergence between the selective interests of pollinating insects and plants, but also diversification of flowers constitutes a feature that in turn may drive diversification of pollinating insects. This virtuous cycle of diversification is what has allowed the formidable radiation of Angiosperms and of pollinating insects.

Within the context of the evolution of conspicuous flowers and the production of scents for insect attraction, one may note that structures resembling corollas are found in some bryophytes such as Splachnaceae and in particular the yellow dung moss, *Splachnum luteum* Hedw., associated with the production of scents attracting flies that will disperse spores (Marino, Raguso, & Goffinet, 2009). However, these structures are not analogous to flowers (e.g., Fig. 36-1 in Schatz, Hossaert-McKey, Kjellberg, 2013). Indeed their function is to allow the transport of spores from the plant to a germination site, in this case faeces. Hence there is no selection on the plant to increase the probability that the flies will visit a second plant, and hence there is no selection for the production of rewards and for diversification of signalling. A potential candidate situation that would have more evolutionary implications similar to those of insect pollination mutualisms is the relationship between Ascomycetes of the genus *Epichloë* and anthomyiid flies of the genus *Botanophila* (Hossaert-McKey, McKey, & Dormont, 2008).

The basic evolutionary principles associated with the evolution of insect pollination are simple. However, there are numerous subtleties in how they are plaid out. For instance flowers producing male gametes are selected to appear similar to flowers at female receptivity. This may explain the frequency of hermaphroditic flowers in insect-pollinated Angiosperms, while wind-pollinated Angiosperms often present unisexual flowers.

While Angiosperms were originally insect pollinated, about 10% of extant Angiosperms are wind pollinated. Anemophily has appeared independently at least 65 times in Angiosperm. This shows that the transition from entomophily to anemophily is easy. But it has little evolutionary future, as witnessed by the limited number of species involved (Friedman & Barrett, 2009). Anemophylous pollination is particularly well represented in temperate climates, most notably in trees and in open habitats.

Entomophilous pollination is particularly well represented in tropical forest understories (a habitat in which anemophilous pollination is difficult because of lack of wind), but also in the emergent trees of the humid tropical forests. Entomophilous pollination is frequently found in the intertropical zone, where local species diversity is the highest (Mittelbach et al., 2007). We may suggest that this high species diversity and its corollary, the low individual density of each species are made possible by insect pollination which allows directional pollen transfer. One may also try to imagine how species will evolve within simplified ecosystems presenting lower plant diversity, and in which insect faunas are also largely reduced (Biesmeijer et al., 2006; Thomann, Imbert, Engstrand, & Cheptou, 2015).

6. THE ENTOMOPHILOUS POLLINATION: A NEVER-ENDING SOURCE OF PROBLEMS FOR PLANTS

While producing conspicuous flowers and particularly flowers presenting large petals, easily visible for insects, can be selected for, it is also a source of weakness. Indeed, petals and styles are particularly sensitive and constitute a point of entry for numerous pathogenic fungi as these tissues do not present the thick cuticle that protects other plant organs (Ngugi & Scherm, 2006). Producing nectar is also producing a culture medium hosting surprisingly large quantities of yeasts (Herrera, de Vega, Canto, & Pozo, 2009). Basically flowers produce nutritive resources that attract pollinating insects, and by essence these resources should be easily digestible, a feature that facilitates the evolution of generalist pollinators. This opposes flowers and leaves which are selected for resistance to phytophagous arthropods leading through density-dependent selection to the production of diversified protection mechanisms and to the selection of specialist phytophagous insects. As a result, phytophagous insects are generally much more specialized than pollinating insects (Fontaine, Thébault, & Dajoz, 2009). A direct consequence of limited resource protection is the risk that a pathogen will develop on the resource. A further problem for the plant is that the pollinating insects may act as efficient vectors of these pathogens. Plants are thus potentially exposed to all the pathogens a pollinating insect may encounter during its peregrinations. Within that perspective, an insect feeding outside flowers on alimentary resources that can be contaminated can be catastrophic for the plant.

Hence a flower that does not rely on pollinators that only feed on plants may be particularly exposed to pathogens as the pollinators search for diverse-feeding resources. Plant species may therefore often be selected to provide food that will satisfy insects that only feed in flowers. This evolutionary process may explain why many pollinating insects are specialists in the sense that they only feed on flowers. However, this process does not protect against sexually transmitted diseases. Indeed, like all living beings having sex, plants are exposed to such diseases (Antonovics, 2005). To ensure seed production, an entomophilous plant must attract sufficient numbers of pollinators. The genetic quality of offspring may even depend on pollination by multiple male parents, as competition among pollen tubes may results in the selection of the most vigorous ones during their growth within the style. However, each additional pollinator visitation increases the risk of transmission of sexually transmitted diseases. Plants may therefore be selected to present an intermediate level of attractiveness to pollinators: sufficient to ensure pollination, but not too strong, to prevent contagion (Shykoff, Bucheli, & Kaltz, 1997). In an exemplary case, in the White Campion, *Silene latifolia* Poir. (Caryophyllaceae), an insect will transport from flower to flower spores of the anther smut fungus, *Microbotryum violaceum* (Pers.) G. Deml & Oberw. 1982 (Ustilaginaceae) that will infect the plant and colonize its tissues. At the next flowering season, the

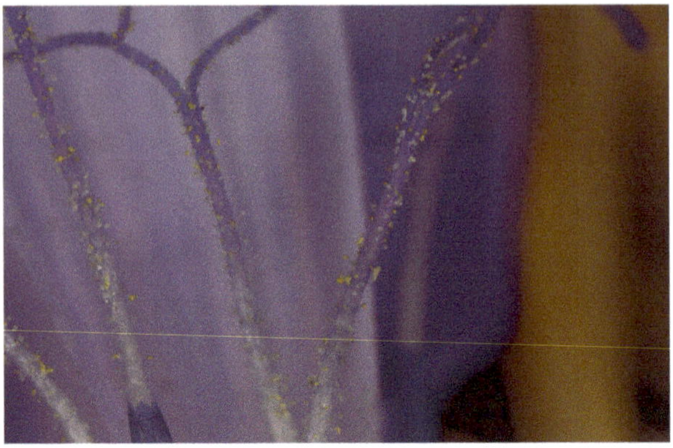

Figure 6 ***Stigmata of the common chicory,*** **Cichorium intybus *L.*** A diversity of pollen are stuck on the stigmas including pollen from *Convolvulus arvensis* L., *Malva sylvestris* L., *Crepis* sp. and also some *C. intybus* pollen.

development of the plant will be highly modified, manipulated by the fungus whose spores will replace pollen (Shykoff & Kaltz, 1998). In other cases, fungi transmitted by pollinating insects will induce the development of plant structures simulating, visually and olfactorily, flowers producing nectar loaded with spores (Raguso & Roy, 1998). In these two cases the plant is castrated.

A similar problem for plants is that they have to capture the pollen that reaches them on a receptive surface, the stigma, which will allow the germination of the pollen grain and subsequent growth of the pollen tube and hence its nutrition. Generally, stigmas are sticky, so that arriving pollen grains will remain on its surface. However as the stigma is sticky it may

Figure 7 ***Flowering*** **Anthoxantum odoratum** ***(Poaceae).*** This is a typical example of large exposed feathery stigmas observed in wind pollinated plants.

also collect dust, pollen grains from other species (Fig. 6), spores of fungi and other pathogens. These different contaminants will on one hand mask the stigmas and for some of them benefit of the pollen tube germination medium to develop as pathogens. For instance in a study of orchids, it was observed that stigmas were contaminated by pollen from other species resulting in substantial reduction of plant reproductive success (Neiland & Wilcock, 1998). The stigmas of wind pollinated plants are particularly exposed (Fig. 7).

7. THE ROLE OF FLOWER MORPHOLOGY IN ENTOMOPHILOUS POLLINATION

Generally the plant must draw a balance between benefits associated with attracting somewhat generalist pollinators and the costs associated with the fact that these pollinators may bring contaminants, including pollen from other plant species. Any trait masking the stigmas and protecting them against contaminants, without limiting arrival of pollen of the same species will be selected for. This masking will be all the more developed as the pollinating insects belong to a circumscribed, morphologically homogeneous group. This will facilitate deposition of pollen on precise parts of the insects which will contact the stigmas on visitation of another flower of the same species. Hence plants manipulate space, through the shape of flowers and the precise location of food rewards attracting the insects, in such a way that insects are automatically loaded with pollen and will automatically deposit pollen on stigmas. Thus, floral morphology plays a central role in entomophilous pollination (van der Cingel, 1995; Faegri & van der Pijl, 1979).

The masking of stigmas reaches its extreme expression in obligate pollination mutualisms, in which pollinator larvae develop in the flowers (or inflorescence) after pollination. This is the case for instance in the association between the globeflower, *Trollius europeaeus* L. (Ranunculaceae) and the flies of genus *Chiastocheta* (Diptera, Anthomyiidae) that ensure its pollination. The evolution of mutualistic pollination in *Trollius* has been accompanied by the closing of the flower, thus protecting stigmas against the visit on nonspecialist insects (Faegri & van der Pijl, 1979; Pompanon, Pettex, & Després, 2006). Similarly Yuccas (Agavaceae) are all pollinated by moths of the genus *Tegeticula* (Lepidoptera, Prodoxidae) whose larvae develop consuming seeds. In this case, the part of the stigma receptive to pollen is deeply sunk into the style. A specific behaviour of the moth is required. The moth pushes pollen down the style allowing pollen germination.

This type of pollination is called active or ethodynamic as the *Tegeticula* have modified mandibular palps that they use to collect pollen in other plants, and later on, after ovipositing in a flower of another *Yucca*, they deposit pollen on the stigma. While the association between *Yucca* and *Tegeticula* is mutualistic and obligate, this has not led to a process of codiversification through co-cladogenesis. A species of *Tegeticula* may pollinate several hosts and over time some *Tegeticula* have switched host. Furthermore, a species pollinated by *Tegeticula*, *Hesperoyucca whipplei* (Torr.) Trel. (Agavaceae), is not a *Yucca* (Smith et al., 2008). In the same way as the first Angiosperms coopted insects pollinating Gymnosperms, this species has coopted the *Yucca* pollination system. Other example, each of the some 800 species of *Ficus* (Moraceae) produces urn-shaped inflorescences, inside which stigmas are protected. Figs are pollinated by specialized insects, namely agaonid wasps (Hymenoptera, Chalcidoidea), whose larvae develop at the expense of ovules (e.g., Fig. 32-1 in Schatz, Proffit, Kjellberg, & Hossaert-McKey, 2013).

Generally, each species of agaonid wasp is associated with a single species of *Ficus*. Nevertheless, as for Yuccas, there has not been strict co-cladogenesis between *Ficus* and agaonid wasps (Cruaud et al., 2012). Furthermore, some *Ficus* species are also pollinated by nonagaonid wasps (Jousselin, Rasplus, & Kjellberg, 2001). In this case, as in the initial diversification of Angiosperms, new insects have been coopted as mutualistic pollinators. Also, in the three cases described previously, no pollinating insect feeds outside the host species. In the case of *Tegeticula* and Agaonidae, adults do not feed at all while adult Chiastochetae feed on *Trollius* pollen. This suggests that the insects protect their offspring larvae by avoiding feeding outside their host, thus limiting sources of contamination.

8. CONSTANTLY EVOLVING INSECT–PLANT INTERACTIONS

The appearance of Angiosperms some 100–120 million years ago has been at the origin of a formidable evolutionary radiation of both plants and insects. This has resulted in the extraordinary diversity of flowers. Indeed flowers are at the centre of plant–insect interactions, and they are the object of the evolution of this fruitful biotic interaction. Convergent evolution has led to the appearance of pollination syndromes and to a diversity of specialization levels in plant–insect interactions ranging from generalist to highly species specific with one species of insect being the sole pollinator a single

Figure 8 Pollen of *Convolvulus arvensis* L. (left) and *Ficus maxima* P. Miller (right). Pollen size varies strongly among species. Generally insect-pollinated plants present large pollen grains as illustrated on the left, while wind-pollinated plants produce small, light pollens. This general rule suffers from exceptions as illustrated here by the small-sized *Ficus* pollen that is dispersed by a specific *Tetrapus* wasp (Agaonidae, Chalcidoidea) breeding in figs.

species of plant as in orchids of genus *Ophrys* (Schatz, Hossaert-McKey, et al., 2013). This diversity of interactions is an important component of biodiversity as it allows the local coexistence of highly diversified sets of species (Fig. 8). Ongoing global change often leads to reduce such diversity and particularly specific interactions. The shift seems to be towards more generalist insect pollination and towards anemophilous pollination. Biesmeijer et al. (2006) show a decline of wild bees and syrphids in parallel with a decline in entomophily in regional floras in Britain and the Netherlands. Will our children witness the demise of the over 100-million-year-old reign of floras and faunas characterized by the association between pollinating insects and entomophilous plants?

REFERENCES

Antonovics, J. (2005). Plant venereal diseases: Insights from a messy metaphor. *New Phytologist, 178*, 71–80. http://dx.doi.org/10.1111/j.1469-8137.2004.01215.x.

Beerling, D. (2007). *The Emerald planet. How plants changed earth's history*. Oxford, New York: Oxford University Press.

Biesmeijer, J. C., Roberts, S. P. M., Reemer, M., Ohlemüller, R., Edwards, M., Peeters, T., … Settele, J. (2006). Parallel declines in pollinators and insect-pollinated plants in Britain and The Netherlands. *Science, 313*, 351–354. http://dx.doi.org/10.1126/science.1127863.

Boltenhagen, E., Dejax, J., & Salard-Cheboldaeff, M. (1985). Evolution de la végétation tropicale africaine du Crétacé à l'Actuel d'après les données de la palynologie. *Comité des Travaux Scientifiques et Historiques, Section Science, Paris, 8*, 165–194.

Carpenter, F. M. (1971). Adaptations among Paleozoic insects. *Proceedings of the First North American paleontological convention, 2*, 1236–1251.

van der Cingel, N. A. (1995). *An atlas of orchid pollination-European orchids*. Rotterdam: Balkema.

Crane, P. R., Friis, E. M., & Pedersen, K. R. (1995). The origin and early diversification of the angiosperms. *Nature, 374*, 27–33. http://dx.doi.org/10.1038/374027a0.

Crepet, W. L. (1995). Timing in the evolution of derived floral characters upper Cretaceous (Turonian) taxa with tricolpate and tricolpate-derived pollen. *Review of Paleobotany and Palynology, 90*, 339–359. http://dx.doi.org/10.1016/0034-6667(95)00091-7.

Crepet, W. L. (2000). Progress in understanding angiosperm history, success and relationships, Darwin's abominably 'perplexing phenomenon'. *Proceedings of the National Academy of Sciences of the United States of America, 97*, 12939–12941. http://dx.doi.org/10.1073/pnas.97.24.12939.

Cruaud, A., Rønsted, N., Chantarasuwan, B., Chou, L. S., Clement, W. L., Couloux, A., … Hossaert-McKey, M. (2012). An extreme case of plant-insect co-diversification: Figs and fig-pollinating wasps. *Systematic Biology, 61*, 1029–1047. http://dx.doi.org/10.1093/sysbio/sys068.

Currano, E. D., Wilf, P., Wing, S. L., Labandeira, C. C., Lovelock, E. C., & Royer, D. L. (2008). Sharply increased insect herbivory during the Paleocene-Eocene thermal maximum. *Proceedings of the National Academy of Sciences of the United States of America, 105*, 1960–1964. http://dx.doi.org/10.1073/pnas.0708646105.

Danieli-Silva, A., Souza, J. M., Donatti, A. J., Campos, E. P., Vicente-Silva, J., Freitas, L., & Varassin, G. (2012). Do pollination syndromes causes modularity and predict interactions in a pollination network in tropical high-altitude grasslands? *Oikos, 121*, 35–43. http://dx.doi.org/10.1111/j.1600-0706.2011.19089.x.

Darwin, C. (1877). *The various contrivances by which orchids are fertilised by insects* (2nd ed.). London: John Murray.

Deuve, T. (2001). Origin of the Hexapoda. *Annales de la Société Entomologique de France, 37*, 304.

Engel, M. S., & Grimaldi, D. A. (2004). New light shed on the oldest insect. *Nature, 427*, 627–630. http://dx.doi.org/10.1038/nature02291.

Faegri, K., & van der Pijl, L. (1979). *The principles of pollination ecology* (3rd ed.). Oxford: Pergamon Press.

Fenster, C. B., Armbruster, W. S., Wilson, P., Dudash, M. R., & Thomson, J. D. (2004). Pollination syndromes and floral specialization. *Annual Review of Ecology, Evolution, and Systematics, 35*, 375–403. http://www.jstor.org/stable/30034121.

Fontaine, C., Thébault, E., & Dajoz, I. (2009). Are insect pollinators more generalist than insect herbivores? *Proceedings of the Royal Society B, 276*, 3027–3033.

Friedman, J., & Barrett, S. C. H. (2009). Wind of change: New insights on the ecology and evolution of pollination and mating in wind-pollinated plants. *Annals of Botany, 103*, 1515–1527. http://dx.doi.org/10.1093/aob/mcp035.

Friis, E. M., & Crepet, W. L. (1987). Time of appearance of floral features. In E. M. Friis, W. G. Chaloner, & P. R. Crane (Eds.), *The origins of angiosperms and their biological consequences*. Cambridge: Cambridge University Press.

Gaunt, M. W., & Miles, M. A. (2002). An insect molecular clock dates the origin of the insects and accords with palaeontological and biogeographic landmarks. *Molecular Biology and Evolution, 19*, 748–761.

Grimaldi, D. (2001). Insect evolutionary history from Handlirsch to Hennig, and beyond. *Journal of Paleontology, 75*, 1152–1160. http://dx.doi.org/10.1017/S0022336000017200.

Grimaldi, D., & Engel, M. S. (2005). *Evolution of the insects*. Cambridge, New York, Melbourne: Cambridge University Press.
Herrera, C. M., de Vega, C., Canto, A., & Pozo, I. M. (2009). Yeasts in floral nectar: A quantitative survey. *Annals of Botany, 103*, 1415–1423. http://dx.doi.org/10.1093/aob/mcp026.
Hossaert-McKey, M., McKey, D., & Dormont, L. (2008). Fungal sex as a private matter: Odour signals in a specialized pollination-like insect-fungus mutualism. *New Phytologist, 178*, 225–227. http://dx.doi.org/10.1111/j.1469-8137.2008.02428.x.
Huang, D. Y., Bechly, G., Nel, P., Engel, M. S., Prokop, J., Azar, D., ... Krogmann, L. (2016). New fossil insect order Permopsocida elucidates major radiation and evolution of suction feeding in hemimetabolous insects (Hexapoda: Acercaria). *Scientific Reports, 6*, 1–9.
Jousselin, E., Rasplus, J.-Y., & Kjellberg, F. (2001). Shift to mutualism in a parasitic lineage of the fig/fig wasp interaction. *Oikos, 94*, 287–294. http://dx.doi.org/10.1034/j.1600-0706.2001.940209.x.
Judd, W. S., Campbell, C. S., Kellogg, E. A., Stevens, P. F., & Donoghue, M. J. (2002). *Plant systematics, a phylogenetic approach*. Sunderland: Sinauer Associates.
Krenn, H. W., Fournel, J., Bauder, J. A.-S., & Hugel, S. (2016). Mouthparts and nectar feeding of the flower visiting cricket *Glomeremus orchidophilus* (Gryllacrididae). *Arthropod Structure & Development, 45*, 221–229.
Labandeira, C. C. (2006). The four phases of plant-arthropod associations in deep time. *Geologica Acta, 4*, 409–438. http://dx.doi.org/10.1344/105.000000344.
Labandeira, C. C., Kvacek, J., & Mostovski, M. B. (2007). Pollination drops, pollen, and insect pollination of Mesozoic gymnosperms. *Taxon, 56*, 663–695.
Labandeira, C. C., & Sepkoski, J. J. J. (1993). Insect diversity in the fossil record. *Science, 261*, 310–315.
Liu, Z.-J., & Wang, X. (2016). Yuhania: A unique angiosperm from the middle Jurassic of inner Mongolia, China. *Historical Biology*. http://dx.doi.org/10.1080/08912963.08912016.01178740.
Maley, J. (1996). The African rainforest. Main characteristics of changes in vegetation and climate from the upper Cretaceous to the quaternary. In I. J. Alexander, M. D. Swaine, & R. Watling (Eds.), *Proceedings of the Royal Society: Vol. 104B. Essays on the ecology of the Guinea-Congo rain forest*. Edinburgh.
Marino, P., Raguso, R., & Goffinet, B. (2009). The ecology and evolution of fly dispersed dung mosses (family Splachnaceae): Manipulating insect behaviour through odour and visual cues. *Symbiosis, 47*, 61–76. http://dx.doi.org/10.1007/BF03182289.
Mayhew, P. J. (2012). Shifts in hexapod diversification and what Haldane could have said. *Proceedings of the Royal Society B: Biological Sciences, 269*, 969–974. http://dx.doi.org/10.1098/rspb.2002.1957.
McKey, D. & Hossaert-McKey, M. (2008). La coévolution entre les plantes et les animaux. In F. Hallé (Vol. Ed.), *Aux origines des Plantes*. Paris: Fayard.
Misof, B., Liu, S., Meusemann, K., Peters, R. S., Donath, A., Mayer, C., ... Niehuis, O. (2014). Phylogenomics resolves the timing and pattern of insect evolution. *Science, 346*, 763–767. http://dx.doi.org/10.1126/science.1257570.
Mittelbach, G. G., Schemske, D. W., Cornell, H. V., Allen, A. P., Brown, J. M., Bush, M. B., ... McCain, C. M. (2007). Evolution and the latitudinal diversity gradient: Speciation, extinction and biogeography. *Ecology Letters, 10*, 315–331. http://dx.doi.org/10.1111/j.1461-0248.2007.01020.x.
Moreau, C. S., Bell, C. D., Vila, R., Archibald, S., & Pierce, N. E. (2006). Phylogeny of the ants: Diversification in the age of angiosperms. *Science, 312*, 101–104. http://dx.doi.org/10.1126/science.1124891.

Neal, P. R., Dafni, A., & Giurfa, M. (1998). Floral symmetry and its role in plant-pollinator systems, terminology, distribution, and hypotheses. *Annual Review of Ecology, Evolution and Systematics, 29*, 345–373. http://www.jstor.org/stable/221712.

Neiland, M. R. M., & Wilcock, C. C. (1998). Fruit set, nectar reward, and rarity in the Orchidaceae. *American Journal of Botany, 85*, 1657–1671.

Nel, A. (1997). The probabilistic inference of unknown data in phylogenetic analysis. *Mémoires du Muséum national d'histoire naturelle, 173*, 305–327.

Nel, P., Azar, D., Prokop, J., Roques, P., Hodebert, G., & Nel, A. (2012). From Carboniferous to recent: wing venation enlightens evolution of thysanopteran lineage. *Journal of Systematic Palaeontology, 10*, 385–399. http://dx.doi.org/10.1080/14772019.2011.598578.

Nel, A., De Ploëg, G., Millet, J., Menier, J.-J., & Waller, A. (2004). The French ambers: A general conspectus and the lowermost Eocene amber deposit of Le Quesnoy in the Paris Basin. *Geologica Acta, 2*, 3–8.

Nel, A., Nel, P., Petrulevičius, J. F., Perrichot, V., Prokop, J., & Azar, D. (2010). The Wagner Parsimony using Morphological Characters: A new method for palaeosynecological studies. *Annales de la Société Entomologique de France, (N.S.), 46*, 276–292.

Nel, A., Petrulevičius, J. F., & Béthoux, O. (2005). Les insectes, témoins privilégiés du Crétacé. *Pour la Science Dossier, 48*, 108–114.

Nel, A., Roques, P., Nel, P., Prokin, A. A., Bourgoin, T., Prokop, J., … Garrouste, R. (2013). The earliest-known holometabolous insects. *Nature, 503*, 257–261. http://dx.doi.org/10.1038/nature12629.

Ngugi, H. K., & Scherm, H. (2006). Biology of flower-infecting fungi. *Annual Review of Phytopathology, 44*, 261–282. http://dx.doi.org/10.1146/annurev.phyto.44.070505.143405.

Ollerton, J., Alarcón, R., Waser, N. M., Price, M. V., Watts, S., Cranmer, L., … Rotenberry, J. (2009). A global test of the pollination syndrome hypothesis. *Annals of Botany, 103*, 1471–1480.

Pellmyr, O. (2002). Pollination by animals. In C. M. Herrera, & O. Pellmyr (Eds.), *Plant animal interactions: An evolutionary approach* (pp. 157–184). Oxford: Blackwell Science.

Peñalver, E., Labandeira, C. C., Barrón, E., Delclòs, X., Nel, P., Nel, A., … Soriano, C. (2012). Thrips pollination of Mesozoic gymnosperms. *Proceedings of the National Academy of Science, 109*, 8623–8628.

Pompanon, F., Pettex, E., & Després, L. (2006). Patterns of resource exploitation in four coexisting globeflower fly species (*Chiastocheta* sp.). *Acta Oecologica, 29*, 233–240.

Raguso, R. A., & Roy, B. A. (1998). 'Floral' scent production by *Puccinia* rust fungi that mimic flowers. *Molecular Ecology, 7*, 1127–1136. http://dx.doi.org/10.1046/j.1365-294x.1998.00426.x.

Regier, J. C., Shultz, J. W., & Kambic, R. E. (2004). Phylogeny of basal hexapod lineages and estimates of divergence times. *Annals of the Entomological Society of America, 97*, 411–419. http://dx.doi.org/10.1603/0013-8746(2004)097[0411:POBHLA]2.0.CO;2.

Ren, D., Labandeira, C. C., Santiago-Blay, J. A., Rasnitsyn, A., Shih, C., Bashkuev, A., … Dilcher, D. (2009). A probable pollination mode before Angiopserms: Eurasian, long-proboscid scorpionflies. *Science, 326*, 840–847. http://dx.doi.org/10.1126/science.1178338.

Rodet, G. (2013). Chapitre 21-Mutualisme entre insectes et plantes, des ennemis réconciliés. In N. Sauvion, P.-A. Calatayud, D. Thiéry, & F. Marion-Poll (Vol. Eds.), *Interactions insectes-plantes* (pp. 303–318). Versailles: Editions Quae, Marseille: IRD Editions.

Ronquist, F., Klopfstein, S., Vihelmsen, L., Schulmeister, S., Murray, D. L., & Rasnitsyn, A. P. (2012). A Total-evidence approach to dating with fossils, applied to the early radiation of the Hymenoptera. *Systematic Biology, 61*(6), 973–999. http://dx.doi.org/10.1093/sysbio/sys058.

Savard, J., Tautz, D., Richards, S., Weinstock, G. M., Gibbs, R. A., Werren, J. H., ... Lercher, M. J. (2006). Phylogenomic analysis reveals bees and wasps (Hymenoptera) at the base of the radiation of Holometabolous insects. *Genome Research, 16*, 1334–1338. http://dx.doi.org/10.1101/gr.5204306.

Schatz, B., Hossaert-McKey, M., & Kjellberg, F. (2013). Histoire évolutive et diversité des interactions entre insectes pollinisateurs et plantes. In N. Sauvion, P.-A. Calatayud, D. Thiéry, & F. Marion-Poll (Vol. Eds.), *Interactions insectes-plantes* (pp. 565–576). Versailles: Editions Quae, Marseille: IRD Editions.

Schatz, B., Proffit, M., Kjellberg, F., & Hossaert-McKey, M. (2013). Un réseau trophique complexe : le cas des figuiers. In N. Sauvion, P.-A. Calatayud, D. Thiéry, & F. Marion-Poll (Vol. Eds.), *Interactions insectes-plantes* (pp. 483–493). Versailles: Editions Quae, Marseille: IRD Editions.

Schlüter, T. (1986). The fossil Planipennia – A review. In J. Gepp, H. Aspöck, & H. Hölzel (Vol. Eds.), *Recent Research in Neuropterology* (pp. 103–111). Graz.

Sereno, P. C., Wilson, J. A., & Conrad, J. L. (2004). New dinosaurs link southern landmasses in the Mid-Cretaceous. *Proceedings of the Royal Society of London B: Biological Sciences, 271*, 1325–1330. http://dx.doi.org/10.1098/rspb.2004.2692.

Shcherbakov, D. E. (2000). Permian faunas of Homoptera (Hemiptera) in relation to phytogeography and the Permo-Triassic crisis. *Paleontological Journal, 34*, S251–S267.

Shykoff, J. A., Bucheli, E., & Kaltz, O. (1997). Anther smut disease in *Dianthus silvester* (Caryophyllaceae): Natural selection on floral traits. *Evolution, 51*, 383–392. http://www.jstor.org/stable/2411110.

Shykoff, J. A., & Kaltz, O. (1998). Phenotypic changes in host plant diseased by *Microbotryum violaceum*: Parasite manipulation, side effects, and trade-offs. *International Journal of Plant Sciences, 159*, 236–243. http://www.jstor.org/stable/2475087.

Smith, C. I., Pellmyr, O., Althoff, D. M., Balcázar-Lara, M., Leebens-Mack, J., & Segraves, K. A. (2008). Pattern and timing of diversification in *Yucca* (Agavaceae): Specialized pollination does not escalate rates of diversification. *Proceedings of the Royal Society of London B: Biological Sciences, 275*, 249–258. http://dx.doi.org/10.1098/rspb.2007.1405.

Terry, M. D., & Whiting, M. (2005). Mantophasmatodea and phylogeny of the lower neopterous insects. *Cladistics, 21*, 240–257. http://dx.doi.org/10.1111/j.1096-0031.2005.00062.x.

Thomann, M., Imbert, E., Engstrand, R. C., & Cheptou, P. O. (2015). Contemporary evolution of plant reproductive strategies under global change is revealed by stored seeds. *Journal of Evolutionary Biology, 28*, 766–778. http://dx.doi.org/10.1111/jeb.12603.

Trautwein, M. D., Wiegmann, B. M., Beutel, R., Kjer, K. M., & Yeates, D. K. (2012). Advances in insect phylogeny at the dawn of the postgenomic Era. *Annual Review of Entomology, 57*, 449–468. http://dx.doi.org/10.1146/annurev-ento-120710-100538.

Whalley, P. (1986). A review of the current fossil evidence of Lepidoptera in the Mesozoic. *Biological Journal of the Linnean Society, 28*, 253–271. http://dx.doi.org/10.1111/j.1095-8312.1986.tb01756.x.

Wheeler, W. C., Whiting, M., Wheeler, Q. D., & Carpenter, J. M. (2001). The phylogeny of the extant hexapod orders. *Cladistics, 17*, 113–169. http://dx.doi.org/10.1006/clad.2000.0147.

Whiting, M. F. (1998). Phylogenetic position of the Strepsiptera: Review of molecular and morphological evidence. *International Journal of Insect Morphology and Embryology, 27*, 53–60. http://dx.doi.org/10.1016/S0020-7322(97)00032-9.

Whiting, M. F. (2002). Phylogeny of the holometabolous insect orders: Molecular evidence. *Zoologica Scripta, 31*, 3–15. http://dx.doi.org/10.1046/j.0300-3256.2001.00093.x.

Wiegmann, B. M., Mitter, C., Regier, J. C., Friedlander, T. P., Wagner, D. M., & Nielsen, E. S. (2000). Nuclear genes resolve mesozoic-aged divergences in the insect order Lepidoptera. *Molecular Phylogenetics and Evolution, 15*, 242–259. http://dx.doi.org/10.1006/mpev.1999.0746.

Wiegmann, B. M., Trautwein, M. D., Kim, J. W., Cassel, B. K., Bertone, M. A., Winterton, S. L., & Yeates, D. K. (2009). Single-copy nuclear genes resolve the phylogeny of the holometabolous insects. *BMC Biology, 7*, 34–50. http://dx.doi.org/10.1186/1741-7007-7-34.

Winterton, S. L., Hardy, N. B., & Wiegmann, B. M. (2010). On wings of lace: Phylogeny and Bayesian divergence time estimates of Neuropterida (Insecta) based on morphological and molecular data. *Systematic Entomology, 35*, 349–378. http://dx.doi.org/10.1111/j.1365-3113.2010.00521.x.

Yeates, D. K., Cameroun, S. L., & Trautwein, M. (2012). A view from the edge of the forest: Recent progress in understanding the relationships of the insect orders. *Australian Journal of Entomology, 51*, 79–87. http://dx.doi.org/10.1111/j.1440-6055.2012.00857.x.

CHAPTER TWO

Evolution of Plant–Insect Interactions: Insights From Macroevolutionary Approaches in Plants and Herbivorous Insects

G.J. Kergoat[1], A.S. Meseguer, E. Jousselin
INRA – CBGP (INRA, IRD, CIRAD, Montpellier SupAgro), Montferrier-sur-Lez, France
[1]Corresponding author: E-mail: gael.kergoat@supagro.inra.fr

Contents

Abstract

In this study, we review recent works in the phylogenetic investigations of plant–insect interactions. Thanks to the development of novel methodological approaches and the ever-increasing availability of informative molecular markers, it is indeed now possible to test more and more complex evolutionary scenarios. Here, we are limiting our review to studies on herbivorous insects (excluding work on the evolution of pollinating insects), and we provide an overview of the variety of approaches employed to answer fundamental questions in plant/insect evolution. More specifically, our review addresses studies that have focused on the following: (1) reconstructing the evolutionary history of the associations with plants; (2) inferring the diversification dynamics of herbivorous insects and (3) studying the biogeographic history of herbivorous insects. Finally, we attempt to decipher whether general trends in the evolution of plant–insect interactions have emerged from these studies and highlight the most promising perspectives in this field.

Advances in Botanical Research, Volume 81
ISSN 0065-2296
http://dx.doi.org/10.1016/bs.abr.2016.09.005

1. INTRODUCTION

With more than 500,000 known species, herbivorous insects represent nearly a quarter of all terrestrial macroscopic biodiversity (Daly, Doyen, & Purcell, 1998; Southwood, 1973; Strong, Lawton, & Southwood, 1984), thus representing a dominant component of the evolutionary forces that have shaped life on the earth. The extraordinary diversification of herbivorous insects is often attributed to their intimate association with land plants and especially with angiosperms (Farrell, 1998; Marvaldi, Sequeira, O'Brien, & Farrell, 2002; Mitter, Farrell, & Wiegmann, 1988). This hypothesis stems from the observation that most herbivorous insects have a very narrow host range, only feeding on one or few (often) related species of plants. The host plants on which the insects feed constitute ideal agents of divergent selection for these organisms. Adaptation towards different host-plant species can potentially lead to ecological specialization of populations and, subsequently, species formation in plant-feeding insects. This idea that associations with plants play in major role in insect evolution is not new, as intimate associations between herbivorous insects and their hosts plants have been highlighted long ago by botanists and entomologists. For instance the pictorial work of M. S. Merian (a scientific illustrator from the 18th century) constitutes a vibrant testimony of the interest that naturalists had for plant–herbivorous insect interactions (Etheridge, 2010). One hundred and fifty years ago the interactions with plants were already suggested to play a significant role in the formation of 'varieties' or 'species' of herbivorous insects (e.g., Walsh, 1867). This ecological speciation scenario has been clearly formulated and demonstrated recently (e.g., Nosil, Crespi, & Sandoval, 2002; Via, Bouck, & Skillman, 2000; see also Matsubayashi, Ohshima, & Nosil, 2010 for a review).

Several review papers have demonstrated that specialization towards their host plants is a dominant feature of herbivorous insects (Bernays, 1998; Bernays & Chapman, 1994; Funk, Filchak, & Feder, 2002; Futuyma, 1991; Futuyma & Moreno, 1988; Jaenike, 1990; Strong et al., 1984; Ward & Spalding, 1993; Winkler & Mitter, 2008). It is estimated that almost 90% of species feed on plants belonging to less than three plant families (Futuyma, 1991). The use of a limited number of host-plant species is not the only feature that characterizes the interactions of insects with their host plants. Most herbivorous insect species also exhibit a strong specialization towards specific plant parts, such as leaves, stems or seeds, with insect species feeding on plant internal parts being usually more specialized (Anderson, 1995; Bernays & Chapman, 1994; Gaston, Reavy, & Valladares, 1992; Marvaldi et al., 2002). The plants also often represent several dimensions of the ecological niche of herbivorous insects, as those generally spend an important part of their life cycle (sometimes from birth to reproduction) on their hosts. Hence, besides being a food source the host plants can also provide an oviposition site, a shelter for hibernation, a place to encounter partners for sexual reproduction or a shelter against potential predators. Therefore colonizing plants and using them as a food source opens up a wide range of ecological opportunities for insects: it is this wealth of ecological niches and their multiple dimensions that are suspected to have fuelled the spectacular diversification of herbivorous insects. The Cretaceous terrestrial revolution (KTR) ca. 125–85 millions years ago (Ma) (Lloyd et al., 2008), during which the floral composition of angiosperms rose from 0% to 80% (Benton, 2010), has been advanced to be a strong factor of insect diversification. Nevertheless, with the exception of few studies (e.g., the study of Labandeira, Dilcher, Davis, & Wagner, 1994 on leaf-mining Lepidoptera), evidence to support this claim is controversial (Labandeira & Sepkoski, 1993), and most recent studies relying on fossil records tend to support the hypothesis that the KTR did not significantly impact the diversification of herbivorous insects (Clapham, Karr, Nicholson, Ross, & Mayhew, 2016; Condamine, Clapham, & Kergoat, 2016; Labandeira, 2005; Nicholson, Ross, & Mayhew, 2014; Sohn, Labandeira, & Davis, 2015; Wang, Zhang, & Jarzembowski, 2013).

These ideas and observations on host plant utilization by insects have further inspired macroevolutionary scenarios in which the colonization of new plant lineages causes bursts of speciation in insect lineages. The most influential of these is probably the scenario put forward 52 years ago by Ehrlich and Raven (1964). They suggested that plants and the insects (in their study butterflies) that feed on them have diversified through successive bursts of diversification. They postulated that plant chemical defences impose strong selective pressures on the insects that feed on them. Once an insect lineage evolves the ability to circumvent these chemical defences, it opens up a new adaptive zone (with fewer competitors), and insects can then quickly diversify on the range of host-plant species exhibiting similar defences (which are often phylogenetically related plants). The scenario of Ehrlich and Raven also postulates that, in response to herbivory, plants evolve new chemical defences that allow them in turn to diversify rapidly in the absence of a strong selective pressure from herbivores. This scenario is now commonly known as 'escape and radiate' (Thompson, 1988, 1994). This theory primarily focuses on the arms race between the evolution of plant chemical defences and the development of counter adaptations exhibited by insect to circumvent them (Fig. 1). Though the ubiquity of the scenario of Ehrlich and Raven remains to be demonstrated (Futuyma & Agrawal, 2009; Suchan & Alvarez, 2015), this work has inspired most of the current studies on the phylogenetic history of plant–insect interactions. This is probably because, as underlined in previous review papers (Janz, 2011; Winkler & Mitter, 2008), it integrates several key concepts: (1) phylogenetic constraints in the evolution of host use; (2) the acquisition of key innovations; (3) coevolution diversification; (4) host-plant mediated speciation and (5) adaptive radiation. Each of these concepts is important for understanding the evolutionary history of herbivorous insects.

Thanks to advances in DNA sequencing technologies, phylogenetic studies have proliferated in the last 3 decades, and the increasing availability of insect and plant phylogenies have provided numerous opportunities to test scenarios of herbivorous insect evolution. Besides clarifying the evolutionary relationships of species, when they are dated (i.e., branch lengths

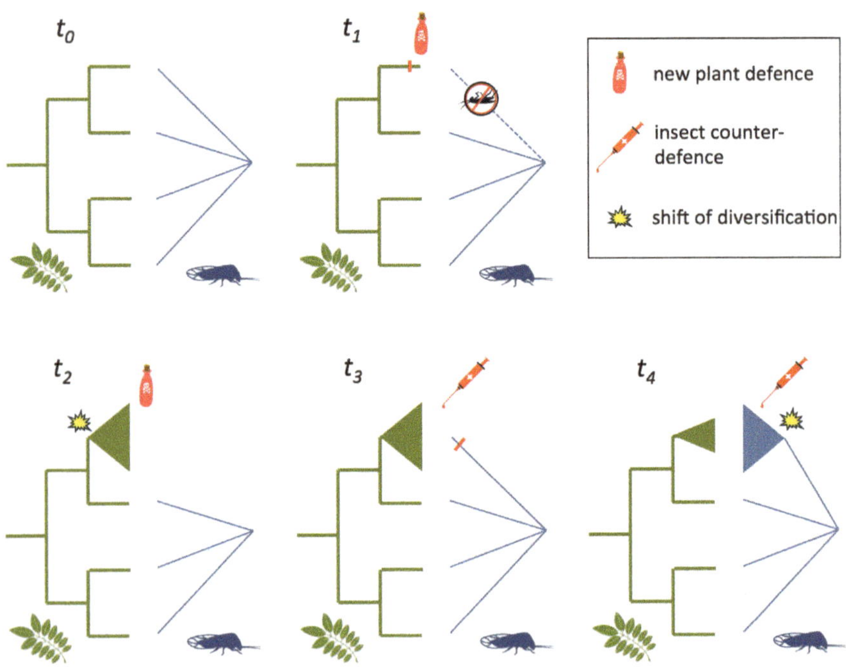

Figure 1 Schematic illustration of the pattern expected under the 'escape and radiate' scenario.

are scaled in temporal units), phylogenetic reconstructions also provide the temporal framework over which their evolutionary history elapsed; they can thus situate diversification events into the paleoenvironmental conditions in which they occurred and also explore variations in the rates of diversification through time (i.e., analyze the acceleration and deceleration in speciation and extinction rates). This has been made possible by recent developments in the estimation of divergence time using relaxed molecular clocks (Drummond, Ho, Phillips, & Rambaut, 2006; Ho & Phillips, 2009; Lartillot, Lepage, & Blanquart, 2009; Pyron, 2010, 2011; Ronquist et al., 2012), historical biogeography (Ree, Moore, Webb & Donoghue, 2005; Ree & Smith, 2008; Ronquist & Sanmartín, 2011) and tools for investigating the dynamics of diversification from the branching patterns of phylogenetic trees (from the pioneering work of Nee, Mooers, & Harvey, 1992 to more advanced methodologies; Condamine, Rolland, & Morlon, 2013; Morlon, 2014; Morlon, Potts, & Plotkin, 2010; Rabosky, 2014; Stadler, 2011). Applications on herbivorous insect phylogenies have shed new lights on the evolutionary processes that have shaped herbivorous insect biodiversity through time.

In this paper, we are going to review recent work in the phylogenetic investigations of plant/insect interactions that have taken advantages of recent methodological developments for testing evolutionary scenarios. We are limiting our review to studies on herbivorous insects (purposefully setting aside work on the evolution of pollinating insects that would necessitate a review in itself), and rather than exhaustively reporting the conclusions of the numerous investigations, we give an overview of the variety of methodological approaches employed to answer fundamental questions in plant/insect evolution. Most studies published so far have focused on (1) reconstructing the evolutionary history of the associations with plants; (2) inferring the diversification dynamics of herbivorous insects and (3) studying the biogeographic history of herbivorous insects. We then attempt to decipher whether general trends in the evolution of plant–insect interactions have emerged from these studies and highlight the most promising perspectives in this field.

2. RECONSTRUCTING THE HISTORY OF THE ASSOCIATIONS WITH PLANTS

2.1 Phylogenetic History of Plant–Insect Interactions and Evolution of Host Range

Phylogenetic trees of herbivorous insect lineages are often used as templates to reconstruct the history of the associations with plants (i.e., infer who ate whom and when). Using information on reported host plants for extant taxa, inferences about host-plant associations throughout the history of the insect groups under investigation are made using ancestral character state reconstructions methods. They are few insect species that are strictly monophagous (i.e., that feed on a single plant species), many insect species feed on a few related plant species, and therefore the associations are often reconstructed at the plant genus level or the plant family level. While the use of Maximum Parsimony optimization dominated the literature until the last decade, Maximum Likelihood optimization and Bayesian inference are now generally used for reconstructing ancestral states (e.g., Pagel, 1999; Pagel & Meade, 2006; Pagel, Meade, & Barker, 2004). Besides trying to give an accurate tale of the major evolutionary transitions in host-plant utilization, one of the main questions addressed by these studies: Are the associations phylogenetically conserved, i.e., do related insects species use related plant species? This question is sometimes addressed by simply describing the reconstructions; most case studies demonstrating host plant conservatism consist in ad hoc verbal scenarios that comment the reconstructions of ancestral associations. Although generally, it could also be statically tested by measuring the 'phylogenetic signal' in host-plant associations using the permutation tail probability test (PTP, Faith & Cranston, 1991, *host associations are randomized onto the phylogeny to tests whether the observed number of host-shifts falls into the null distribution*) or some index of phylogenetic signal such as the lambda Λ of Pagel (1999) that has been adapted to fit with discrete characters. Most studies demonstrate nonrandom associations with host-plant groups throughout the history of insects across a large diversity of insect lineages; see Winkler and Mitter (2008) for a compilation of studies until 2007 and more recent examples for some genera of Lepidoptera (Condamine, Sperling, Wahlberg, Rasplus, & Kergoat, 2012; Nylin, Slove, & Janz, 2014; Toussaint et al., 2012), Coleoptera (Kergoat, Delobel, Le Ru, & Silvain, 2008; Kergoat et al., 2015), Hymenoptera (Leppanen, Altenhofer, Liston, & Nyman, 2012; Stone et al., 2009) and Hemiptera (Meseguer, Coeur d'acier, Genson, & Jousselin, 2015; Meseguer, Lobo, Ree, Beerling, & Sanmartín, 2015; Ouvrard, Chalise, & Percy, 2015), suggesting that closely related species tend to feed on similar plant species.

The associations of herbivorous insects with their host plants are generally so conserved in some cases, that the hypothesis of cospeciation (i.e., simultaneous speciation events in two associated lineages generating similar branching patterns in their phylogenetic trees) has been put forward in the literature (Funk, Futuyma, Orti, & Meyer, 1995; Yokoyama, 1995). Robust cospeciation analyses rely on evaluating the congruence of phylogenetic trees, both in branching patterns and timing of speciation events (see Doyon, Ranwez, Daubin, & Berry, 2011; de Vienne et al., 2013; Martínez-Aquino, 2016 for reviews on methodological approaches). Event-based methods such as reconciliation analyses (Doyon et al., 2011) can even produce codiversification scenarios that depict when and where host shifts have occurred. Using these methods the only plant–insects associations showing significant signals of cospeciation are species-specific interactions in which the herbivorous insects also act as obligatory pollinators of their host plants (e.g., fig trees and their pollinators: Cruaud et al., 2012; yuccas and yucca moths: Althoff, Segraves, Smith, Leebens-Mack, & Pellmyr, 2012). This is not surprising since, for the diversification of an interspecific interaction to be governed by cospeciation, this interaction must be specific for both partners, and both associates should be highly dependent on one other for their reproduction, so that any vicariance events (i.e., geographical separation) affecting the populations of one of the partners should simultaneously affect the other. This is indeed the case in obligate plant–pollinator interactions. In plant–herbivorous insect, the insect is the only one that is dependent of its partner. Consequently a pattern of phylogenetic tracking is more likely than cospeciation (Althoff, Segraves, & Johnson, 2014). In this scenario an herbivorous insect species, after colonizing a host-plant species, will switch to phylogenetically related hosts, specializing on the new host plants and speciate in the process. This should ultimately produce an insect phylogeny with a similar topology to the one of the host lineage, but much more recent on time (Fig. 2).

Support for phylogenetic tracking is controversial. The study of Mitter, Farrell, and Futuyma (1991), reviewing the studies conducted until then showed that the parallelism of the phylogenies of insects and their plants was rare (only one study out of the 14 surveyed provided evidence for such a parallelism). However, since then, the study of Percy, Page, and Cronk (2004) on psyllids clearly shows a strong congruence between the phylogenetic tree of these insects and that of their host plants. The study of Wilson et al. (2012) on geometrid moths also shows some parallelism between insect and plant phylogenies. However, both studies suggest delayed speciation in insects. Molecular dating studies recovered contrasted patterns, either suggesting that the radiation of herbivorous insect groups lagged behind the ones of their host plants (e.g., Gómez-Zurita, Hunt, Kopliku, & Vogler, 2007; Lopez-Vaamonde et al., 2006; McLeish, Miller, & Mound, 2013; McKenna, Sequeira, Marvaldi, & Farrell, 2009; Meseguer, Coeur

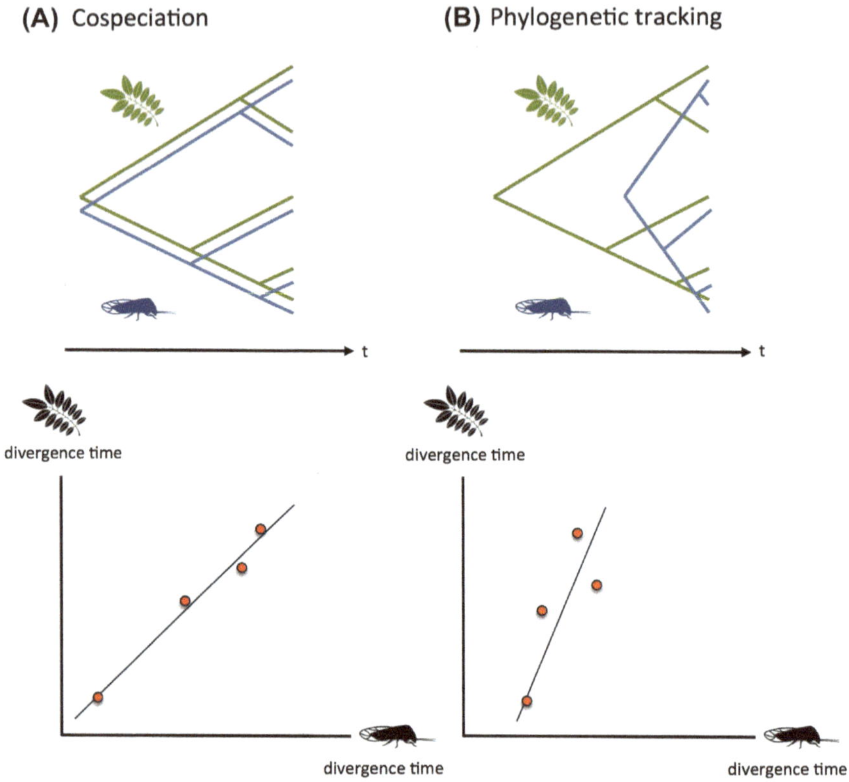

Figure 2 Schematic illustration of the patterns expected under a strict cospeciation pattern (A) or resulting from phylogenetic tracking (B).

d'acier, et al., 2015; Meseguer, Lobo, et al., 2015) or followed it more closely (e.g., Ahrens, Schwarzer, & Vogler, 2014; Cardinal & Danforth, 2013; Kergoat et al., 2011, 2015; Song et al., 2015). All these studies were carried out without testing for topological similarities between plants and insect phylogenies, probably because a mere observation of patterns of species specificity and host association was enough to show that this hypothesis did not hold.

Although the phylogenetic conservatism of host association in the diversification of herbivorous insects is widely demonstrated, the evolutionary processes that have generated this pattern and their influence on insect diversification remain unclear. For instance, host conservatism can be interpreted as fitting the predictions of the 'escape and radiate' hypothesis, suggesting that 'host-plant associations' have driven speciation. On the other hand, adaptation towards restricted ranges of host plant has been suggested to lead to evolutionary dead ends limiting further species diversification (Moran, 1988). Phylogenetic studies that infer the frequency of host-plant shifts in relation to speciation events (see Nyman, Vikberg, Smith, & Boeve, 2010 for an example in Nematinae and Jousselin et al., 2013 for a study on conifer-feeding aphids) can give a better estimation of the contribution of host-plant association in the speciation processes. Both studies, along with the synthesis of Winkler and Mitter (2008), suggest that less than 50% of the speciation events in herbivorous insects are congruent with a shift in trophic niche, moderating the importance of 'host-plant mediated speciation scenario' in insects.

Additional insights into the processes underlying host-plant conservatism to test if it can represent an evolutionary dead end can be gained from analyzing evolutionary transitions in host-plant breadth. While most herbivorous insect species feed on a few related host plants, some species are able to feed on a large number of host plants that belong to very distinct lineages. These species are referred to as 'polyphagous' or 'generalist' species. Several studies have investigated how these species were distributed in the phylogeny of insects and whether transitions from being a specialist to a generalist were equally likely than the reverse transitions. Among the first studies that statistically compared the transition rates between specialization and generalization, the work of Nosil (2002) and Nosil and Mooers (2005) based on 15 insect phylogenies suggested that directional evolution towards increased specialization was more likely, but also indicated that specialization did not always represent an evolutionary dead end, a generalist species sometimes having specialist ancestors. This result has been confirmed in other insect lineages (e.g., aphids: Jousselin, Genson, & Coeur d'acier, 2010; Coleoptera: Kelley & Farrell, 1998; Morse & Farrell, 2005; Lepidoptera: Hardy & Otto, 2014; Menken, Boomsma, & van Nieukerken, 2010; Nylin et al., 2014). Phylogenetic comparative studies that aimed at identifying insect characters, geographic or climatic circumstances that are associated with changes in host breadth have also been conducted. Nyman, Farrell, Zinovjev, and Vikberg (2006) and Menken et al. (2010) in leaf-mining groups showed that internal feeders were more specialized than external feeders. Life cycle variation in aphids (Jousselin et al., 2010; Ward, 1991) influences the degree of insects specialization; and latitude (correlated with host-plant richness) also has been shown to influence host breath within insect clades (see Hardy, Peterson, & Normark, 2015 for a study of scale insects). All these studies confirm that host-plant associations and the number of hosts that can be used are not solely the results of dietary specialization (Forister, Dyer, Singer, Stireman, & Lill, 2012). In cases where host plants represent several dimensions of the insects ecological niche (the host plant is also a mating site, a shelter against predators), more specialized interactions evolve.

2.2 Diversification Dynamics of Herbivorous Insects

Diversification rates have become some of the most important metrics in macroecological and macroevolutionary studies. They have been increasingly used to explore the factors that explain the exceptional diversity of specific groups of insects (e.g., Farrell & Sequeira, 2004; Hunt et al., 2007; Kergoat et al., 2014; Moreau & Bell, 2013; Wahlberg et al., 2013) or insects as a whole (Condamine et al., 2016; Rainford, Hofreiter, Nicholson, & Mayhew, 2014; Rainford & Mayhew, 2015; Wiens, Lapoint, & Whiteman, 2015). As underlined in the introduction, it has been suggested that diversification rates are more elevated in herbivorous insects than in other groups, implying that association with plants could have fuelled their diversification (Farrell, 1998; Farrell & Mitter, 1990; Strong et al., 1984). At first, this hypothesis has been tested by comparing the species richness of herbivorous and nonherbivorous sister clades in several lineages of Lepidoptera and Coleoptera (Farrell, 1998; Mitter et al., 1991, 1988); these studies have all suggested that herbivorous clades were more diverse than their nonherbivorous counterparts. However, Hunt et al. (2007), using a more comprehensive beetle phylogeny that the one used by Farrell (1998), and Rainford and Mayhew (2015), using a hexapod phylogeny, both adopting a sister clade comparison, refuted this hypothesis. Recently Wiens et al. (2015) have revisited this question using phylogenetic comparative methods (i.e., *Phylogenetic Generalized Least Square Regressions* that test for the relationship between diversification rates and the proportion of herbivorous insect species in a clade) and comprehensive phylogenies for several insect orders. Their analyses show that herbivory had a significant impact on diversification rates among insect orders, but this relationship does not hold in all orders (e.g., in Coleoptera, Hymenoptera and Orthoptera, herbivorous clades do not diversify more rapidly than nonherbivorous clades). They also demonstrate, along with Rainford et al. (2014), that other characters (e.g., wings, holometaboly) spur insect diversification and suggest that studies at finer evolutionary scales (i.e., using species-level phylogenies) will be necessary to deepen our understanding of the relationship between evolutionary transitions towards herbivory and diversification. Most studies have identified shifts in diversification rates and made ad hoc explanations on the processes that have generated them. But analytical methods modelling trait evolution, diversification and the influence of the former on the latter (FitzJohn, 2012; Maddison, Midford, & Otto, 2007) could be used to explore how the evolution of herbivorous traits has influenced diversification dynamics. They could be powerful tools to test whether shifts towards herbivory increase diversification, but as underlined by Wiens et al. (2015), they require well resolved species-level phylogenies, which are not always available.

Another way to provide evidence for the positive impact of host-plant association on insect diversification is to explore concomitant bursts of diversification in insects and plants. Such analyses could be performed using methods that allow detecting changes of diversification rates, assess key innovation and diversity-dependence patterns (e.g., Rabosky, 2014). Recently, Condamine et al. (2016) explored global patterns of insect diversification using diversification models and, contrary to the results suggested by the study of Rainford et al. (2014) and Wiens et al. (2015), their analyses did not recover significant shifts in diversification rate associated with the origin of wings (in Pterygota) or holometaboly. Interestingly, all inferred diversification rate shifts, but one, either largely predate (six shifts out of eight) or postdate the KTR, suggesting that the rise of angiosperms did not have an immediate positive impact on insect diversification within major insect groups.

Some of the trait dependant diversification analyses cited above have been successfully used to test the hypothesis that major host shifts spur species formation by offering a new adaptive zone. For instance, Pena and Espeland (2015) found that a shift to Solanaceae in the tribe Ithomiini (Nymphalidae) was correlated with an increase in diversification rates. Surprisingly, to our knowledge, no one has investigated whether insect lineages associated with very diverse plant lineages show higher diversification rates than lineages associated with species poor plant lineages. Such results would suggest that the diversification of herbivorous insects is fuelled by the number of ecological niches offered by their host plants.

Further studies have then investigated how changes in diet breadth have influenced diversification dynamics of herbivorous insects. Indeed, if specialization towards a restricted number of host plant species are a major pathway towards the formation of new species in herbivorous insects, then evolutionary transitions in diet breaths have the capacity to provoke significant shifts in speciation rates. Janz, Nylin et al. have elaborated a scenario in which expansions in diet breadth followed by specialization constantly fuel the diversification of herbivorous insects. They called it the 'oscillation hypothesis': transitions towards a generalist diet generally open a new adaptive zone that favours the capture of new host plants, these expansions are also correlated with an increase in geographic range size (Janz, Nyblom, & Nylin, 2001; Janz & Nylin, 2008; Janz, Nylin, & Wahlberg, 2006; Nylin & Janz, 2009). The specialization of some populations on newly captured host plants and population fragmentation lead to the formation of new species. This hypothesis can be seen as an updated version of the 'escape and radiate' scenario of Ehrlich and Raven; it is both more precise (the speciation processes, involving host-plant adaptation and geographic isolation, are clearly laid out) and applicable to a broader range of interactions (as it is not focused on the evolution of plant defences). The predictions of this hypothesis in terms of diversification dynamics are controverted. Hardy and Otto (2014) have predicted that under this scenario, generalist species should have higher diversification rates than specialists. This prediction, using butterflies phylogenies and trait-dependent diversification models, was not met, and the authors suggested a new hypothesis called '*the musical chairs*', where the speciation of herbivorous insects is mostly driven by host-plant switches in specialist species. Hamm and Fordyce (2015) in Nymphalidae also found that the transitions in host breadth were decoupled from diversification rates. However as, underlined by Janz, Braga, Wahlberg, and Nylin (2016), the transient nature of host breadth makes it very difficult to detect generalist species in phylogenies, which render further tests on their influence on diversification rates difficult. They also caution against the use of trait-dependant diversification methods that are very sensitive to model specifications. One of the predictions of the *oscillation hypothesis* that could be tested is that clades showing high lability in host-plant breadth should have higher diversification rates (Hardy & Otto, 2014; Janz et al., 2016). To assess the generality of the oscillation model, tests outside butterflies' families should be conducted. Aphids that encompass very specific species as well as polyphagous species and show evolutionary lability in host breadth (Jousselin et al., 2010; Moran, 1992; Peccoud et al., 2010) are good candidates for such a test.

Finally, several studies have looked at changes in diversification rates associated with environmental changes. Indeed, landscape transformation associated with large-scale climatic changes is widely believed to have affected the diversification of major groups of herbivorous insects (McLeish, Chapman, & Schwarz, 2007; Pena & Wahlberg, 2008; Winkler, Mitter, & Scheffer, 2009). In all these studies, as underlined by the authors, such changes have often caused drastic biotic turnover in ecosystems, and their impact on herbivorous insect diversification might have been a cascading result of their impact on host plant communities (Nyman, Linder, Pena, Malm, & Wahlberg, 2012).

2.3 Historical Biogeography of Herbivorous Insects

Thanks to the development of powerful parametric methods, it is now feasible to use dated phylogenies to trace back the evolution of ancestral areas of distribution for both herbivorous insects and their host plants (Ronquist & Sanmartín, 2011). Numerous biogeographical studies have been carried out independently in plants (e.g., Baker & Couvreur, 2013; Bouchenak-Khelladi, Maurin, Hurter, & van der Bank, 2010; Buerki et al., 2011; Couvreur et al., 2011; Schaefer et al., 2012) and phytophagous insect groups (e.g., Condamine, Silva-Brandao, Kergoat, & Sperling, 2012; Condamine, Toussaint, et al., 2013; De-Silva, Elias, Willmott, Mallet, & Day, 2015; Toussaint & Balke, 2016). They have allowed addressing numerous questions related to island radiations (Condamine, Sperling, & Kergoat, 2013; Lewis et al., 2015), long distance dispersals (e.g., Kergoat et al., 2012; Rota, Pena, & Miller, 2016), latitudinal gradients of diversity (Condamine, Sperling, et al., 2012) and impact of climate changes on insect distribution (Meseguer, Coeur d'acier, et al., 2015; Meseguer, Lobo, et al., 2015; Vila et al., 2011).

Few studies have interpreted the biographic history of groups of herbivorous insects in light of changes in the distribution of their host plants. Yet, biogeographic patterns of herbivorous insects could be directly linked to the biogeographic history of their host plants. Plants do not only constitute a source of nutrients for the insects but also represent different dimensions of the insect's niche. The geographic evolution of herbivorous insects could thus follow the one of their hosts: expansion of plant species to new geographic areas could favour range increases in their associated insects, while the fragmentation of the host plant range might promote range contractions in the distribution of insects as well. For example, one of the most conspicuous patterns among herbivorous insects in the Northern Hemisphere is the existence of related lineages distributed in widely separate continents, such as Asia and North America (Sanmartín, Enghoff, & Ronquist, 2001). Intercontinental disjunctions were originally reported in plants (Wen, 1999) and generally explained by the fragmentation of an ancient forest, the '*mixed mesophytic*' forest; this forest once extended across the Holarctic but was probably divided by climate cooling at the end of the Cenozoic (Xiang, Soltis, Soltis,

Manchester, & Crawford, 2000). The fragmentation of the host plant range has been suggested to explain intercontinental disjunctions in insects such as aphids (von Dohlen, Kurosu, & Aoki, 2002; Kim, Lee, & Jang, 2011; Ren et al., 2013). Comparing the age of the disjunctions between insects and associated plants could allow to test this hypothesis, estimating whether vicariance events are simultaneous, or if, conversely, similar distributions are the result of biogeographical pseudocongruence – when two or more lineages display the same biogeographic pattern but with different temporal origins (Donoghue & Moore, 2003; Sanmartín, 2014). The statistical inference of biogeographic processes could reveal if current insect intercontinental disjunctions are indeed the result of vicariance (this is expected when the fragmentation of the host plant range drove the fragmentation of the insect distribution), or more recent dispersal (Meseguer, Coeur d'acier, et al., 2015; Meseguer, Lobo, et al., 2015).

It is however very difficult to disentangle the relative role of host plant association and environmental changes (i.e., climate, geographic barriers) in shaping insect species distribution and diversity patterns, since environmental changes affect simultaneously both plants and insects. The only way to decipher the relative importance of each category of factor is to conduct comparative analyses of plant and associated insect biogeographic histories: i.e., formally test whether insects follow their host plants in their dispersion or shifts to available plants in their environment.

3. CONCLUSION AND PERSPECTIVES

3.1 General Trends

With reference to phylogenetic-based studies in plant–insect interactions, this review has allowed us to highlight some general trends in herbivorous insect diversification:

1. insects and their hosts plants have not cospeciated, but we generally observe phylogenetic conservatism of host use, with insect diversification generally lagging behind the radiation of their hosts;
2. feeding diet is highly labile, and specialization is not a dead end;
3. the link between herbivory and increase in speciation is not clearly established; impacts of evolutionary transitions in host breadth and diversification dynamics are also discussed with some controversy on the methods that can reliably detect such links (Janz et al., 2016) and

4. distinguishing the relative contribution of biotic and abiotic factors and at which time scale they influence diversification patterns remains a challenging task.

3.2 Word of Caution

Phylogenetic reconstructions provide a powerful framework for understanding plant–insect evolutionary dynamics. In the last 2 decades progresses have been made to make sense of phylogenetic trees using statistical tools. Phylogenetic analyses have been made relatively accessible by the development of efficient algorithms for tree reconstruction and their implementation in user-friendly software (e.g., BEAST; Drummond, Suchard, Xie, & Rambaut, 2012) and R packages (e.g., phytools; Revell, 2013). As emphasized in this chapter, phylogenetic reconstructions provide powerful tools to test evolutionary hypotheses; they have made their way into evolutionary ecology and are no longer only used to refine taxonomic classifications. However one must keep in mind that phylogenetic trees only represent a hypothesis of the 'true' evolutionary relationships of species. Hence phylogenetic methods that aim at testing hypotheses about factors underlying species diversification (or any hypothesis tests about ancestral character states) must rely on robust reconstructions.

Yet, when the phylogenetic reconstruction itself is not the focus of a research study but just a tool to conduct further analyses, some important aspects of conducting robust phylogenetic analyses are sometimes ignored. To publish in any molecular systematic journals (e.g., Molecular Phylogenetics and Evolution; Systematic Biology, Systematic Entomology), phylogenetic reconstructions must follow several standards such as using multiple DNA markers, relying on extensive species sampling in the lineages under investigation, and drawing conclusions from well-supported nodes only. We think that similar standards have to be enforced for all studies in evolutionary ecology using phylogenetic comparative analyses, in order to avoid methodological biases that will likely distort macroevolutionary interpretations.

3.3 Perspective – Integrating Fossil and Phylogenetic Evidence Into Macroevolutionary Studies on Insect–Plant Evolution

For many years, macroevolutionary scenarios of insect–plant evolution have been investigated in two different domains. Palaeontologists have used direct evidence from the fossil record to study origination and extinction rates in insect and their associated plants and infer the evolution of the interactions (see chapter: Plant-Insect Interactions: A Paleontological and an Evolutionary Perspective by Schatz, Sauvion, Kjellberg, & Nel, 2017). Meanwhile, phylogeneticists have most often used phylogenetic trees of herbivorous insects as templates to reconstruct the long-term history of the associations with plants. While both approaches have aimed at answering similar questions, they have adopted different perspectives; phylogeneticists mostly have tried to understand processes leading to present diversity patterns, while palaeontologists have focused on the taxa that existed in the past. Fossils can provide direct evidence of when, where and how the ancestors lived. Morphological studies of fossils (e.g., insect mouthparts or activity traces — generally consumption — in the plants) have revealed the diet and trophic behaviour of extinct insects, as well as major changes in host-plant use, such as the major floristic turnovers that occurred at the Triassic—Jurassic transition (McLoughlin, Martin, & Beattie, 2015) or during the KTR (Labandeira, 2006, 2007; Labandeira & Currano, 2013). Insect fossil studies have also allowed detecting major originations and extinctions of insect and plant groups (Grimaldi & Engel, 2005; Labandeira & Sepkoski, 1993). Speciation, extinction and preservation rates may be determined from the analysis of the fossil record (Silvestro, Salamin, & Schnitzler, 2014; Silvestro, Schnitzler, Liow, Antonelli, & Salamin, 2014), which has been recently used to explore the diversification dynamics of insects (Condamine et al., 2016) and vascular plants (Silvestro, Cascales-Minana, Bacon, & Antonelli, 2015). Unfortunately, by its own nature the fossil record is incomplete, and many lineages do not have good fossil representatives. This is the case for insects, which are relatively scarce in the fossil record, given their present diversity levels (Grimaldi & Engel, 2005; Le Gall & Nel, 2013; Nicholson, Mayhew, & Ross, 2015; Sohn et al., 2015). Further, fossil studies are prone to sampling biases (Alroy et al., 2001, 2008) and generally limited to the generic or familiar level because of the difficulty and, sometimes, impossibility to assign fossils to species taxa. This concerns numerous (herbivorous) insect groups such as aphids, beetles or moths, which often display high levels of evolutionary convergence in morphological traits. While the limitations of the sole use of fossils for macroevolutionary studies have been long recognized, the weakness associated with the exclusive use of molecular phylogenies to uncover long-term patterns of species evolution has just began to be explored. The phylogenetic approach uses phylogenies of extant taxa and present observations to infer changes in the past; for example, the combination of phylogenetic relationships and morphological/ecological traits, such as host-plant association, is often used to infer phenotypic evolution and ancestral host associations (as discussed earlier). However, present diversity represents just a small fraction of the diversity that once existed in the past. In case of elevated extinction and/or fast evolution, it is possible that many evolutionary events occurred in the past without leaving a trace (Sanmartín & Meseguer, 2016). It then turns that some macroevolutionary scenarios, such as directional evolutionary trends (Slater, Harmon, & Alfaro, 2012), complete host change or elevate rates of extinction associated with a particular host plant, would be impossible or hardly impossible to detect with phylogenetic data alone.

Though Palaeontology and Phylogenetics have been traditionally separated by different conceptual and methodological frameworks, both disciplines are complementary (Davis, Nicholson, Saunders, & Mayhew, 2011; Mayhew, 2007), and their combination could minimize some of the biased associated to each particular data set. Up to date a standard practice in phylogenetic analysis is to use fossils to provide calibration points, usually expressed as minimum ages for the clades the fossils belong to. As explained earlier, dated phylogenies are often used to test diversification and biogeographic hypotheses. However, fossils can also provide direct information for the reconstruction of phylogenies, biogeographic scenarios, phenotypic evolution, diversification patterns and paleoenvironments. New approaches have emerged in the last years to directly combine molecular and fossil datasets (Pennell & Harmon, 2013). The most powerful one is probably the application of statistical comparative methods to phylogenetic data including extant and fossil taxa, which has been termed a 'total evidence approach' – similar approaches could be applied to paleotrees including only fossil taxa (Bapst, 2012). This often relies on the inference of phylogenetic relationships between extant and extinct taxa, using a combination of molecular (generally available for extant species only) and morphological data (Ronquist et al., 2012). Insect phylogenies including extant and fossil data have been also constructed based on morphological data alone (e.g., Solodovnikov, Yue, Tarasov, & Ren, 2013). Treating fossils as terminal taxa could benefit the inference of divergence times (Pyron, 2011), producing more precise age estimations (Ronquist et al., 2012; Wood, Matzke, Gillespie, & Griswold, 2013; Vea & Grimaldi, 2016), removing the subjectivity of assigning prior distributions on node calibrations based on minimum age constrains (Parham et al., 2012). Total evidence trees could also be used to reconstruct biogeographic scenarios taking into account the distribution of extinct lineages, which allows placing ancestors in areas where present taxa do not exist any longer (Betancur-R, Ortí, & Pyron,

2015; Wood et al., 2013). While these approaches have been proven to be very powerful, they are generally difficult to implement because of the lack of genetic data and the paucity of morphological characters preserved in fossils. An alternative solution would be the inclusion of fossil constraints in comparative analyses based on phylogenetic trees of living species (Slater & Harmon, 2013); the data from fossils could be used as informative priors on the character state at internal nodes of the phylogeny (Mao et al., 2012; Meseguer, Coeur d'acier, et al., 2015; Meseguer, Lobo, et al., 2015; Slater et al., 2012). Fossil-informed phylogenetic approaches are especially interesting when fossil characteristics (i.e., distribution, morphology) are no longer represented across extant taxa (Betancour-R et al., 2015; Meseguer, Coeur d'acier, et al., 2015; Meseguer, Lobo, et al., 2015; Wood et al., 2013). One could envision very different types of useful information to extract from the fossil record: fossil ages, occurrence locations, climatic tolerances, feeding plants (e.g., the appearance of insects in amber could be related with the plant producing this amber), feeding habits (i.e., mode and type), fossil morphology, community assemblages, etc. Although this field of research is still in its early stages a complete macroevolutionary research program will require the integration of different sources of evidence (extant and extinct) to produce more realistic reconstructions of the past plant—insect interactions (e.g., Fritz et al., 2013).

3.4 Other Perspectives

Thanks to recent improvements in DNA bar coding and sequencing technologies, it is now possible to determine feeding preferences of herbivorous insects using DNA isolated from insect bodies (e.g., Garcia-Robledo, Erickson, Staines, Erwin, & Kress, 2013; Jurado-Rivera, Vogler, Reid, Petitpierre, & Gómez-Zurita, 2009; Kajtoch, Kubisz, Heise, Mazur, & Babik, 2015; Navarro, Jurado-Rivera, Gómez-Zurita, Lyal, & Vogler, 2010). These recent developments will be hugely beneficial for researches on evolution of host-plant associations (especially studies involving their reconstruction), as the identification of herbivorous insect diets has always been a challenge.

Another interesting perspective is to reconcile microevolutionary studies (where host plants mediate speciation) and macroevolutionary views on insect diversification, as there are not many studies that precisely estimate if host shifts relate to speciation events. Such studies could combine an experimental evolution research framework (*sensu* Kawecki et al., 2012) with phylogenetic frameworks. They are conceivable in some well-studied herbivorous insect groups, such as pest species of the genus *Callosobruchus* (Coleoptera,

Bruchinae). This genus has already been the focus of numerous studies involving experimental evolution (e.g., Fricke & Arnqvist, 2007; Messina & Durham, 2013; Messina & Johnson, 2014; Messina, Jones, Mendenhall, & Muller, 2009), whose results have been discussed in the light of phylogenetic reconstructions (Tuda, Kagoshima, Toquenaga, & Arnqvist, 2014).

Additional insights into adaptive mechanisms will also be useful. In the era of high-throughput sequencing genomics, numerous studies have been carried out on the genomics of adaptation of herbivorous insects to their host plants (see Simon et al., 2015 for a review). However, up to now only a few studies have been able to put genomic data into phylogenetic perspectives. One of the most inspiring studies here is the work carried out by Edger et al. (2015) on the evolutionary arm race between Pierinae butterflies and their Brassicales host plants. The authors have showed that shifts in diversification rates within the plants and their insect predators are associated with an evolutionary arm race implying gradual changes in plant chemical defences and insect molecular counter adaptations. Using transcriptome and whole genome sequencing, they have identified the genomic mechanisms (gene and genome duplications) explaining the evolution of the biosynthetic pathways associated with this arm race (evolution of glucosinolates defences in the Brassicales and evolution of the nitrile-specifier family in the Pierinae). In another study, Celorio-Mancera et al. (2016) compared tissue-specific transcriptomes from *Vanessa cardui* (Nymphalidae) caterpillars that were reared on six distinct host plants. Interestingly they found that the tissue-specific patterns of caterpillar gene expression are better explained by the evolutionary history of insects' adaptation to the plants than by plant phylogeny itself.

All these recent results illustrate the interest of conducting multidisciplinary approaches to study the evolution of plant–insect interactions within an historical (i.e., phylogenetic) framework.

REFERENCES

Ahrens, D., Schwarzer, J., & Vogler, A. P. (2014). The evolution of scarab beetles tracks the sequential rise of angiosperms and mammals. *Proceedings of the Royal Society B, 281*. http://dx.doi.org/10.1098/rspb.2014.1470.

Alroy, J., Aberhan, M., Bottjer, D. J., Foote, M., Fürsich, F. T., Harries, P. J., … Visaggi, C. C. (2008). Phanerozoic trends in the global diversity of marine invertebrates. *Science, 321*, 97–100.

Alroy, J., Marshall, C. R., Bambach, R. K., Bezusko, K., Foote, M., Fürsich, F. T., … Webber, A. (2001). Effects of sampling standardization on estimates of Phanerozoic marine diversification. *Proceedings of the National Academy of Sciences of the United States of America, 98*, 6261–6266.

Althoff, D. M., Segraves, K. A., & Johnson, M. T. J. (2014). Testing for coevolutionary diversification: Linking pattern with process. *Trends in Ecology & Evolution, 29*, 82–89.
Althoff, D. M., Segraves, K. A., Smith, C. I., Leebens-Mack, J., & Pellmyr, O. (2012). Geographic isolation trumps coevolution as a driver of yucca and yucca moth diversification. *Molecular Phylogenetics and Evolution, 62*, 898–906.
Anderson, R. S. (1995). An evolutionary perspective of diversity in Curculionoidea. *Memoirs of the Entomological Society of Washington, 14*, 103–114.
Baker, W. J., & Couvreur, T. L. P. (2013). Global biogeography and diversification of palms sheds light on the evolution of tropical lineages. I. Historical biogeography. *Journal of Biogeography, 40*, 274–285.
Bapst, D. W. (2012). Paleotree: An R package for paleontological and phylogenetic analyses of evolution. *Methods in Ecology and Evolution, 3*, 803–807.
Benton, M. J. (2010). The origins of modern biodiversity on land. *Philosophical Transactions of the Royal Society B, 365*, 3667–3679.
Bernays, E. A. (1998). Evolution of feeding behavior in insect herbivores. *Bioscience, 48*, 35–44.
Bernays, E. A., & Chapman, R. F. (1994). *Host-plant selection by phytophagous insects*. London: Chapman & Hall.
Betancur-R, R., Ortí, G., & Pyron, A. R. (2015). Fossil-based comparative analyses reveal ancient marine ancestry erased by extinction in ray-finned fishes. *Ecology Letters, 18*, 441–450.
Bouchenak-Khelladi, Y., Maurin, O., Hurter, J., & van der Bank, M. (2010). The evolutionary history and biogeography of Mimosoideae (Leguminosae): An emphasis on African acacias. *Molecular Phylogenetics and Evolution, 57*, 495–508.
Buerki, S., Forest, F., Alvarez, N., Nylander, J. A. A., Arrigo, N., & Sanmartín, I. (2011). An evaluation of new parsimony-based versus parametric inference methods in biogeography: A case study using the globally distributed plant family Sapindaceae. *Journal of Biogeography, 38*, 531–550.
Cardinal, S., & Danforth, B. N. (2013). Bees diversified in the age of eudicots. *Proceedings of the Royal Society B, 280*. http://dx.doi.org/10.1098/rspb.2012.2686.
Celorio-Mancera, M. P., Wheat, C. W., Huss, M., Vezzi, F., Neethiraj, R., Reimegard, J., … Janz, N. (2016). Evolutionary history of host use, rather than plant phylogeny, determines gene expression in a generalist butterfly. *BMC Evolutionary Biology, 16*, 59.
Clapham, M. E., Karr, J. A., Nicholson, D. B., Ross, A. J., & Mayhew, P. J. (2016). Ancient origin of high taxonomic richness among insects. *Proceedings of the Royal Society B, 283*. http://dx.doi.org/10.1098/rspb.2015.2476.
Condamine, F. L., Clapham, M. E., & Kergoat, G. J. (2016). Global patterns of insect diversification: Towards a reconciliation of fossil and molecular evidence? *Scientific Reports, 6*, 19208.
Condamine, F. L., Rolland, J., & Morlon, H. (2013). Macroevolutionary perspectives to environmental change. *Ecology Letters, 16*, 72–85.
Condamine, F. L., Silva-Brandao, K., Kergoat, G. J., & Sperling, F. A. H. (2012). Biogeographic and diversification patterns of Neotropical Troidini butterflies (Papilionidae) support a museum model of diversity dynamics for Amazonia. *BMC Evolutionary Biology, 12*, 82.
Condamine, F. L., Sperling, F. A. H., & Kergoat, G. J. (2013). Global biogeographic patterns of swallowtails' diversification resolve their use of alternative colonization routes in Northern and Southern hemispheres. *Journal of Biogeography, 40*, 9–23.
Condamine, F. L., Sperling, F. A. H., Wahlberg, N., Rasplus, J.-Y., & Kergoat, G. J. (2012). What causes latitudinal gradients in species diversity? Evolutionary processes and ecological constraints on swallowtail biodiversity. *Ecology Letters, 15*, 267–277.

Condamine, F. L., Toussaint, E. F. A., Cotton, A. M., Sperling, F. A. H., Genson, G., & Kergoat, G. J. (2013). Fine-scale biogeographical and temporal diversification processes of peacock swallowtails (*Papilio* subgenus *Achillides*) in the Indo-Australian Archipelago. *Cladistics, 29*, 88–111.

Couvreur, T. L. P., Pirie, M. D., Chatrou, L. W., Saunders, R. M. K., Su, Y. C. F., Richarson, J. E., … Erkens, R. H. J. (2011). Early evolutionary history of the flowering plant family Annonaceae: Steady diversification and boreotropical geodispersal. *Journal of Biogeography, 38*, 664–680.

Cruaud, A., Ronsted, N., Chantarasuwan, B., Chou, L. S., Clement, W. L., Couloux, A., … Savolainen, V. (2012). An extreme case of plant-insect codiversification: Figs and fig-pollinating wasps. *Systematic Biology, 61*, 1029–1047.

Daly, H. V., Doyen, J. T., & Purcell, A. H., III (1998). *Introduction to insect biology and diversity* (2nd ed.). Oxford, NY: Oxford University Press.

Davis, R. B., Nicholson, D. B., Saunders, E. L. R., & Mayhew, P. J. (2011). Fossil gaps inferred from phylogenies alter the apparent nature of diversification in dragonflies and their relatives. *BMC Evolutionary Biology, 11*, 252.

De-Silva, D. L., Elias, M., Willmott, K., Mallet, J., & Day, J. J. (2015). Diversification of the clearwing butterflies with the rise of the Andes. *Journal of Biogeography, 43*, 44–58.

von Dohlen, C. D., Kurosu, U., & Aoki, S. (2002). Phylogenetics and evolution of the eastern Asian-eastern North American disjunct aphid tribe, Hormaphidini (*Hemiptera: Aphididae*). *Molecular Phylogenetics and Evolution, 23*, 257–267.

Donoghue, M. J., & Moore, B. R. (2003). Towards an integrative historical biogeography. *Integrative and Comparative Biology, 43*, 261–270.

Doyon, J.-P., Ranwez, V., Daubin, V., & Berry, V. (2011). Models, algorithms and programs for phylogeny reconciliation. *Briefings in Bioinformatics, 12*, 392–400.

Drummond, A. J., Ho, S. Y. W., Phillips, M. J., & Rambaut, A. (2006). Relaxed phylogenetics and dating with confidence. *PLoS Biology, 4*, e88.

Drummond, A. J., Suchard, M. A., Xie, D., & Rambaut, A. (2012). Bayesian phylogenetics with BEAUti and the BEAST 1.7. *Molecular Biology and Evolution, 29*, 1969–1973.

Edger, P. P., Heidel-Fischer, H. M., Bekaert, M., Rota, J., Glöckner, G., Plats, A. E., … Wheat, C. W. (2015). The butterfly plant arms-race escalated by gene and genome duplications. *Proceedings of the National Academy of Sciences of the United States of America, 112*, 8362–8366.

Ehrlich, P. R., & Raven, P. H. (1964). Butterflies and plants: A study in coevolution. *Evolution, 18*, 586–608.

Etheridge, K. (2010). Maria Sibylla Merian and the metamorphosis of natural history. *Endeavour, 35*, 15–21.

Faith, D. P., & Cranston, P. S. (1991). Could a cladogram this short have arisen by chance alone? On permutation tests for cladistic structure. *Cladistics, 7*, 1–28.

Farrell, B., & Mitter, C. (1990). Phylogenesis of insect/plant interactions: have *Phyllobrotica* leaf beetles (Chrysomelidae) and the Lamiales diversified in parallel? *Evolution, 44*, 1389–1403.

Farrell, B. D. (1998). "Inordinate fondness" explained: Why are there so many beetles? *Science, 281*, 555–559.

Farrell, B. D., & Sequeira, A. S. (2004). Evolutionary rates in the adaptive radiation of beetles on plants. *Evolution, 58*, 1984–2001.

FitzJohn, R. G. (2012). Diversitree: Comparative phylogenetic analyses of diversification in R. *Methods in Ecology and Evolution, 3*, 1084–1092.

Forister, M. L., Dyer, L. A., Singer, M. S., Stireman, J. O., 3rd, & Lill, J. T. (2012). Revisiting the evolution of ecological specialization, with emphasis on insect-plant interactions. *Ecology, 93*, 981–991.

Fricke, C., & Arnqvist, G. (2007). Rapid adaptation to a novel host in a seed beetle (*Callosobruchus maculatus*): The role of sexual selection. *Evolution, 61*, 440–454.

Fritz, S. A., Schnitzler, J., Eronen, J. T., Hof, C., Böhning-Gaese, K., & Graham, C. H. (2013). Diversity in time and space: Wanted dead and alive. *Trends in Ecology & Evolution, 28*, 509–516.

Funk, D. J., Filchak, K. E., & Feder, J. L. (2002). Herbivorous insects: Model systems for the comparative study of speciation ecology. *Genetica, 116*, 251–267.

Funk, D. J., Futuyma, D. J., Orti, G., & Meyer, A. (1995). A history of host association and evolutionary diversification for *Ophraella* (Coleoptera: Chrysomelidae): New evidence form mitochondrial DNA. *Evolution, 49*, 1008–1017.

Futuyma, D. J. (1991). Evolution of host specificity in herbivorous insects: Genetic, ecological, and phylogenetic aspects. In P. W. Price, T. M. Lewinsohn, G. W. Fernandes, & W. W. Benson (Eds.), *Plant-animal interactions: Evolutionary ecology in tropical and temperate regions* (pp. 431–454). New York: Wiley, NY, USA.

Futuyma, D. J., & Agrawal, A. A. (2009). Macroevolution and the biological diversity of plants and herbivores. *Proceedings of the National Academy of Sciences of the United States of America, 106*, 18054–18061.

Futuyma, D. J., & Moreno, G. (1988). The evolution of ecological specialization. *Annual Review of Ecology, Evolution and Systematics, 19*, 207–233.

Garcia-Robledo, C., Erickson, D. L., Staines, C. K., Erwin, T. L., & Kress, W. J. (2013). Tropical plant-herbivores networks: Reconstructing species interactions using DNA barcodes. *PLoS One, 8*, e52967.

Gaston, K. J., Reavy, D., & Valladares, G. R. (1992). Intimacy and fidelity: Internal and external feeding by the British Microlepidoptera. *Ecological Entomology, 17*, 86–88.

Gómez-Zurita, J., Hunt, T., Kopliku, F., & Vogler, A. (2007). Recalibrated tree of leaf beetles (Chrysomelidae) indicates independent diversification of angiosperms and their insect herbivores. *PLoS One, 2*, e360.

Grimaldi, D., & Engel, M. S. (2005). *Evolution of the insects*. New York, NY: Cambridge Univ. Press.

Hamm, C. A., & Fordyce, J. A. (2015). Patterns of host plant utilization and diversification in the brush-footed butterflies. *Evolution, 69*, 589–601.

Hardy, N. B., & Otto, S. P. (2014). Specialization and generalization in the diversification of phytophagous insects: Tests of the musical chairs and oscillation hypotheses. *Proceedings of the Royal Society B, 281*, 1795.

Hardy, N. B., Peterson, D. A., & Normark, B. B. (2015). Scale insect host ranges are broader in the tropics. *Biology Letters, 11*. http://dx.doi.org/10.1098/rsbl.2015.0924.

Ho, S. Y. W., & Phillips, M. J. (2009). Accounting for calibration uncertainty in phylogenetic estimation of evolutionary divergence times. *Systematic Biology, 58*, 367–380.

Hunt, T., Bergsten, J., Levkanicova, Z., Papadopoulou, A., John, O. S., Wild, R., … Vogler, A. P. (2007). A comprehensive phylogeny of beetles reveals the evolutionary origins of a superradiation. *Science, 318*, 1913–1916.

Jaenike, J. (1990). Host specialization in phytophagous insects. *Annual Review of Ecology, Evolution and Systematics, 21*, 243–273.

Janz, N. (2011). Ehrlich and Raven revisited: Mechanisms underlying codiversification of plants and enemies. *Annual Review of Ecology, Evolution and Systematics, 42*, 71–89.

Janz, N., Braga, M. P., Wahlberg, N., & Nylin, S. (2016). On oscillations and flutterings—A reply to Hamm and Fordyce. *Evolution, 70*, 1150–1155.

Janz, N., Nyblom, K., & Nylin, S. (2001). Evolutionary dynamics of host-plant specialization: a case study of the tribe Nymphalini. *Evolution, 55*, 783–796.

Janz, N., & Nylin, S. (2008). The oscillation hypothesis of host-plant range and speciation. In D. Tilman (Ed.), *The evolutionary biology of herbivorous insects: Specialization, speciation and radiation* (pp. 203–215). University of California Press.

Janz, N., Nylin, S., & Wahlberg, N. (2006). Diversity begets diversity: Host expansions and the diversification of plant-feeding insects. *BMC Evolutionary Biology, 6*, 4.

Jousselin, E., Cruaud, A., Genson, G., Chevenet, F., Foottit, R. G., & d'acier, A. C. (2013). Is ecological speciation a major trend in aphids? Insights from a molecular phylogeny of the conifer-feeding genus *Cinara*. *Frontiers in Zoology, 10*, 56.

Jousselin, E., Genson, G., & Coeur d'acier, A. (2010). Evolutionary lability of a complex life cycle in the aphid genus *Brachycaudus*. *BMC Evolutionary Biology, 10*, 295.

Jurado-Rivera, J. A., Vogler, A. P., Reid, C. A., Petitpierre, E., & Gómez-Zurita, J. (2009). DNA barcoding insect-host plant associations. *Proceedings of the Royal Society B, 276*, 639–648.

Kajtoch, L., Kubisz, D., Heise, W., Mazur, M. A., & Babik, W. (2015). Plant–herbivorous beetle networks: Molecular characterization of trophic ecology within a threatened steppic environment. *Molecular Ecology, 24*, 4023–4038.

Kawecki, T. J., Lenski, R. E., Ebert, D., Hollis, B., Olivieri, I., & Whitlock, M. C. (2012). Experimental evolution. *Trends in Ecology & Evolution, 27*, 547–560.

Kelley, S. T., & Farrell, B. D. (1998). Is specialization a dead end? The phylogeny of host use in *Dendroctonus* bark beetles (Scolytidae). *Evolution, 52*, 1731–1743.

Kergoat, G. J., Bouchard, P., Clamens, A.-L., Abbate, J. L., Jourdan, H., Zahab, R., & Condamine, F. L. (2014). Cretaceous environmental changes led to high turnovers and decreases in diversification rates in a hyperdiverse non-phytophagous beetle family. *BMC Evolutionary Biology, 14*, 220.

Kergoat, G. J., Delobel, A., Le Ru, B., & Silvain, J.-F. (2008). Seed-beetles in the age of the molecule: Recent advances on systematics and host-plant association patterns. In P. Jolivet, J. Santiago-Blay, & M. Schmitt (Eds.), *Researches on Chrysomelidae* (Vol. 1, pp. 59–86). Leiden: Brill.

Kergoat, G. J., Le Ru, B. P., Genson, G., Cruaud, C., Couloux, A., & Delobel, A. (2011). Phylogenetics, species boundaries and timing of resource tracking in a highly specialized group of seed beetles (Coleoptera: Chrysomelidae: Bruchinae). *Molecular Phylogenetics and Evolution, 59*, 746–760.

Kergoat, G. J., Le Ru, B. P., Sadeghi, S. E., Tuda, M., Reid, C. A. M., György, Z., … Delobel, A. (2015). Evolution of *Spermophagus* seed beetles (Coleoptera, Bruchinae, Amblycerini) indicates both synchronous and delayed colonizations of host plants. *Molecular Phylogenetics and Evolution, 89*, 91–103.

Kergoat, G. J., Prowell, D. P., Le Ru, B. P., Mitchell, A., Dumas, P., Clamens, A.-L., … Silvain, J. F. (2012). Disentangling dispersal and vicariance patterns in armyworms: Evolution and historical biogeography of the pest genus *Spodoptera* (Lepidoptera: Noctuidae). *Molecular Phylogenetics and Evolution, 65*, 855–870.

Kim, H., Lee, S., & Jang, Y. (2011). Macroevolutionary patterns in the Aphidini aphids (Hemiptera: Aphididae): Diversification, host association, and biogeographic origins. *PLoS One, 6*, e24749.

Labandeira, C. C. (2005). The fossil record of insect extinction: New approaches and future directions. *American Entomologist, 51*, 14–29.

Labandeira, C. C. (2006). The four phases of plant-arthropod associations in deep time. *Geologica Acta, 4*, 409–438.

Labandeira, C. C. (2007). The origin of herbivory on land: The initial pattern of live tissue consumption by arthropods. *Insect Science, 14*, 259–274.

Labandeira, C. C., & Currano, E. D. (2013). The fossil record of plant-insect dynamics. *Annual Review of Earth Planetary Science, 41*, 287–311.

Labandeira, C. C., Dilcher, D. L., Davis, D. R., & Wagner, D. L. (1994). Ninety-seven million years of angiosperm-insect association: Paleobiological insights into the meaning of coevolution. *Proceedings of the National Academy of Sciences of the United States of America, 91*, 12278–12282.

Labandeira, C. C., & Sepkoski, J. J. (1993). Insect diversity in the fossil record. *Science, 261*, 310–315.
Lartillot, N., Lepage, T., & Blanquart, S. (2009). PhyloBayes 3: A Bayesian software package for phylogenetic reconstruction and molecular dating. *Bioinformatics, 25*, 2286–2288.
Le Gall, P., & Nel, A. (2013). Paléoentomologie des insectes associés aux plantes. In N. Sauvion, P.-A. Calatayud, D. Thiéry, & F. Marion-Pol (Eds.), *Interactions insectes-plantes* (pp. 23–31). Versailles: Quae.
Leppanen, S. A., Altenhofer, E., Liston, A. D., & Nyman, T. (2012). Phylogenetics and evolution of host-plant use in leaf-mining sawflies (Hymenoptera: Tenthredinidae: Heterarthrinae). *Molecular Phylogenetics and Evolution, 64*, 331–341.
Lewis, D. S., Sperling, F. A. H., Nakahara, S., Cotton, A. M., Kawahara, A. Y., & Condamine, F. L. (2015). Role of Caribbean islands in the diversification and biogeography of Neotropical *Heraclides* swallowtails. *Cladistics, 31*, 291–314.
Lloyd, G. T., Davis, K. E., Pisani, D., Tarver, J. E., Ruta, M., Sakamoto, M., … Benton, M. J. (2008). Dinosaurs and the Cretaceous terrestrial revolution. *Proceedings of the Royal Society B, 275*, 2483–2490.
Lopez-Vaamonde, C., Wikström, N., Labandeira, C., Godfray, H. C. J., Goodman, S. J., & Cook, J. M. (2006). Fossil-calibrated molecular phylogenies reveal that leaf-mining moths radiated millions of years after their host plants. *Journal of Evolutionary Biology, 19*, 1314–1326.
Maddison, W. P., Midford, P. E., & Otto, S. P. (2007). Estimating a binary character's effect on speciation and extinction. *Systematic Biology, 56*, 701–710.
Mao, K., Milne, R. I., Zhang, L., Peng, Y., Liu, J., Thomas, P., … Renner, S. S. (2012). Distribution of living Cupressaceae reflects the breakup of Pangea. *Proceedings of the National Academy of Sciences of the United States of America, 109*, 7793–7798.
Martínez-Aquino, A. (2016). Phylogenetic framework for coevolutionary studies: A compass for exploring jungles of tangled trees. *Current Zoology, 62*, 393–403.
Marvaldi, A. E., Sequeira, A. S., O'Brien, C. W., & Farrell, B. D. (2002). Molecular and morphological phylogenetics of weevils (Coleoptera, Curculionoidea): Do niche shifts accompany diversification? *Systematic Biology, 51*, 761–785.
Matsubayashi, K. W., Ohshima, I., & Nosil, P. (2010). Ecological speciation in phytophagous insects. *Entomologia Experimentalis et Applicata, 134*, 1–27.
Mayhew, P. J. (2007). Why are there so many insect species? Perspectives from fossils and phylogenies. *Biological Reviews of the Cambridge Philosophical Society, 82*, 425–454.
McKenna, D. D., Sequeira, A. S., Marvaldi, A. E., & Farrell, B. D. (2009). Temporal lags and overlap in the diversification of weevils and flowering plants. *Proceedings of the National Academy of Sciences of the United States of America, 106*, 7083–7088.
McLeish, M. J., Chapman, T. W., & Schwarz, M. P. (2007). Host-driven diversification of gall-inducing *Acacia* thrips and the aridification of Australia. *BMC Biology, 5*, 3.
McLeish, M. J., Miller, J. T., & Mound, L. A. (2013). Delayed colonisation of *Acacia* by thrips and the timing of host-conservatism and behavioural specialisation. *BMC Evolutionary Biology, 13*, 188.
McLoughlin, S., Martin, S. K., & Beattie, R. (2015). The record of Australian Jurassic plant-arthropod interactions. *Gondwana Research, 27*, 940–959.
Menken, S. B. J., Boomsma, J. J., & van Nieukerken, E. J. (2010). Large-scale evolutionary patterns of host plant associations in the Lepidoptera. *Evolution, 64*, 1098–1119.
Meseguer, A. S., Coeur d'acier, A., Genson, G., & Jousselin, E. (2015). Unravelling the historical biogeography and diversification dynamics of a highly diverse conifer-feeding aphid genus. *Journal of Biogeography, 42*, 1482–1492.
Meseguer, A. S., Lobo, J. M., Ree, R. H., Beerling, D. J., & Sanmartín, I. (2015). Integrating fossils, phylogenies, and niche models into biogeography to reveal ancient

evolutionary history: The case of *Hypericum* (Hypericaceae). *Systematic Biology, 64*, 215–232.
Messina, F. J., & Durham, S. L. (2013). Adaptation to a novel host by a seed beetle (Coleoptera: Chrysomelidae: Bruchinae): Effect of source population. *Environmental Entomology, 42*, 733–742.
Messina, F. J., & Johnson, D. (2014). Specificity of adaptation to a novel host plant by a seed beetle. *Entomologia Experimentalis et Applicata, 153*, 231–239.
Messina, F. J., Jones, J. C., Mendenhall, M., & Muller, A. (2009). Genetic modification of host acceptance by a seed beetle, *Callosobruchus maculatus* (Coleoptera: Bruchidae). *Annals of the Entomological Society of America, 102*, 181–188.
Mitter, C., Farrell, B., & Futuyma, D. J. (1991). Phylogenetic studies of insect-plant interactions: Insights into the genesis of diversity. *Trends in Ecology & Evolution, 6*, 290–293.
Mitter, C., Farrell, B. D., & Wiegmann, B. (1988). The phylogenetic study of adaptive zones: Has phytophagy promoted insect diversification? *The American Naturalist, 132*, 107–128.
Moran, N. A. (1988). The evolution of host alternation in aphids: Evidence that specialization is a dead end. *The American Naturalist, 132*, 681–706.
Moran, N. A. (1992). The evolution of aphid life cycles. *Annual Review of Entomology, 37*, 321–348.
Moreau, C. S., & Bell, C. D. (2013). Testing the museum versus cradle tropical biological diversity hypothesis: Phylogeny, diversification, and ancestral biogeographic range evolution of the ants. *Evolution, 67*, 2240–2257.
Morlon, H. (2014). Phylogenetic approaches for studying diversification. *Ecology Letters, 17*, 508–525.
Morlon, H., Potts, M. D., & Plotkin, J. B. (2010). Inferring the dynamics of diversification: A coalescent approach. *PLoS Biology, 8*, e1000493.
Morse, G. E., & Farrell, B. D. (2005). Ecological and evolutionary diversification of the seed beetle genus *Stator* (Coleoptera: Chrysomelidae: Bruchinae). *Evolution, 59*, 1315–1333.
Navarro, S. P., Jurado-Rivera, J., Gómez-Zurita, J., Lyal, C. H. C., & Vogler, A. (2010). DNA profiling of host-herbivore interactions in tropical forests. *Ecological Entomology, 35*, 18–32.
Nee, S., Mooers, A., & Harvey, P. H. (1992). Tempo and mode of evolution revealed from molecular phylogenies. *Proceedings of the National Academy of Sciences of the United States of America, 89*, 8322–8326.
Nicholson, D. B., Mayhew, P. J., & Ross, A. J. (2015). Changes to the fossil record of insects through fifteen years of discovery. *PLoS One, 10*, e0128554.
Nicholson, D. B., Ross, A. J., & Mayhew, P. J. (2014). Fossil evidence for key innovations in the evolution of insect diversity. *Proceedings of the Royal Society B, 281*, 20141823.
Nosil, P. (2002). Transition rates between specialization and generalization in phytophagous insects. *Evolution, 56*, 1701–1706.
Nosil, P., Crespi, B. J., & Sandoval, C. P. (2002). Host-plant adaptation drives the parallel evolution of reproductive isolation. *Nature, 417*, 440–443.
Nosil, P., & Mooers, A. O. (2005). Testing hypotheses about ecological specialization using phylogenetic trees. *Evolution, 59*, 2256–2263.
Nylin, S., & Janz, N. (2009). Butterfly host plant range: An example of plasticity as a promoter of speciation? *Evolutionary Ecology, 23*, 137–146.
Nylin, S., Slove, J., & Janz, N. (2014). Host plant utilization, host range oscillations and diversification in nymphalid butterflies: A phylogenetic investigation. *Evolution, 68*, 105–124.
Nyman, T., Farrell, B. D., Zinovjev, A. G., & Vikberg, V. (2006). Larval habits, host-plant associations, and speciation in nematine sawflies (Hymenoptera, Tenthredinidae). *Evolution, 60*, 1622–1637.
Nyman, T., Linder, H. P., Pena, C., Malm, T., & Wahlberg, N. (2012). Climate-driven diversity dynamics in plants and plant-feeding insects. *Ecology Letters, 15*, 889–898.

Nyman, T., Vikberg, V., Smith, D. R., & Boeve, J. L. (2010). How common is ecological speciation in plant-feeding insects? A 'Higher' Nematinae perspective. *BMC Evolutionary Biology, 10*, 266.

Ouvrard, D., Chalise, P., & Percy, D. M. (2015). Host-plant leaps versus host-plant shuffle: A global survey reveals contrasting patterns in an oligophagous insect group (Hemiptera, Psylloidea). *Systematics and Biodiversity, 13*, 1–21.

Pagel, M. D. (1999). The maximum likelihood approach to reconstructing ancestral character states of discrete characters on phylogenies. *Systematic Biology, 48*, 612–622.

Pagel, M., & Meade, A. (2006). Bayesian analysis of correlated evolution of discrete characters by reversible-jump Markov chain Monte Carlo. *The American Naturalist, 167*, 808–825.

Pagel, M. D., Meade, A., & Barker, D. (2004). Bayesian estimation of ancestral character states on phylogenies. *Systematic Biology, 53*, 673–684.

Parham, J. F., Donoghue, P. C. J., Bell, C. J., Calway, T. D., Head, J. J., Holroyd, P. A., … Benton, M. J. (2012). Best practices for justifying fossil calibrations. *Systematic Biology, 61*, 346–359.

Peccoud, J., Simon, J.-C., von Dohlen, C., Coeur d'acier, A., Plantegenest, M., Vanlerberghe-Masutti, F., & Jousselin, E. (2010). Evolutionary history of aphid-plant associations and their role in aphid diversification. *Comptes Rendus Biologies, 333*, 474–487.

Pena, C., & Espeland, M. (2015). Diversity dynamics in Nymphalidae butterflies: effect of phylogenetic uncertainty on diversification rate shift estimates. *PLoS One, 10*, e0120928.

Pena, C., & Wahlberg, N. (2008). Prehistorical climate change increased diversification of a group of butterflies. *Biology Letters, 4*, 274–278.

Pennell, M. W., & Harmon, L. J. (2013). An integrative view of phylogenetic comparative methods: Connections to population genetics, community ecology, and paleobiology. *Annals of the New York Academy of Sciences, 1289*, 90–105.

Percy, D. M., Page, R. D., & Cronk, Q. C. B. (2004). Plant-insect interactions: Double-dating associated insect and plant lineages reveals asynchronous radiations. *Systematic Biology, 53*, 120–127.

Pyron, R. A. (2010). A likelihood method for assessing molecular divergence time estimates and the placement of fossil calibrations. *Systematic Biology, 59*, 185–194.

Pyron, R. A. (2011). Divergence time estimation using fossils as terminal taxa and the origins of Lissamphibia. *Systematic Biology, 60*, 466–481.

Rabosky, D. L. (2014). Automatic detection of key innovations, rate shifts, and diversity-dependence on phylogenetic trees. *PLoS One, 9*, e89543.

Rainford, J. L., Hofreiter, M., Nicholson, D. B., & Mayhew, P. J. (2014). Phylogenetic distribution of extant richness suggests metamorphosis is a key innovation driving diversification in insects. *PLoS One, 9*, e109085.

Rainford, J. L., & Mayhew, P. (2015). Diet evolution and clade richness in Hexapoda: A phylogenetic study of higher taxa. *The American Naturalist, 186*, 777–791.

Ree, R. H., Moore, B. R., Webb, C. O., & Donoghue, M. J. (2005). A likelihood framework for inferring the evolution of geographic range on phylogenetic trees. *Evolution, 59*, 2299–2311.

Ree, R. H., & Smith, S. A. (2008). Maximum likelihood inference of geographic range evolution by dispersal, local extinction, and cladogenesis. *Systematic Biology, 57*, 4–14.

Ren, Z. M., Zhong, Y., Kurosu, U., Aoki, S., Ma, E. B., von Dohlen, C. D., & Wen, J. (2013). Historical biogeography of eastern Asian-eastern North American disjunct Melaphidina aphids (Hemiptera: Aphididae: Eriosomatinae) on *Rhus* hosts (Anacardiaceae). *Molecular Phylogenetics and Evolution, 69*, 1146–1158.

Revell, L. J. (2013). Phytools: An R package for phylogenetic comparative biology (and other things). *Methods in Ecology and Evolution, 3*, 217–223.

Ronquist, F., Klopfstein, S., Vilhelmsen, L., Schulmeister, S., Murray, D. L., & Rasnitsyn, A. P. (2012). A total-evidence approach to dating with fossils, applied to the early radiation of the Hymenoptera. *Systematic Biology, 61*, 973–999.
Ronquist, F., & Sanmartín, I. (2011). Phylogenetic methods in biogeography. *Annual Review of Ecology, Evolution and Systematics, 42*, 441–464.
Rota, J., Pena, C., & Miller, S. E. (2016). The importance of long-distance dispersal and establishment events in small insects: Historical biogeography of metalmark moths (Lepidoptera, Choreutidae). *Journal of Biogeography, 43*, 1254–1265.
Sanmartín, I. (2014). Biogeography. In P. Vargas, & R. Zardoya (Eds.), *The tree of life* (pp. 156–166). Sunderland: Sinauer Associates.
Sanmartín, I., Enghoff, H., & Ronquist, F. (2001). Patterns of animal dispersal, vicariance and diversification in the Holarctic. *Biological Journal of the Linnean Society, 73*, 345–390.
Sanmartín, I., & Meseguer, A. S. (2016). Extinction in phylogenetics and biogeography: From lineages to the assembly of regional biotas. *Frontiers in Genetics, 7*, 35.
Schaefer, H., Hechenleitner, P., Santos-Guerra, A., de Sequeira, M. M., Pennington, R. T., Kenicer, G., & Carine, M. A. (2012). Systematics, biogeography, and character evolution of the legume tribe Fabeae with special focus on the middle-Atlantic island lineages. *BMC Evolutionary Biology, 12*, 250.
Schatz, B., Sauvion, N., Kjellberg, F., & Nel, A. (2017). Plant–insect interactions: a palaeontological and an evolutionary perspective. In N. Sauvion, D. Thiery, & P. A. Calatayud (Eds.), *Insect-Plant Interactions in a Crop Protection Perspective* (Vol. 81, pp. 1–53).
Silvestro, D., Cascales-Minana, B., Bacon, C. D., & Antonelli, A. (2015). Revisiting the origin and diversification of vascular plants through a comprehensive Bayesian analysis of the fossil record. *New Phytologist, 207*, 425–436.
Silvestro, D., Salamin, N., & Schnitzler, J. (2014). PyRate: A new program to estimate speciation and extinction rates from incomplete fossil record. *Methods in Ecology and Evolution, 5*, 1126–1131.
Silvestro, D., Schnitzler, J., Liow, L. H., Antonelli, A., & Salamin, N. (2014). Bayesian estimation of speciation and extinction from incomplete fossil occurrence data. *Systematic Biology, 63*, 349–367.
Simon, J.-C., d'Alençon, E., Guy, E., Jacquin-Joly, E., Jaquiéry, J., Nouhaud, P., … Streiff, R. (2015). Genomics of adaptation to host-plants in herbivorous insects. *Briefing in Functional Genomics, 14*, 413–423.
Slater, G. J., & Harmon, L. J. (2013). Unifying fossils and phylogenies for comparative analyses of diversification and trait evolution. *Methods in Ecology and Evolution, 4*, 699–702.
Slater, G. J., Harmon, L. J., & Alfaro, M. E. (2012). Integrating fossils with molecular phylogenies improves inference of trait evolution. *Evolution, 66*, 3931–3944.
Sohn, J.-C., Labandeira, C. C., & Davis, D. R. (2015). The fossil record and taphonomy of butterflies and moths (Insecta, Lepidoptera): Implications for evolutionary diversity and divergence-time estimates. *BMC Evolutionary Biology, 15*, 12.
Solodovnikov, A., Yue, Y., Tarasov, S., & Ren, D. (2013). Extinct and extant rove beetles meet in the matrix: Early Cretaceous fossils shed light on the evolution of a hyperdiverse insect lineage (Coleoptera: Staphylinidae: Staphylininae). *Cladistics, 29*, 360–403.
Song, H., Amédégnato, C., Cigliano, M. M., Desutter-Grandcolas, L., Heads, S. W., Huang, Y., … Whiting, M. F. (2015). 300 million years of diversification: Elucidating the patterns of orthopteran evolution based on comprehensive taxon and gene sampling. *Cladistics, 31*, 621–651.
Southwood, T. R. E. (1973). The insect/plant relationship — An evolutionary perspective. *Symposium of the Royal Entomological Society London, 6*, 3–30.
Stadler, T. (2011). Mammalian phylogeny reveals recent diversification rate shifts. *Proceedings of the National Academy of Sciences of the United States of America, 108*, 6187–6192.

Stone, G. N., Hernandez-Lopez, A., Nicholls, J. A., di Pierro, E., Pujade-Villar, J., Melika, G., & Cook, J. M. (2009). Extreme host plant conservatism during at least 20 million years of host plant pursuit by oak gallwasps. *Evolution, 63*, 854–869.
Strong, D. R., Lawton, J. H., & Southwood, T. R. E. (1984). *Insects on plants: Community patterns and mechanisms*. Cambridge, Massachusset: Harvard University Press.
Suchan, T., & Alvarez, N. (2015). Fifty years after Ehrlich and Raven, is there support for plant–insect coevolution as a major driver of species diversification? *Entomologia Experimentalis et Applicata, 157*, 98–112.
Thompson, J. N. (1988). Evolutionary genetics of oviposition preference in swallowtail butterflies. *Evolution, 42*, 1223–1234.
Thompson, J. N. (1994). *The coevolutionary process*. Chicago: Chicago University Press.
Toussaint, E. F. A., & Balke, M. (2016). Historical biogeography of *Polyura* butterflies in the oriental Palaeotropics: Trans-archipelagic routes and South Pacific island hopping. *Journal of Biogeography, 43*, 1560–1572.
Toussaint, E. F. A., Condamine, F. L., Kergoat, G. J., Silvain, J.-F., Capdevielle-Dulac, C., Barbut, J., & Le Ru, B. P. (2012). Palaeoenvironmental shifts drove the adaptive radiation of a noctuid stemborer tribe (Lepidoptera, Noctuidae, Apameini) in the Miocene. *PLoS One, 7*, e41377.
Tuda, M., Kagoshima, K., Toquenaga, Y., & Arnqvist, G. (2014). Global genetic differentiation in a cosmopolitan pest of stored beans: Effects of geography, host-plant usage and anthropogenic factors. *PLoS One, 9*, e106268.
Vea, I. M., & Grimaldi, D. A. (2016). Putting scales into evolutionary time: The divergence of major scale insect lineages (Hemiptera) predates the radiation of modern angiosperm hosts. *Scientific Reports, 6*, 23487.
Via, S., Bouck, A. C., & Skillman, S. (2000). Reproductive isolation between divergent races of pea aphids on two hosts. II. Selection against migrants and hybrids in the parental environments. *Evolution, 54*, 1626–1637.
de Vienne, D. M., Refregier, G., Lopez-Villavicencio, M., Tellier, A., Hood, M. E., & Giraud, T. (2013). Cospeciation vs host-shift speciation: Methods for testing, evidence from natural associations and relation to coevolution. *New Phytologist, 198*, 347–385.
Vila, R., Bell, C. D., Macniven, R., Goldman-Huertas, B., Ree, R. H., Marshall, C. R., & Pierce, N. E. (2011). Phylogeny and palaeoecology of *Polyommatus* blue butterflies show Beringia was a climate-regulated gateway to the New World. *Proceedings of the Royal Society B, 278*, 2737–2744.
Wahlberg, N., Wheat, C. W., & Peña, C. (2013). Timing and patterns in the taxonomic diversification of Lepidoptera (butterflies and moths). *PLoS One, 8*, e80875.
Walsh, B. D. (1867). The apple-worm and the apple maggot. *The Journal of Horticultural Science and Biotechnology, 2*, 338–343.
Wang, B., Zhang, H., & Jarzembowski, E. A. (2013). Early Cretaceous angiosperms and beetle evolution. *Frontiers in Plant Science, 4*, 360.
Ward, S. A. (1991). Reproduction and host selection by aphids: The importance of 'rendevous' hosts. In W. J. Bailey, & J. Ridsdill-Smith (Eds.), *Reproductive behaviour of Insects: Individuals and populations* (pp. 202–226). London: Chapman & Hall.
Ward, L., & Spalding, D. F. (1993). Phytophagous British insects and mites and their food plant families: Total numbers and polyphagy. *Biological Journal of the Linnean Society, 49*, 257–276.
Wen, J. (1999). Evolution of eastern Asian and eastern North American disjunct distributions in flowering plants. *Annual Review in Ecology, Evolution and Systematics, 30*, 421–455.
Wiens, J. J., Lapoint, R. T., & Whiteman, N. K. (2015). Herbivory increases diversification across insect clades. *Nature Communications, 6*, 8370.

Wilson, J. S., Forister, M. L., Dyer, L. A., O'Connor, J. M., Burls, K., Feldman, C. R., ... Young, B. (2012). Host conservatism, host shifts and diversification across three trophic levels in two Neotropical forests. *Journal of Evolutionary Biology, 25*, 532—546.
Winkler, I. S., & Mitter, C. (2008). The phylogenetic dimension of insect/plant interactions: A summary of recent evidence. In K. Tilmon (Ed.), *Specialization, speciation, and radiation: The evolutionary biology of herbivorous insects* (pp. 240—263). Berkeley: University of California Press.
Winkler, I. S., Mitter, C., & Scheffer, S. J. (2009). Repeated climate-linked host shifts have promoted diversification in a temperate clade of leaf-mining flies. *Proceedings of the National Academy of Sciences of the United States of America, 106*, 18103—18108.
Wood, H. M., Matzke, N. J., Gillespie, R. G., & Griswold, C. E. (2013). Treating fossils as terminal taxa in divergence time estimation reveals ancient vicariance patterns in the palpimanoid spiders. *Systematic Biology, 62*, 264—284.
Xiang, Q. Y., Soltis, D. E., Soltis, P. S., Manchester, S. R., & Crawford, D. J. (2000). Timing the eastern Asian-eastern North American floristic disjunction: Molecular clock corroborates paleontological estimates. *Molecular Phylogenetics and Evolution, 15*, 462—472.
Yokoyama, J. (1995). Insect-plant coevolution and cospeciation. In R. Arai, M. Kato, & Y. Doi (Eds.), *Biodiversity and evolution. Vol. Science Museum Foundation* (pp. 115—130). Tokyo.

CHAPTER THREE

From Plant Exploitation to Mutualism

F. Lieutier*[,1], K. Bermudez-Torres[§], J. Cook[¶], M.O. Harris[||], L. Legal[#], A. Sallé*, B. Schatz, D. Giron[§§]**

*INRA, Laboratoire de Biologie des Ligneux et des Grandes Cultures, Université d'Orléans, Orléans, France
[§]National Polytechnical Institute, Yautepec, Mexico
[¶]Western Sydney University, NSW, Australia
[||]North Dakota State University, Fargo, ND, United States
[#]ECOLAB, UMR 5245 (INP, UPS, CNRS), Ecologie Fonctionnelle et Environnement, Toulouse, France
**CEFE (Centre d'Ecologie Fonctionnelle et Evolutive), UMR 5175 CNRS, Université de Montpellier – Université Paul-Valéry Montpellier – EPHE, Montpellier, France
[§§]IRBI (Insect Biology Research Institute) – UMR 7261 CNRS/Université François-Rabelais de Tours, Tours, France
[1]Corresponding author: E-mail: francois.lieutier@univ-orleans.fr

Contents

Advances in Botanical Research, Volume 81
ISSN 0065-2296
http://dx.doi.org/10.1016/bs.abr.2016.10.001

Abstract

Phytophagous insects have developed mechanisms of various complexity levels to utilize plants in spite of the barriers that plants have built to resist aggressions. Plant exploitation, the simplest level, is the use of plant defence chemicals for the benefit of insects. It is illustrated by the use of plant toxins for defence against predators. The energetic cost of that defence strategy is discussed according to the toxicity of the chemicals and the necessity of protecting the herbivore, and the modes of action on predators are presented. Furthermore, manipulation of the plant can reorient the plant metabolism to satisfy insect needs. Drastic remodelling of the host plant can occur, from ultrastructure to anatomy levels, with alteration of both its nutritional quality and secondary metabolism. The mechanisms involved are being investigated. Outcomes concern optimization of the nutritional value of the host plant and protection from adverse abiotic and biotic (natural enemies, competition) conditions. Cooperation with conspecifics or microorganisms often interferes. At the highest level of complexity, mutualism is the result of a compromise between insect and plant where each partner benefits from the association. Pollination is a typical example. Pollinators vary from generalists to specialists and belong to a community of insect linked to a community of plants. In the fig—fig wasp mutualism, the various mechanisms involved in situations of monoecy and dioecy are discussed, as well as the existence of coadaptations and cospeciations. The chapter ends with a presentation of research perspectives for improving crop productivity.

1. INTRODUCTION

Living plants are source of food for phytophagous insects furnishing them with most of the nutrients required for their development and reproduction. However, plants may lack certain essential compound. Moreover, they possess a wide array of defensive strategies, including physical structures, a large diversity of secondary metabolites and proteins which can interfere with the colonization and development of phytophagous insects. Nevertheless, insects have evolved to cope with these physical and chemical barriers and have developed mechanisms/strategies to avoid and even to use them. The present chapter covers diverse situations of utilization of plant tissues and compounds by insects, representing various levels of relations between insects and their living host plant, including various 'insect strategies'.

The simplest level is plant exploitation. It is more than simple plant utilization and can be defined as a misappropriation of plant defence chemicals for the benefit of insects. A more elaborated level is plant manipulation, where the plant natural mechanisms are reoriented so that the plant system becomes manipulated to serve the insect needs. This can be realized without or with modification of the plant genetic program. At the highest level of plant utilization, a compromise has been established between insects and their host plants, which benefits both partners in a mutual agreement called mutualism. A large array of insect feeding behaviours (from browsers to galling and wood-borers), as well as of plant categories (herbs, trees) can be concerned and are illustrated in the chapter. Finally, applied perspectives are tentatively presented.

2. DEFENCE AGAINST PREDATORS

Insect predation is a complex world involving multiple interactions. Among the phytophagous species, many examples illustrate extreme sophistication in the use of secondary metabolites for their *defence* (Agrawal et al., 2015; Mithöfer et al., 2012; Schuman & Baldwin, 2016), their adaptive processes of evolutionary history-dependent among insect families (Stam et al., 2014) and also in their relationship to host plants (Gols, 2014). Though the term *defence* implies broad protection against a threat, it can be split into active and passive defence.

While many Hymenoptera such as wasps, bees and ants are capable to defend themselves actively with poisonous stings or bites, most insects have developed antipredator behaviours/strategies (Clark & Faeth, 1997; Vulinec, 1990), which is to accumulate or transform active/toxic substances from host plants. However, in few cases, insects can produce de novo antipredatory substances (Pasteels, Duffey, & Rowell-Rahier, 1990). Some intermediate cases are illustrated by some bugs or some beetles which project chemical substances on their predators (Krall, Bartelt, Lewis, & Whitman, 1999). In this part, we will only present the passive chemical antipredation strategies.

2.1 Sequestration or Metabolism?

There are two types of passive chemical antipredation. The first is represented by insects feeding on very poisonous plants (Euphorbiaceae, Asclepiadaceae, Solanaceae) but without accumulating these secondary compounds (Hundsdörfer, Tshibangu, Wetterauer, & Wink, 2005). The flow of toxic compounds in the haemolymph or digestive tract is sufficient to make insects unpalatable to predators. In general, only a stage of the insect (mainly larvae) is protected. In the second, and more common type, the insect accumulates and often concentrates a toxic plant substance. This kind of antipredation strategy is known as *sequestration*, which includes, the selective uptake, transport, modification, storage and deployment of plant secondary metabolites for the insect's own defence (Heckel, 2014). Several studies indicate a relationship between sequestration and the degree of phytophagy specialization, showing that specialists sequester more efficiently than generalist insects (Dobler, 2001; Dyer, 1995; Lampert & Bowers, 2010; Lampert, Dyer, & Bowers, 2014). Mode and sites of accumulation are generally poorly known (Duffey, 1980; Petschenka & Anurag, 2016). Some authors assume the existence of cellular compartments comparable to plant vacuoles (Frick & Wink, 1995), while others hypothesize the existence of a balance between toxicity of chemical compounds and metabolic resistance of the insects. In addition, some insects are specialized in storing chemical compounds in glandular systems (Bowers, 1990). In this type of accumulation, there are three possible processes:

- The insect can sequester a plant substance that is toxic to predators without being affected by it, suggesting that the insect is resistant to the toxin (Scott, Liu, & Wen, 1998).
- More frequently, the insect can sequester a little or nontoxic precursor of a toxic substance. When attacked by a predator the precursor passes from an amorphous to a toxic state, often by air oxidation, similarly to action mode of some phytoalexines in plants (Nishida & Fukami, 1990).
- In few cases, the insect accumulates a chemical precursor from the host plant but performs one or more metabolic steps that transform this substance into a toxin against which it is resistant (Von Nickisch-Rosenegk & Wink, 1993; Wink & Legal, 2001; Wink, Legal, & von Nickisch-Rosenegk, 1998).

2.1.1 Example of Metabolism Excretion

The study by Hundsdörfer et al. (2005) on the Spurge Hawk-moth, *Hyles euphorbiae* (Linnaeus, 1758) (Lepidoptera, Sphingidae), a common species in Europe, is a good illustration of such a situation. The sphingid moths are large nocturnal moths whose caterpillars often have bright colours and have the particularity to feed on very toxic plants such as Euphorbiaceae and Solanaceae (Bernays & Janzen, 1988). These bright colours associated with toxic compounds are called aposematism because they can inform the predators on the toxicity of potential prey. Certain chemical compounds found in Euphorbiaceae are phorbol esters which are among the most potent natural toxins known (Goel, Makkar, Francis, & Becker, 2007). Experiments were performed with caterpillars of *H. euphorbiae* to determine if phorbol ester compounds were sequestered and used as substances for defence against predators (Hundsdörfer et al., 2005). Approximately 80% of the 12-tetradecanoyl-phorbol-13-acetate (TPA) was used for metabolism and the remaining 20% eliminated in the faeces. When TPA was directly injected into caterpillars, almost all of the TPA was found circulating into haemolymph and intestinal tracts. When attacked by a predator, caterpillars protect themselves by regurgitation of their food bowl filled with fragments of euphorbias and therefore still rich in TPA. Aposematic colour is well associated with a hazard for the predator, but it does not automatically mean a phenomenon of sequestration by prey, in the above case just a regular flux of toxin in the intestinal tract is sufficient to provide protection.

2.1.2 Example of Toxic Compounds Sequestration

Brassicaceae are rich in glucosinolates which are toxic compounds or at least antipalatable for many phytophagous. Larvae of the cabbage sawfly, *Athalia rosae* (Linnaeus, 1758) (Hymenoptera, Tenthredinidae), are rarely attacked by predators. It is not known if this species metabolizes (slow circulation strategy, see former case) or sequesters toxic compounds but the larvae, when attacked by a predator, produce small drops of haemolymph excreting it to the integument (Müller et al., 2001). This type of defence is called 'reflex bleeding'. It has been demonstrated that the larvae incorporate and concentrate glucosinolates produced by the host plant (Müller et al., 2001).

2.1.3 Example of Precursor Sequestration and De Novo Biosynthesis

This example is borrowed from the papers by Laurent, Braekman, Daloze, and Pasteels (2003) and Laurent, Dooms, et al. (2003). Among Chrysomelidae, there is a strong dependency on the chemistry of host plants. In many cases, the compounds (pyrolizidic alkaloids, curcurbitacines) come directly from the host plant with little or no changes. Some de novo synthesis have been described, and it is worth noting that the biosynthesis pathways of the various compounds produced by insects are very close or even identical to the pathways found in plants (Wink, 2016).

Some very recent advances suggest that horizontal gene transfer (HGT) may be at the origin of such similarities, with some complete genetic systems almost identical between bacteria/fungi/plants but also insects. These HGTs may be much more common than formerly suspected and represent a new and promising research theme (Wink, 2016).

Only few insects are able to perform for a same defence compound, both de novo biosynthesizes and sequestration from their food plant. The Burnet moth *Zygaena filipendulae* (Linnaeus, 1758) (Lepidoptera, Zygaenidae) is one of these rare examples. Its larvae alternate between de novo biosynthesis and sequestration of cyanide glucosides based on the ingested amount of plant-derived cyanide glycosides from the food-plant *Lotus corniculatus* L. (Fabaceae). Thereby, the de novo biosynthesis is associated with an apparent higher cost or reduced fitness compared to the sequestration from the food-plant (Fürstenberg-Hägg et al., 2014).

2.2 Energy Cost of Chemical Defence

The insect using plant toxics to defend itself against other insects faces a paradox. It is indeed itself basically susceptible to these substances since they act mostly on insect's basal metabolism. It must thus acquire at least a partial resistance to them. Various processes can be involved but all have an energy cost. A balance between the antipredation chemical and efficacy against predators is therefore subject to a negative trade-off. Several strategies aiming at minimizing the cost of this chemical defence are recognized.

- The energetically cheapest strategy is to accumulate products of low toxicity (Pasteels, Theuring, Witte, & Hartmann, 2003). A precursor of the plant toxin is generally drained by the Malpighian tubes of the digestive tract (Tsoupras, Luu, Hetru, Muller, & Hoffmann, 1983). Subsequently, in all known cases, the precursor can stay in the haemolymph or be accumulated in specialized glands. When staying in the haemolymph, the toxin will be produced by an open air oxidation of the precursor during an injury of the insect (Pasteels et al., 1990). A similar mechanism also exists in some cases when the precursor is stored in glands, but one (or a few) additive metabolic step can be necessary in other cases to transform it into a toxin within the glandular tissues (Von Nickisch-Rosenegk & Wink, 1993).

- A more costly strategy is the direct sequestration of the plant toxin itself by the insect. Many hypotheses have been advanced for the storing of such substances. Organized granules are often visible in cells and some authors suspect the sequestration is stored in pseudovacuoles (Wink & Roberts, 1998). Thereby, the toxin is not directly in contact with the potential metabolic targets.
- Finally, some insects have an impressive armada of detoxification enzymes such as esterases and cytochrome P450, which allow a steady stream of toxins in the haemolymph without affecting the metabolism of the insect (Schuler, 1996). A continuous regulated enzyme activity must be balanced with the concentration of ingested toxins. In such cases, the insects are living on the most toxic plants and the host plant specificity is generally very high (Scott & Wen, 2001).

2.3 Modes of Action of Secondary Metabolites on Predators

According to its mode of action, the chemical defence of insects from predators are divided into three broad categories:

- Toxicity: The chemicals can irritate, injure, poison or drug the predators.
- Antiappetence: The defence is based on antipalatable compounds for predators, mainly products that are bitter for the predator's palate.
- Adherence and immobilisation: Products that adhere to the predators and immobilize them are involved.

2.3.1 Toxicity

We can subdivide the actions of the various families of molecules into two broad categories:

- Selective actions of defence molecules that act specifically on a chemical target by changing its configuration and metabolic function.
- Nonselective molecules that cause generalized metabolic disorders in predators.

In the first category, the defence molecules can have multiple targets depending on whether they have an amino group. The positive charged nitrogen can interact with the negative charged groups from glutamic or aspartic acids of proteins. In this way, defence compounds modulate the three-dimensional structure of proteins, inducing a loss or reduction of their metabolic activity, thus modifying the predator metabolism. Within the nonpolar substances without nitrogen, for example, terpenes, an association with lipophilic molecules interferes (Wink, 2016 for review and references therein). In the second category, toxic molecules act mostly by changes in the permeability of cell membranes. Biomembranes are amphipathic consisting of a double layer of phospholipids which is impervious to a variety of external molecules. Within this structure, several types of proteins allow a selective passage of some molecules such as ion channels, transporters and receptors. Several types of toxic molecules such as terpenes, alkaloids and steroids may interact with these proteins, thus modifying the membrane permeability.

An example is given by Eisner et al. (2000), Peschke and Eisner (1987), Palmeira and Wallace (1996), Guo, Reigan, Siegel, and Ross (2008) and Cosby et al. (1976). The bombardier beetle, *Metrius contractus* (Eschscholtz, 1829) (Coleoptera, Carabidae), is capable of projecting on the predators a mixture of products, among them 1, 4-benzoquinone. This product has the advantage of being less volatile and more stable than 2-methyl-1, 4-benzoquinone that is the most common compound produced in other species of bombardiers. These authors have showed that not only the quinone but also the mix of hydrocarbons plays a role in defence against predators. These products were tested and they were shown to be repellent against the ant *Monomorium pharaonis* (Linnaeus, 1758) (Hymenoptera, Formicidae) and irritating on the cockroach *Periplaneta americana* (Linnaeus, 1758) (Blattaria, Blattidae). Using the mitochondria of rat liver, the same authors have showed that several anthraquinones induce a formation of nonspecific pores in the (cell or mitochondria) membrane. In addition, these products inhibit the action of nicotinamide adenine dinucleotide (NADH)-ubiquinone oxidoreductase. Also, Cosby et al. (1976) have demonstrated that naphthoquinones inhibit cancer cell growth. These quinones induce the oxidation of NADH, and one of the results is the blocking of the conversion of thymine preventing DNA synthesis.

2.3.2 Antiappetence

Insects have developed another defence strategy which consists in producing ‘unpleasant’ compounds that are repellent to the predators’ palate. These compounds interact on the predator’s chemical receptors and more specifically on deterrence receptors, informing of the unpalatable characteristic of the prey. Such compounds are found in various types of molecules: terpenes, alkaloids and quinones. A review on the different ways for a predator to recognize antipalatable preys was done by Glendinning (2007).

Eisner, Goetz, Hil, Smedley, and Meinwald (1997), González, Hare, and Eisner (1999), and González, Schroeder, Meinwald, and Eisner (1999) have shown that species of the genus *Photuris* (Coleoptera, Lampyridae) (called 'femmes fatales') attract males of another species of glow worms (of the genus *Photinus*) to consume them. The mode of attraction is to mimic sex pheromones. *Photuris* females not only consume males of other species but also assimilate their prey defence compounds, which consist of steroids (lucibufagines) that the predatory *Photuris* cannot produce. The *Photuris* females that have fed off *Photinus* males are repulsive for spiders of the genus *Phidippus* (Salticidae). Moreover, lucibufagine together with betaine is transferred by *Photuris* females to their eggs, inducing their 'protection'. This mixture is also antipalatable for larvae of the ladybird *Harmonia axyridis* (Pallas, 1773) (Coleoptera, Coccinellidae) and ants of the species *Leptothorax longispinosus* (Roger, 1863) (Hymenoptera, Formicidae), but less effective against the Common Earwig, *Forficula auricularia* (Linnaeus, 1758) (Dermaptera, Forficulidae).

2.3.3 Adherent Products

Adherent products act in a mechanical manner to immobilize temporarily or permanently the predator. A variant of this strategy is to 'paste' the predators' mandible or (sensilla) receptors. In some cases, these products (typically proteins, terpene resins and mixtures of hydrocarbons or long-chain waxes) are mixed with low molecular weight molecules that have the function of antiappetence (Betz & Kölsch, 2004).

Chen, Henderson, and Laine (1999), Quintana et al. (2003), and Dettner, Scheuerlein, Fabian, Schulz, and Francke (1996) have shown that, when termite soldiers *Coptotermes formosanus* (Shiraki, 1909) (Isoptera, Rhinotermitidae) are attacked by predators, they release a mixture of sticky products through their mandibles. This glue is composed of n-alkanes, mucopolysaccharides and long-chain fatty acids, mainly lignoceric and hexacosanoic acids. Other species of termites among the genus *Reticulitermes* (Isoptera, Rhinotermitidae) produce a sticky mixture mainly composed of terpenes: monoterpenes such as α-pinene, β-pinene and limonene and a wide variety of sesquiterpenes. In the case of the collembolan, *Tetrodontophora bielanensis* (Waga, 1842) (Entognatha, Collembola), a sticky mixture is produced which induces disorientation and cleaning behaviour in the beetle *Nebria brevicollis* (Fabricius, 1792) (Coleoptera, Carabidae), allowing enough time for the collembolan to jump away.

2.4 Diversity of Chemical Defence Types

2.4.1 Defence Against Insect Predators

Only a small proportion of insects is specialized in predation of other insects (Table 1) but it is found in the largest insect orders such as Odonata, Heteroptera, Coleoptera and Diptera (e.g., asilides). Few specialist predators exist in Hymenoptera; they are mostly specialized in parasitism. Finally, a few Lepidoptera caterpillars are also entomophagous.

Table 1 Main Insects and Insect Predators

Order	Family	Name	Prey
Coleoptera	Cantharidae	Several genera	Small nectar-feeding
Coleoptera	Carabidae	*Pasimachus* sp.	Terrestrial insects
Coleoptera	Carabidae	*Cicindela* sp.	Terrestrial insects
Coleoptera	Coccinellidae	Several genera	Mites, aphids and mealy bugs
Coleoptera	Melyridae	Especially *Collops* sp.	Terrestrial insects
Diptera	Asiliidae	Several genera	Flying insects
Diptera	Cecidomyiidae	Especially *Aphidoletes* sp.	Especially aphids
Diptera	Dolichopodidae	Several kind	Small insects of wetlands
Diptera	Syrphidae	Several genera	Especially aphids
Dyctioptera	Mantidae	Several genera	Terrestrial insects
Heteroptera	Anthocoridae	Especially *Orius* sp.	Especially thrips
Heteroptera	Gerridae	Several genera	Aquatic insects
Heteroptera	Nabidae	Especially *Nabis* sp.	Phytophagous insects
Heteroptera	Pentatomidae	Several genera	Phytophagous insects
Heteroptera	Phymatidae	Especially *Phymata* sp.	Pollinators, bees
Heteroptera	Reduviidae	Several genera	Aquatic insects
Hymenoptera	Crabonidae	Especially *Eucercis* sp.	Flying insects
Lepidoptera	Lycaenidae	*Feniseca, Maculinea*	Aphids, ants
Neuroptera	Chrysopidae	*Chrysoperla* sp.	Phytophagous, aphids
Neuroptera	Corydalidae	Several genera	Aquatic and terrestrial insects
Neuroptera	Hemerobiidae	Several genera	Aphids
Neuroptera	Mantispidae	Several genera	Terrestrial insects
Neuroptera	Myrmeleotidae	Several genera	Especially ants
Neuroptera	Raphiidae	Especially *Raphidia* sp.	Mostly beetles
Odonata	All		Aquatic and terrestrial insects
Orthoptera	Tettigonidae	Several genera	Flying insects
Plecoptera	Mainly Perlodidae	Several genera	Aquatic insects

Most insect predators are generalists (Table 1). However, some are more specialized and feed off chemically protected insects. This is the case for beetles of the Coccinelidae family and the genus *Pasimachus* (Carabidae), and some Diptera such as the Syrphidae and the Cecidomyiidae. These predatory insects are part of a trophic network and in turn may become potential preys for top predators (mostly birds). A selective advantage of this specialization is the sequestration or, at least, the use of defence compounds of the preys, the toxic compounds of the plant passing to the predator via phytophagous feeding. An extreme case occurs when two predators are in competition. For example, the spotted ladybug beetle, *Coleomegilla maculata* (De Geer, 1775) (Coleoptera, Coccinellidae), and the chrysopid, *Chrysoperla rufilabris* (Burmeister, 1839) (Neuroptera, Chrysopidae) are predating on the same (chemically protected) aphid. When populations of the two predators are high, intraguild predation occurs between the two predators. Thereby first and second instars of the Coccinellidae are actively captured and consumed by chrysopid larvae (Lucas, Coderre, & Brodeur, 1997) (Fig. 1).

2.4.2 Defence Against Noninsect Predators

Apart from insects, several organisms are insectivorous. Among them are arachnids, which are all predators of insects, many birds, reptiles, amphibians and some mammals. The case of bats will not be treated here, even though they are the main predators of nocturnal insects. Unlike predator insects, arachnids do not seem to sequester or synthesize defence substances. Arachnid venoms are mostly proteic and are not derived from chemical compounds of the host plants of their prey. Within arachnids, we can observe two broad strategies:

- The first is to build traps such as nets that are typically unselective. The main question is what to do with toxic preys which are caught in these traps. Our results and those described in the literature show that in model species such as *Nephila clavipes* (Leach, 1815) (Arachnida, Araneae), prey selection is done according to two main criteria: the size (compared to that of the spider) (Hénaut, Delmé, Legal, Williams, 2005) and the chemical defences (Lucas-Silva & Trigo, 2002). When some toxic preys are trapped [especially Lepidoptera Danainae, Ithomiinae, Heliconinae (all Nymphalidae)], they are placed by the spider in a sort of 'trash can' without consuming them. This 'garbage' is often the food supply of other spiders (referred to as kleptoparasites) such as Argyrodes (Hénaut et al., 2005), but also some lizards which, while feeding on this 'trash', can become prey of *N. clavipes*.

Figure 1 How a praying *Mantis* can use a toxic nutrient source coming from a chemically protected insect? How to hunt a predator which detoxifies defence substances from an initial prey? *Baronia brevicornis* (Lepidotera, Papilionidae) is the most 'primitive' species in the world. It is endemic to the Mexico. This butterfly lives on a highly toxic Acacia (*Acacia cochliacantha,* Fabaceae) very rich in cyanogenic glucosinolates. This chemical defence (probably by sequestration) is ineffective against the predatory bugs (Heteroptera, Pentatomidae, *Brochymena* or *Parabrochymena* sp.). Although mantises are poisoned by this type of caterpillars, they are fond of their predators (which metabolize cyanogenic compounds). Indirectly, the praying *Mantis* can thus feed off not palatable preys. (A) Caterpillar, (B) Caterpillar eaten by a Pentatomide bug, (C) Mantis predates the same bug species, (D) Mantis eats the Pentatomide. *Photos: L. Legal and J. Albre, Sierra de Huautla, Morelos, Center Mexico.*

- Arachnids that hunt actively are more diverse. Some consume chemically protected insect species while others seem to avoid them. Most of time, diurnal species that using their vision appear to avoid toxic prey (Skow & Jakob, 2006), while the nocturnal species that locate their prey by vibration and/or chemical detection appears to be indifferent to the toxicity of the prey (Dor, Machkour-M'Rabet, Legal, Williams, & Hénaut, 2008).

Some birds specialize on one of the most toxic butterfly species, *Danaus plexippus* (Linnaeus, 1758) (Lepidoptera, Nymphalidae), also called the Monarch, an emblematic species in North and Central America. Every year, they migrate from Canada to Mexico (Brower, 1995; Zhu, H. et al., 2008). The mixture of cardiac glycosides and alkaloids from its host plants (milkweed, Asclepiadaceae) makes this butterfly very toxic (Malcolm & Brower, 1989; Oyeyele & Zalucki, 2008). Only traces of the toxins are accumulated in the wings at adult stage. Therefore, the continuous fluxes of toxins during the larval stage and from nectar of toxic plants eaten by adults are required to guarantee the defence. The toxins act on sodium channels which regulate the ouabaine levels (Holzinger & Wink, 1996). Two species of birds [*Icterus galbula* (Linnaeus, 1758) (Icterinae) and *Pheucticus melanocephalus* (Swainson, 1827) (Cardinalinae)] manage to pass the chemical barrier of defence of the butterflies. In the State of Michoacán/Mexico, it was estimated that 60% of predation of the monarch was carried out by these two species (Fink & Brower, 1981). The question is how these birds overpass the chemical defence of the monarch. Their learning behaviour through observation is exceptional. During migration, the rate of toxins in the insect body goes down when density of host plant availability decreases (Fink & Brower, 1981), and the birds learn to predate butterflies preferentially in places where the toxic plant is rare. In addition, the birds eat only the abdomen of the butterflies, where most defence substances have already been metabolized by the insect. Finally, the two bird species are partially resistant to cardiac glycosides and alkaloids of the insect. This case represents an extreme situation that combines coevolution and learning behaviour of the predators.

3. HOST PLANT MANIPULATION

Manipulation of plant development by insects results in improved nutritional value of plant tissues and/or nutrient access for insects. These changes have been interpreted as adaptations that address the mismatch between the nutrient status of unmodified host-plant tissues and the herbivore's requirements (Giron, Huguet, Stone, & Body, 2016).

3.1 Insect-Induced Effects on Plants

3.1.1 Remodelling From Ultrastructure to Anatomy

Gall-inducing insects are iconic examples of plant manipulation, with spectacular and complex tissue reorganization, sometimes resulting in new visually apparent plant organs within which the insect feeds and grows (Mani, 1964; Price, Waring, & Fernandes, 1986; Shorthouse, Wool, & Raman, 2005). In response to stimuli from the ovipositing mother and/or her offspring, host tissues usually dedifferentiate and gall development often involves a combination of cell division and growth (Carneiro, Oliveira, &

Isaias, 2014; Giron et al., 2016; Suzuki, Bedetti, & Isaias, 2015). Changes in cell ultrastructure also occur. For example, larval-induced nutritive cells in galls of the Hessian fly, *Mayetiola destructor* (Say, 1817) (Diptera, Cecidomyiidae), exhibit an increase in cytoplasmic staining and numbers of cellular organelles, along with development of numerous fragmented small vacuoles and endo-reduplicated or hypertrophied nuclei and nucleoli (Harris et al., 2006). Additionally, nutritive cells often have thin walls, presumably due to their developmental stage as well as insect-induced inhibition of cell-wall fortification and expansion. Thin walls facilitate the breakdown that releases cell contents to larvae of cynipid gall wasps (Hymenoptera) and cecidomyiid gall midges (Diptera) (Bronner, 1992; Harris et al., 2006; Rohfritsch, 1992). Most likely, however, harvest of the contents of nutritive cells by insects results from a combination of increased permeability of plant cell walls, enlarged plasmodesmata, high turgor pressure in the nutritive cells, mechanical action of larval mouthparts and possibly secretion and injection of proteases into cells by gall-inducing insects.

Modified plant tissues also are often supported by changes in vascular tissues which facilitate nutrient translocation towards the insect's feeding site (Rohfritsch, 1992; Stone, Schönrogge, Atkinson, Bellido, & Pujade-Villar, 2002; Wool, Aloni, Ben-Zvi, & Wollberg, 1999). Gall tissues act as a strong resource sink for photo-assimilates and can be associated with massive changes in plant growth, metabolism and investment (Giron et al., 2016; see Section 3.3.1).

3.1.2 Interaction With Primary Metabolism

Plants are often considered suboptimal food for phytophagous insects (Schoonhoven, Van Loon, & Dicke, 2005). Host manipulation offers the opportunity to alter the nutritional quality of the host plant by creating additional or novel feeding tissues, upregulating nutrient synthesis in situ or modifying source—sink relationships (Giron, Kaiser, Imbault, & Casas, 2007; Kaiser, Huguet, Casas, Commin, & Giron, 2010; Schwachtje & Baldwin, 2008; Stone & Schönrogge, 2003).

The ability to alter the physiological state of plant tissues, particularly of the cells adjacent to the feeding site, has been well described for a range of plant-manipulating arthropods (e.g., Abrahamson & Weis, 1987; Dardeau et al., 2015; Giron, Frago, Glevarec, Pieterse, & Dicke, 2013; Harris et al., 2006; Hartley, 1998; Larson & Whitham, 1991; Nabity, Haus, Berenbaum, & DeLucia, 2013). Gall inducers, but also some leaf miners such as *Phyllonorycter blancardella* (Fabricius, 1781) (Lepidoptera,

Gracillariidae) are among the best examples of how plants can be manipulated to improve the nutrition they supply to insects (Giron et al., 2016). Induced nutritive tissues contain high levels of nutrients, including minerals, lipids, proteins, amino acids, sugars and/or starch (Bronner, 1992; Stuart, Chen, Shukle, & Harris, 2012). For example, some gall-inducing insects increase plant amino acid content through alteration of their synthesis and/or transport (Gündüz & Douglas, 2009; Zhu L. et al., 2008). Active transport of sugars towards the insect's feeding site is also frequently observed due, at least partially, to increased invertase activity, and glucose in excess can be transformed into lipids (Nabity et al., 2013; Rehill & Schultz, 2003).

3.1.3 Interaction With Secondary Metabolites

Besides offering imperfect nutrition, plants are suboptimal hosts because of a dazzling array of secondary metabolites, many of which harm insects (Schoonhoven et al., 2005). Insects have a number of strategies for dealing with these metabolites, including excretion, sequestration and degradation. As a fourth strategy, direct modulation of plant metabolism as a way to circumvent plant defences at source, has also been reported for several gall-inducing insect lineages, leaf miners and caterpillars that build leaf shelters (Lill & Marquis, 2007). Phenolic compounds, for example, are substantially lower in tissues close to the insect feeding site but accumulate at the periphery of host plant tissues infested by the woolly poplar aphid, *Phloeomyzus passerinii* (Signoret, 1875), some *Pontania* sawflies and of cynipid wasp galls (Dardeau, Deprost, et al., 2014; Ikai & Hijii, 2007; Nyman & Julkunen-Tiitto, 2000). Similarly, *M. destructor* and *P. blancardella* have also been shown to cope with plant defences by inhibiting phenylpropanoid pathways (Liu et al., 2007; Giron unpublished). A fifth strategy is to manipulate the plant, through mass attack, to overwhelm its production of secondary metabolites. This cooperative strategy is presented in Section 3.2.3.

Finally, phytophagous insects can also interfere with the plant's defence signalling pathways (reviewed in Guiguet et al., 2016). Interference of the corn earworm, *Helicoverpa zea* (Boddie, 1850) (Lepidoptera, Noctuidae) with early danger signalling by the plant suppresses glandular trichome production and inhibits expression of defensive genes regulated by jasmonic acid (JA) and ethylene (ET) pathways (Wu, Peiffer, Luthe, & Felton, 2012). Release of bacteria in the oral secretions of the Colorado potato beetle, *Leptinotarsa decemlineata* (Say, 1824) (Coleoptera, Chrysomelidae), results in activation of a plant microbial defence response through induction of the salicylic acid (SA) pathway (Chung et al., 2013). This benefits the caterpillar by leading, through negative cross talk, to downregulation of the JA-responsive antiherbivore response.

3.2 Mechanisms Involved in Plant Manipulation

3.2.1 Effectors

Oral secretions of insect herbivores are important recognition cues that can be used by plants as 'elicitors' of induced defences. Oral secretions also have important functions for herbivores as effector proteins having various functions, including, suppression of plant defence, alteration of plant development and manipulation of plant resources (Chung et al., 2013; Consales et al., 2012; Giron et al., 2016; Guiguet et al., 2016; Hogenhout & Bos, 2011). Information about the effectors used by herbivorous insects is just beginning to emerge (e.g., Giron et al., 2016; Zhao et al., 2016). In-depth functional molecular approaches are now required to determine the mechanisms that contribute to plant manipulation and to characterise a greater number of candidate effectors in a larger array of plant-manipulating species, possibly leading to identification of convergent mechanisms.

To date, the Hessian fly is the only gall-inducing insect with a sequenced genome as well as a researched salivary gland transcriptome and proteome (Chen et al., 2008; Zhao et al., 2015). It is also the only gall-inducing insect for which there are candidate genes having both elicitor and effector functions (Aggarwal et al., 2014; Zhao et al., 2015, 2016). Elicitor function, which benefits the plant, is better understood than effector function, which benefits the attacking larva. Hessian fly salivary proteins can function as elicitors because they can be detected by a plant surveillance system mediated by *resistance* (*R*) genes. Each *R* gene product detects the product of a particular Hessian fly *Avirulence* (Avr) gene in what is known as the 'gene-for-gene' interaction (Harris et al., 2003, 2015; Stuart et al., 2012). Detection triggers induced downstream defence responses, which kill the larva. These can be avoided if the larva has a mutation in its matching *Avirulence* gene (Aggarwal et al., 2014; Zhao et al., 2015, 2016). Presumably these mutations spread through populations under selection pressure from the *R* gene. Gene-for-gene interactions have also been documented for two other gall-inducing insects, the wheat midge, *Sitodiplosis mosellana* (Gehin, 1857) and the Asian rice gall midge, *Orseolia oryzae* (Wood-Mason, 1889) (Diptera, Cecidomyiidae), and also for several aphids that are not gall inducers (Harris et al., 2003, 2015). It is expected that plant parasites that exhibit gene-for-gene interactions (including microbes, insects and nematodes) produce effectors to suppress the plant's basal immune systems and manipulate the plant to deliver better or more food (Jones & Dangl, 2006). It is not known how proteins encoded by the four Hessian fly candidate effector genes contribute to these two goals, although interference with downstream signalling and phytohormones seems likely (Aggarwal et al., 2014; Zhao et al., 2015, 2016). The Hessian fly's genome sequence points to the importance of effectors (Zhao et al., 2015). One-eighth of genes encode putative effectors. One of the effector gene families is the largest known arthropod gene family. Members have structural similarities with both bacterial effectors and the E3 ubiquitin ligases of plants. The authors proposed that Hessian fly effectors have proliferated as a result of an arms race with plant immune systems mediated by *Resistance* genes.

Emerging evidence suggests that plant-feeding insects use salivary effectors to interfere with 'early danger signals' of their hosts, deploying apyrases, calreticulins and peroxiredoxins in saliva (reviewed in Guiguet et al., 2016). Given links between these molecules and plant growth and defence, we propose that these effectors interfere with phytohormone signalling, and therefore have a special importance for plant-manipulating insects. More specifically, insect-derived apyrases act by degrading extracellular ATP released by the plant during wounding and insect feeding (Guiguet et al., 2016). Apyrases and/or an ATPase activity have been identified in the salivary secretions of several insect herbivores (reviewed in Guiguet et al., 2016).

3.2.2 Plant Growth Regulators

Experimental data support the role of phytohormones in plant manipulation (Bartlett & Connor, 2014; Giron et al., 2016; Tooker & Helms, 2014). Because auxins and cytokinins (CKs) lie at the very core of molecular mechanisms controlling the balance between the rate of cell division and differentiation, they have long been hypothesized to be involved in insect-induced plant manipulation. However, it is likely that plant defensive hormones such as jasmonic acid (JA), ethylene (ET), SA and abscisic acid (ABA) are also involved (Bartlett & Connor, 2014; Tooker & Helms, 2014; Zhang et al., 2016).

Insect salivary secretions and accessory oviposition gland substances applied to the plant can act directly on the plant's hormone biosynthesis, degradation, transport or signalling pathways to alter phytohormonal balance (Giron et al., 2016). Many organisms are also known to be able to synthesize these phytohormones de novo (especially auxins and CKs), and insect-associated symbionts may play a key role in the production/delivery of phytohormones (Giron et al., 2013, 2016; Kaiser et al., 2010; Zhang et al., 2016).

In many plants, herbivory stimulates the production of JA and ET while other organisms stimulate the production of SA (Erb, Meldau, & Howe, 2012). Several lines of evidence suggest that plant-manipulating insects can counteract defences mediated by JA and SA. Hormones such as auxins and CKs can influence plant defensive responses (Erb et al., 2012; Giron et al., 2013). This hypothesis is consistent with data showing that several gall inducers and leaf miners induce increased levels of auxins or CKs in galls and mines, but fail to induce higher levels of JA or SA (Giron et al., 2016; Zhang et al. 2016).

3.2.3 *Behaviour and Cooperation With Conspecifics*

Actual intraspecific cooperation (i.e., excluding dose-dependence effects) for the purpose of plant manipulation is only known in the particular case of bark beetles (Scolytinae). For most bark beetle species, establishment on a host tree requires mass attacks to overcome tree resistance, and there is a critical attack density threshold below which all attacks fail and above which they succeed (Berryman, 1976). The main mechanism of conifer defence against bark beetles is a hypersensitive reaction, which rapidly develops in the phloem and the outer sapwood adjacent to attack sites (e.g., Berryman, 1972; Reid, Whitney, & Watson, 1967). Tissues are invaded by terpenes and phenols, with this leading localized death, which renders tissues unsuitable for beetles. Because beetle attacks are generally completed in just a few days, the tree must rapidly synthesize large quantities of secondary compounds simultaneously at all attack points to achieve the concentrations needed to stop the aggressors. However, synthesis of such compounds is energy demanding. Each tree only has a limited quantity of energy. An important element of the beetle population strategy is thus to manipulate tree defensive activities by increasing the energy demand for syntheses through increasing the number of attacks (Berryman, 1972; Raffa & Berryman, 1983; Wood, 1982). Above a certain attack density, the tree becomes unable to build efficient defences at all aggression points, and the critical threshold is reached (Christiansen, Waring, & Berryman, 1987). This 'exhausting tree defences' or 'intraspecific cooperative' strategy (Lieutier, 2004) is typically enabled via aggregation pheromones emitted by pioneer beetles, allowing a rapid gathering of a large number of conspecifics (Wood, 1982). After its defences are exhausted, the tree, no longer able to stop beetle brood development or invasion by the various types of organisms introduced by the beetle, finally dies.

The energy that the tree is able to rapidly mobilize depends on its genotype and physiological status. The critical attack density threshold thus depends on tree genotype and physiological conditions, as well as beetle aggressiveness. Threshold values range from 50 to 850 attacks per m^2 of bark (Lieutier, Yart, & Sallé, 2009). Whether such intraspecific cooperation is at play in other plant-manipulating systems awaits further investigations. There are other species whose colonization is characterized by coordinated attacks by large numbers of individuals, e.g., the horse-chestnut leaf miner, *Cameraria ohridella* (Deschka & Dimic, 1986) (Lepidoptera, Gracillariidae) (Lees, Lopez-Vaamonde, & Augustin, 2011).

3.2.4 Cooperation With Other Organisms

Microorganisms have been shown to be important 'hidden players' in insect—plant interactions (e.g., Gutzwiller, Dedeine, Kaiser, Giron, & Lopez-Vaamonde, 2015; Sugio, Dubreuil, Giron, & Simon, 2015; see chapter: Influence of Microbial Symbionts on Plant-Insect Interactions by Giron et al., 2017 for details) and can affect many insect traits, including their ability to manipulate plant physiology for their own benefit (Kaiser et al. 2010; Sugio et al. 2015). Insect symbionts can directly or indirectly affect the plant by interfering with plant signal transduction pathways or by altering plant primary and secondary metabolism (Body, Kaiser, Dubreuil, Casas, & Giron, 2013; Giron et al. 2013; Sugio et al. 2015; Zhu, Poelman, Dicke, 2014).

In many conifer bark beetle species that attack live trees, the remarkable intraspecific cooperation that leads to attack success (see Section 3.2.3) is made even much more efficient through cooperation with fungi, which also elicit the tree's defences (Lieutier et al., 2009). In fact, a complex community of organisms is present in the beetle galleries, including several species of fungi, yeasts, bacteria, mites and nematodes (Hofstetter, Dinkins-Bookwalter, Davis, & Klepzig, 2015). These organisms interact among each other as well as with beetles and the tree. The role of fungi in stimulating the tree's defences during bark beetle attack seems now recognized in several bark beetle—fungi associations (Hofstetter et al., 2015; Lieutier et al., 2009). Information regarding other groups is, however, very limited.

3.2.4.1 Fungal Contribution to Manipulating Tree Defences

Most bark beetle species are associated with and carry spores of Ophiostomatoid fungal species mainly of the *Ophiostoma*, *Ceratocystis*, *Ceratocystiopsis* and *Grosmannia* genera and the related anamorph *Leptographium* (Kirisits, 2004;

Six, 2012). In addition to the contribution of fungi to exhausting tree defences, beetles can benefit from improved nutrition through mycophagy or modification of phloem substrate (e.g., Klepzig & Six, 2004). For fungi, the obvious benefit is the transportation that beetles provide. The added stimulation of the tree's defences provided by fungi can be measured by comparing, on a same tree, the development of the hypersensitive reaction around beetle galleries with or without fungi (Lieutier, Garcia, Yart, & Romary, 1995). The tree devotes more energy to defence when fungi are present. An immediate benefit for the beetle is a lowering of the critical attack density threshold. This can mean the difference between colonization success and failure, as is the case of the southern pine beetle, *Dendroctonus frontalis* (Zimmermann, 1868) (Coleoptera, Curculionidae) (Bridges, Nettleton, & Conner, 1985).

3.2.4.2 Modulation of the Fungal Contribution to Manipulating Tree Defences

Efficiency of the beetle—fungus association to stimulate the tree's defences varies depending on fungal species, as well as isolates of the same species (Lieutier, Yart, Ye, Sauvard, & Gallois, 2004). For some bark beetle species, their fungal associates participate in the stimulation of tree defences but also can invade the phloem, making it unsuitable for larvae (Klepzig & Six, 2004; Paine, Raffa, & Harrington, 1997; Six & Paine, 1998; among others). Why these fungal species have not been selected against may be explained by a trade-off between the benefits the insect gains from the fungus helping to counter plant defence versus the losses the insect suffers because of the reduced availability of larval feeding sites (Lieutier et al., 2009). In the case of *D. frontalis*, complex interactions among fungal species can occur. For instance, an *Entomocorticium* sp. is able to stop the development of *Ophiostoma minus*, a species involved in host defence stimulation. Consequently, *O. minus* cannot reach the phloem used by larvae and do not compromise their survival (e.g., Klepzig & Six, 2004).

Experimental observations strongly suggest that bacteria may interfere, positively or negatively, in tree defence manipulation by ophiostomatoid fungi. A diverse set of bacteria inhabits beetle galleries. *In vitro* bioassays have showed that some are able to stimulate or inhibit fungal growth, alone or in combination with host tree compounds (Adams, Currie, Cardoza, Klepzig, & Raffa, 2009). For example, the mycelial growth of *Grosmannia clavigera*, a fungus associated with the mountain pine beetle, *Dendroctonus ponderosae* (Hopkins, 1902) (Coleoptera, Curculionidae), known to strongly stimulate the hypersensitive reaction of pine, is favoured by a combination of *Pseudomonas* and racemic alpha-pinene, whereas the mycelial growth of *Ophiostoma ips*, an associate of several *Ips* species, appears less favoured. Fungal reproduction can also be stimulated by bacteria, especially in the case of beetles that are adapted to successfully attack living trees having vigorous defences (Adams et al., 2009). In the *D. frontalis* system, bacteria are able to produce selective antibiotics that strongly inhibit *O. minus* (Scott et al., 2008).

3.3 Outcomes of the Interaction

3.3.1 Consequences for Plant Vigour and Survival

In most cases plant-manipulating insects are considered to be parasites, that is, they harm but do not kill their hosts (Stone & Schönrogge, 2003). However, in some systems, the host plant may reap benefits from the modifications of its tissues. For instance, eucalypts colonized by *Leptocybe invasa* (Fisher & La Salle, 2004) (Hymenoptera, Chalcidoidea) have an increased tolerance to cold (Rocha et al., 2013). The interaction can also secondarily evolve into a mutualistic relationship (see Section 4.2 for example).

While manipulation is typically expected to have a limited impact on the vigour or survival of the plant, there are clearly interactions where the plant's survival and fitness is drastically affected. The strategy of defence exhaustion of bark beetles is an extreme situation, which generally results in host plant death inasmuch as the plant is no longer able to control herbivores and pathogens (see Section 3.2.3). Gall induction can also have drastic effects on plant survival (e.g., Dardeau, Deprost, et al., 2014; Harris et al., 2015). When attacked by Hessian fly larvae, seedling plants could die, presumably because the creation of the larval-induced nutritive tissue 'starves' younger leaves of resources needed for cell growth (Anderson & Harris, 2006; Harris et al., 2006). A similar process probably impedes bud break of poplars previously infested by *P. passerinii* (Dardeau et al., 2015; Sallé unpublished). The differentiation of novel organs, or misshaping of preexisting ones, can dramatically alter the metabolism and anatomy of host plant tissues and compromise their initial function. For instance, galls induced in root systems and/or stems can interfere with uptake of nutrients and conduction of water or sap, resulting in partial or total plant stunting (Brown, Glenn, & Wisniewski, 1991; Granett, Walker, Kocsis, & Omer, 2001). Likewise, host-plant fitness can be sharply reduced by galls affecting the differentiation of reproductive organs and preventing seed production (e.g., Harris et al., 2003; Post, Kleinjan, Hoffmann, & Impson, 2010). Modification of within-plant allocation of carbohydrates and nutrients can lead to similar outcomes. Plant-manipulating insects often turn their feeding sites into physiological sinks, drawing nutrients and photo-assimilates from surrounding tissues (see Section 3.1). The manipulated tissues therefore compete with natural plant sinks, like buds or catkins (Larson & Whitham, 1991, 1997). This can delay flowering and reduce seed viability, affect growth patterns or survival of plant modules and even modify the architecture of the host-plant (e.g., Kurzfeld-Zexer, Wool, & Inbar, 2010; Marini-Filho & Fernandes, 2012).

3.3.2 Costs and Benefits for the Insects

Plant manipulation can result in optimization of the nutritional value of host-plant tissues (Hartley & Lawton, 1992; Price, Fernandes, & Waring, 1987): this is the *nutrition hypothesis*. Such benefits have been described for gall inducers (Giron et al., 2016) and other plant-manipulating insects, such as aphids inducing chlorotic lesions (Sandström, Telang, & Moran, 2000), caterpillars that build leaf shelters (Lill & Marquis, 2007) and leaf miners (Kaiser et al., 2010). This is generally achieved through an optimization in nutrients and/or secondary metabolites (see Section 3.1) and enhances larval development, fecundity and survival (e.g., Dardeau, Pointeau, et al., 2014; Kaiser et al., 2010; Lill & Marquis, 2007). For the leaf miner *P. blancardella*, manipulation also maintains nutrient-rich green tissues and creates an enhanced nutritional microenvironment in leaves which are otherwise senescing (Body et al., 2013; Giron et al., 2007). Behavioural benefits may also be reaped from plant manipulation, and higher acceptance rates, and optimized probing and feeding behaviours have been reported from insects feeding on manipulated tissues (Dardeau, Pointeau, et al., 2014; Kaiser et al., 2010). These benefits, together with an increased development rate, could shorten the exposure of plant-manipulating insects to adverse biotic and abiotic factors (Dardeau, Pointeau, et al., 2014).

Plant manipulation could also shelter insects from adverse abiotic conditions, particularly desiccation (Price et al., 1987; Stone & Schönrogge, 2003): this is the *microenvironment hypothesis*, which is relevant for partially or totally enclosing galls, leaf mines, and leaf shelters (Lill & Marquis, 2007; Pincebourde & Casas, 2006; Stone & Schönrogge, 2003). Abiotic conditions within plant 'shelters' can differ sharply from ambient conditions in terms of temperature, relative humidity, solar radiation and wind exposure (Lill &

Marquis, 2007; Price et al., 1987; Stone & Schönrogge, 2003). Insects develop in moister conditions than they would in ambient air and are therefore protected from drought. Leaf temperature modulation by leaf miners, through a manipulation of leaf absorbance and stomatal conductance, can also increase the development of caterpillars and shorten their exposure to parasitoids (Pincebourde & Casas, 2006).

Manipulation may also protect plant-manipulating insects from natural enemies, decreasing detection by predators, preventing attacks and/or recruiting protecting allies (Lill & Marquis, 2007; Price et al., 1987; Stone & Schönrogge, 2003): this is the *enemy hypothesis*. Intraspecific variation in some gall traits such as increased hardness and thickness of gall tissues, external coatings with hairs and recruitment of ant guards with nectar secretions can be associated with increased survival of gall inducers (Stone & Schönrogge, 2003).

Finally, plant manipulation can also offer an opportunity to *escape from intraguild competition* and/or *limit exploitation of the same host plant* by other herbivores and pathogens (Giron et al., 2016). For example, host defence exhaustion through intraspecific cooperation and association with Ophiostomatoid fungi allows the establishment of insects in vigorous trees, while most bark beetle species are only able to colonize dead or decaying trees. However, mass attacks, especially those that occur on vigorous trees for which the critical threshold of attack density is high, may result in intense intraspecific larval competition impacting the productivity and the quality of offspring (Sallé & Raffa, 2007). In addition, cooperation with microorganisms can also result in interspecific competition between offspring and particular inoculated fungi (see Section 3.2.4.2).

3.3.3 Modulation of Plant Manipulation

3.3.3.1 Host-Related Factors Including Host Resistance

The physiological status of the host plant can have different outcomes on plant–insect interactions, depending on the manipulation strategy considered. In the bark beetle strategy of host defence exhaustion, the critical threshold of attack density depends on the amount of energy that can be mobilized by the host tree. Consequently, this threshold, and therefore the host resistance level, positively correlates with host vigour (Lieutier et al., 2009; Raffa & Berryman, 1983). On the contrary, the abundance and performance of gall-inducing insects often correlates with the size of the manipulated plant module, or more generally with host vigour (Koricheva, Larsson, & Haukioja, 1998; Quiring, Flaherty,

Johns, & Morrison, 2006). In this regard, gall inducers fit the predictions of the plant vigour hypothesis (Price, 1991; Quiring et al., 2006). There are several exceptions to this general pattern though, and negative, parabolic or nonsignificant relationships between plant module size and the abundance and performance of gall-inducing insects have also been observed (Quiring et al., 2006). An alternative hypothesis, the *optimal module size hypothesis*, predicts that gall-inducer performance will be enhanced on intermediate-sized plant modules (Björkman, 1998; McKinnon, Quiring, & Bauce, 1999). Fast-growing and large plant modules may alter the performance of gall inducers because of higher competition among plant and insect-induced sinks. In addition, in such plant modules the induction stimulus could be excessively diluted, which would compromise the success of manipulation (Björkman, 1998; McKinnon et al., 1999).

Host-manipulating insects also maintain complex interactions with the endophytic fungi inhabiting the tissues of their host plants. The profile of fungal endophyte communities is modified within manipulated tissues because of their anatomical, biochemical and physiological remodelling (Lawson, Christian, & Abbot, 2014). Depending on the systems investigated, the endophytic fungi may have positive, negligible or antagonistic effects on the development and survival of leaf miners or gall inducers and may also alter the spatial and temporal distribution of the insects on their host plant (Raman et al., 2012).

3.3.3.2 Site-Related Factors

Environmental factors affecting host growth and vigour can modulate the ability of an insect to manipulate its host. When manipulation requires a reconfiguration of host metabolism or anatomy, host-manipulating insects are predicted to be adversely affected by environmental constraints affecting plant growth, such as water deficit, which could impair the initiation and growth of manipulated tissues (e.g., Koricheva et al., 1998). Nonetheless, environmental constraints could also enhance the manipulation success in resistant host plants, in situations where the constraints interfere with resistance mechanisms, while still allowing a reconfiguration of primary metabolism by the insect (e.g., Björkman, 1998). This emphasizes that both constraint intensity and host genotype can affect the outcome of an environmental constraint on plant manipulation. Similarly, depending on the study system, environmental conditions promoting plant growth and nitrogen content in plant tissues can have very different effects on the success of plant manipulation, which might be further modulated by interactions with the host plant genotype (e.g., Dardeau et al., 2015; McKinnon et al., 1999). When manipulation is favoured by reduced host vigour, as for host defence exhaustion, environmental constraints are generally predicted to promote the establishment and development of insects (e.g., Koricheva et al., 1998). However, in agreement with the growth differentiation balance hypothesis (Herms & Mattson, 1992), constraint intensity might modulate these predictions since a mild water shortage can enhance host resistance (Sallé, Ye, Yart, & Lieutier, 2008).

4. MUTUALISM: A RESULT OF COEVOLUTION? GENETIC AND FUNCTIONAL ASPECTS

4.1 Pollination Mutualisms

While insect–plant interactions often involve antagonism, because the insects eat the plants, there are also important cases of mutualism, in which both insect and plant benefit from the interaction, e.g., in seed dispersion or leaf/bud protection by ants. However, a much wider category of mutualism, of considerable ecological and evolutionary significance, involves insect pollination of plants. Perhaps 90% of angiosperms are pollinated by animals, with insects playing the largest role. Insects typically benefit from the interaction by gaining food from either pollen itself and/or nectar. The plants benefit because insects vector their pollen to conspecifics and bring the pollen of conspecifics to their ovaries. Another form of interaction that might be regarded as indirect mutualism involves emission of volatile organic compounds by plants under attack by herbivores, leading to recruitment of parasitoids that attack the herbivores.

4.1.1 From Generalists to Specialists

Insect pollination has evolved many times and takes many different forms, involving more than 20,000 species of bees, wasps, flies, beetles, moths and butterflies and thrips. The best-known pollinators are probably honey bees, *Apis mellifera* (Linnaeus, 1758) (Hymenoptera, Apidae), whose activities are crucial to human agriculture. Both wild and managed pollinators have a globally significant role in crop pollination, although their relative contributions differ according to crop and location (IPBES, 2016). In addition, beekeepers manage hives to provide pollination services to commercial growers, e.g., the massive almond crop in California. Honeybees are highly generalist (polylectic) pollinators and may visit a wide variety of plant species, although a particular hive at a given time and place may be far more specialized. Wild bees in general, with perhaps 10,000 species globally, are a hugely important group of pollinators. Many species are polylectic, such as the honeybee, but other oligolectic species visit only a restricted range of plants.

Finally, some pollinators may be strongly or strictly associated with a single plant species. One system involves the *Ophrys* orchids in the Mediterranean region that mimic female bees, using both chemical and visual cues, deceiving male bees into acting as pollen vectors (Schatz, Hossaert-McKey, & Kjellberg, 2013). This pollinator attraction without nectar is a clear case of manipulation and useful to keep in mind when considering other cases of mutualism, because it emphasizes that the evolutionary interests of insect and plant are separate and may not be well aligned. Floral morphologies can also select for pollinator specialization. Deep corollas and long spurs of flowers oblige insect visitors to be equipped respectively by a long tongue or by a trunk (only in Lepidoptera) to locate nectar and then become pollinators.

We have established that insect pollinators vary from host plant generalists to specialists. An active research field in the last decade has involved the study of pollination networks, i.e., networks linking a community of plants to a community of insect pollinators at a particular site. This can help reveal many interesting aspects, including the number of different insect species visiting a given plant species and the number of plants visited by a given pollinator species. Such networks can also be summarized statistically in terms of aspects such as linkage and connectance, as has been done extensively for trophic food webs. As a result, they allow to identify species acting as 'hubs' by being central in the organization of such pollination networks while other species are more peripheral (Bascompte & Jordano, 2007).

In cases where there is low plant–pollinator partner specificity, the associations may be diffuse and involve potentially complex mixtures of costs and benefits. For example, it is beneficial to plants for insects to carry pollen to or from another conspecific plant. However, if the insect caries pollen to or from a heterospecific plant, this wastes pollen/ovules or produces hybrid offspring (Neiland & Wilcock, 1999). In addition, pollinators can also vector plant sexual diseases such as fungi or various pathogens. Variations in the quality and quantity of resource provided by plants, and in the activity and capacity of insects to transport pollen, lead to a great diversity of pollination situations, linked to various costs/benefits and partner specificities.

In cases with high partner specificity, a given insect species always interacts with a given plant species, which provides the opportunity for strong plant—insect coevolution with species-specific costs/benefits and mechanisms. In cases such as the sexually deceptive orchid, there may be strong evolutionary forces but the interaction is not a mutualism, since the plant exploits the insect without reward. However, a major class of plant—pollinator interactions with high partner specificity is characterized by mutualism; these are obligate pollination mutualisms (OPMs) involving insects whose larvae are seed predators.

4.1.2 Overview of Obligate Pollination Mutualisms

OPMs are highly specialized species interactions in which both the plant and the pollinator depend upon the partner species for successful reproduction and, for the insects, offspring survival. The most famous case involves fig trees (*Ficus*) and fig wasps (see below). However, another well-known example involves yucca plants and moths, and a dramatic recent discovery in 2003 was the existence of OPMs involving hundreds of plants in the family Phyllanthaceae (especially the genus *Glochidion*) and small moths in the genus *Epicephala*. Table 2 lists known cases of OPMs but it is likely that the novel systems remain to be discovered, even in other plant families.

In general the insect pollinators involved in OPM's transport pollen between male and female flowers of a single plant species. This high specificity is obviously beneficial to the host plant in reducing the production of hybrid offspring and wastage of pollen or ovules through crosses with gametes from other plant species. In some OPMs (e.g., the *Yucca* and *Glochidion* systems) the insects have evolved specialized behaviours to affect active pollination. This is in stark contrast to most insect pollination, which occurs passively as a by-product of insect foraging. OPM pollinators typically lay eggs within at least some of the ovules that they fertilize and pollinator offspring develop by feeding on one or more of the plant seeds, leading to the term 'nursery pollination systems'.

The pollinator offspring develops only in the host plant seeds and the pollen is transported only by adults of the same insect species. Consequently, there is obligate reproductive interdependence of the partners. However, as in other mutualisms, their evolutionary interests are not perfectly aligned and partnership brings both costs and benefits; some plants, as in *Yucca* and *Glochidion,* may selectively abort flowers or inflorescences when too many pollinator eggs are laid in them. An obvious avenue for conflict is the number of seeds that pollinator offspring eat and a considerable amount

Table 2 Obligate Pollination Mutualisms Involving Insects and Plants, With an Indication of the Approximate Number of Plant Species

Plants (Family, Genus)	No. of Species	Pollinators (Family, Order)	References
Agavaceae: *Yucca*, *Senita*	Ca. 50	Yucca moths (Lepidoptera: Proxodidae)	Pellmyr and Huth (1994), Svensson et al. (2005), and Svensson, Pellmyr, and Raguso (2006)
Araceae: *Peltandra*	Ca. 50	*Elachipetra* (Diptera: Chloropidae)	Patt et al. (1992)
Arecaceae: *Chamaerops*	Ca. 200	*Derelomus* (Coleoptera: Curculionidae)	Dufaÿ, Hossaert-McKey, and Anstett (2003, 2004)
Caryophyllaceae: *Silene*, *Dianthus* and *Saponaria*	Ca. 50	*Hadena* (Lepidoptera: Noctuidae)	Jürgens, Witt, and Gottsberger (2002, 2003), Dötterl, Burkhardt, Jürgens, and Mosandl (2007), Dötterl et al. (2006), and Gimenez-Benavides, Dötterl, Jürgens, Escudero, and Iriondo (2007)
Eupomatiaceae: *Eupomatia*	1	*Elleschodas* (Coleoptera: Curculionidae)	Bergström et al. (1991)
Moraceae: *Ficus*	>750	Fig wasps (Hymenoptera: Agaonidae)	Chen and Song (2008), Proffit et al. (2009), and Hossaert-McKey et al. (2010)
Phyllanthaceae: *Breynia*, *Glochidion* and *Phyllanthus*	Ca. 500	*Epicephala* (Lepidoptera: Gracillariidae)	Kato, Takimura, and Kawakita (2003) and Okamoto, Kawakita, and Kato (2007)
Ranunculaceae: *Trollius*	1	*Chiastochaeta* flies (Diptera: Anthomyiidae)	Ibanez et al. (2010) and Lemke and Porembski (2013)
Zamiaceae: *Zamia* and *Macrozamia*	Ca. 10	*Cycadothrips* (Thysanoptera: Aeolothripidae), *Tranes* (Coleoptera: Curculionidae)	Terry et al. (2004) and Terry, Walter, Moore, Roemer, and Hull (2007)

of research has been carried out in OPM systems towards understanding this fundamental problem (Dufaÿ & Anstett, 2003; Jaeger, Till-Bottraud, & Després, 2000).

In these OPMs, the chemical message emitted by the host plant plays a role in maintaining species specificity with the pollinator, thanks to quantitative and qualitative differences in relative proportions of both major and minor volatile organic compounds emitted by host plants (Raguso, 2008). Moreover, scents also provide information on developmental stage, particularly on whether the plant is receptive and ready to be pollinated (Hossaert-McKey, Soler, Schatz, & Proffit, 2010). The host plant scent signal must be specific to attract its specific and obligate partner and permit the required tight coordination and synchronization of their phenologies. Scent also supports various forms of intraspecific variation (dioecy, phenological stages, postpollination) and may be also used by the third actors to exploit these mutualisms. In the system with most data available on floral scent (figs and fig wasps), the message responsible for pollinator attraction is usually constituted by a species-specific blend of compounds, generally dominated by one or a few common terpenoids, the identity of which differs among sympatric species (Chen et al. 2009; Proffit et al. 2009).

4.2 The Fig/Pollinator Mutualism

Figs and fig wasps comprise the most famous OPM; indeed, knowledge of the life cycle of the fig—wasp interaction is essential for the traditional cultural practice of caprification with the common edible fig *Ficus carica* L. This involves removing small branches from trees with fruits emitting wasps and securing them to branches of other trees bearing receptive figs awaiting pollinators. Cultivation of *F. carica* in the Mediterranean region represents one of the earliest forms of agriculture, dating back perhaps 4000 years, but different species of figs (*F. pumila*) have also long been cultivated in China.

4.2.1 Origin, Diversity, Biogeography

Recent research has led to independent dated molecular phylogenies for fig trees and their associated pollinating wasps (family Agaonidae), suggesting a single origin of the mutualism approximately 75 million years ago (Cruaud et al., 2012), followed by radiation to produce more than 750 fig species, with a somewhat larger number of pollinator species. Globally, figs and fig wasps are found throughout the tropics and some other warmer parts of the world. The genus *Ficus* is of considerable importance in rainforests throughout the tropics and one of few genera of plants to be of high ecological and evolutionary significance on all tropical landmasses (Harrison, 2005). Rainforest figs have a number of different growth forms, including hemi-epiphytes or 'stranglers', but also free-standing trees, banyans, creepers and understory shrubs. Figs are also important in many savannah habitats, where two important life strategies are rock splitters (lithophytes) typically growing on boulders on rocky outcrops, and free-standing trees that grow along the banks of seasonal watercourses.

Figs are classified in one large genus *Ficus* (subdivided into 6 subgenera and about 20 sections), whereas their pollinators have family status as the Agaonidae with about 20 genera. To a first approximation, pollinator genera match with *Ficus* sections. For example, section *Americana* is pollinated by wasps in the genus *Pegoscapus* and section *Malvanthera* by *Pleistodontes* wasps, which reflects their underlying history of cospeciation. Nevertheless, taxonomic and phylogenetic analyses also provide evidence for some recent and even ancient host shifts by wasps to different fig taxa. In addition, some fig lineages in Africa, support two or more pollinator that each seem to be coradiating with the figs (McLeish & van Noort, 2012).

Fig sections and their associated wasp genera provide a useful taxonomic level for considering the biodiversity and biogeography of the mutualism. For example, only two fig sections occur naturally in the New World—the large *Americana* with >100 species and smaller *Pharmacosycea* with about 40 species. The former has active pollinators and a diversity of plant species, while the latter is a small radiation of free-standing trees associated with passive pollinators. Diversity is higher in the Palaeotropics and especially in Australasia with >500 species. Most *Ficus* sections have restricted geographic distributions, often in only one continent, but a few like *Urostigma* are distributed widely in the Old World.

4.2.2 Monoecy, Dioecy and How to Protect Seeds

An important aspect of fig diversity is the distinction between monoecious and functionally dioecious species (Fig. 3). In monoecious species all trees produce the same kind of figs, but dioecious species have two kinds of tree (functionally female or male) that specialize in producing either seeds, or pollen and wasps. In dioecious figs, when female wasps enter the fruits of male trees they are able to lay eggs in the flowers by pushing their ovipositors down the short styles of the individualized female flowers and placing the egg in the plant ovule. The wasp larva hatches and feeds on the plant tissue produced by the single fig ovule, completing its whole development inside one flower. Typically only a few female wasps enter each fig to lay

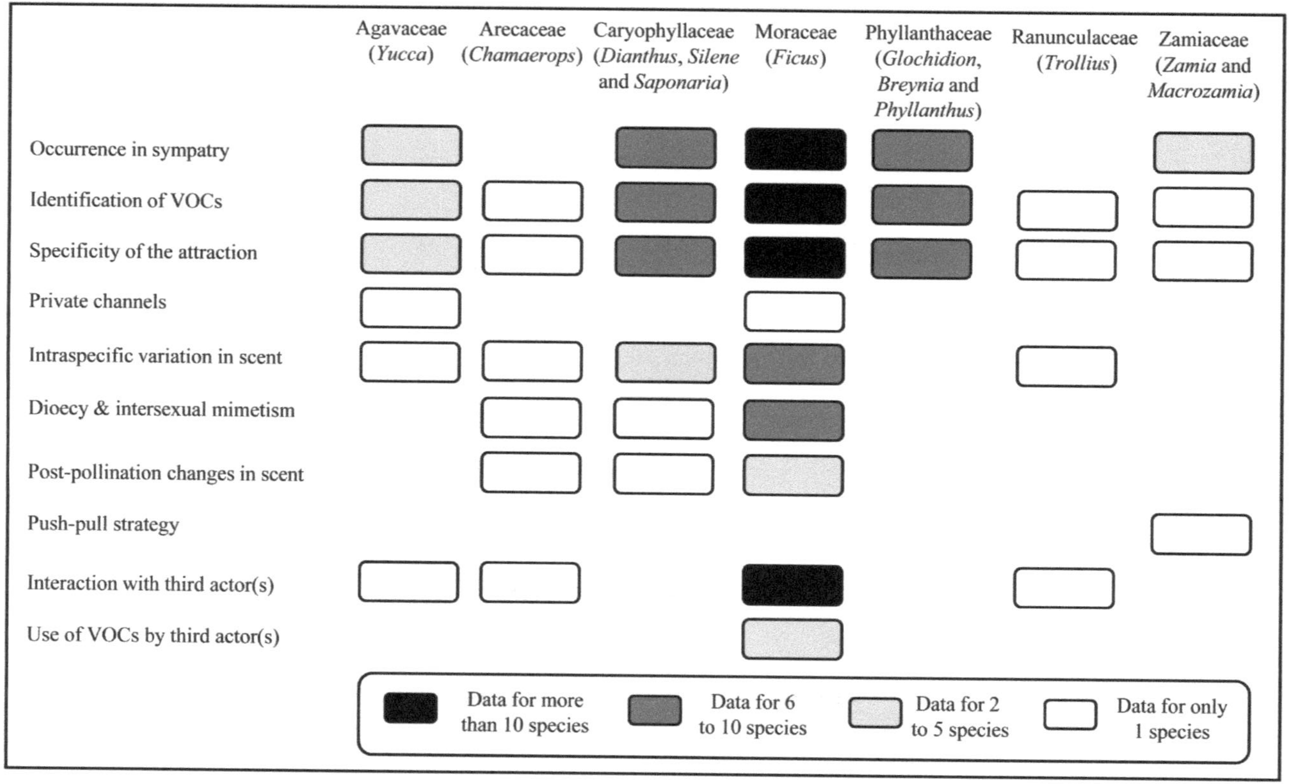

Figure 2 Diagram summarizing knowledge of nursery pollination interactions in which chemical mediation has been described. *Updated from Hossaert-McKey, M., Soler, C., Schatz, B., & Proffit, M. (2010). Floral scents: their role in nursery pollination mutualism.* Chemoecology, 20, *75–88. http://dx.doi.org/10.1007/s00049-010-0043-5.*

eggs but tens to hundreds of wasp offspring (dependent on species) develop in each fig. When they reach maturity they hatch and mate inside the fig before the females disperse to find new receptive figs to enter. At the time when the wasps are hatching individualized male flowers are also maturing inside the fig. Some fig wasps species are active pollinators—the females deliberately use brushes (coxal combs) on their legs to collect pollen into thoracic pollen pockets. In others, whose host figs liberate pollen inside the fig, there is no such behaviour and pollen is carried passively on the body, as in other insects.

After mating and collecting pollen, adult female wasps disperse through holes in the fig wall chewed by the wingless male wasps, which die soon afterwards. Female wasps now search for and enter receptive figs. If they enter a fig on a male tree, the life cycle repeats as described above. However, if they enter a fig on a female tree, the wasps are able to pollinate the flowers but unable to reproduce themselves, because the flowers have much longer styles, preventing them from laying eggs in the necessary place for successful offspring development. Facing the underlying conflict over the reproductive

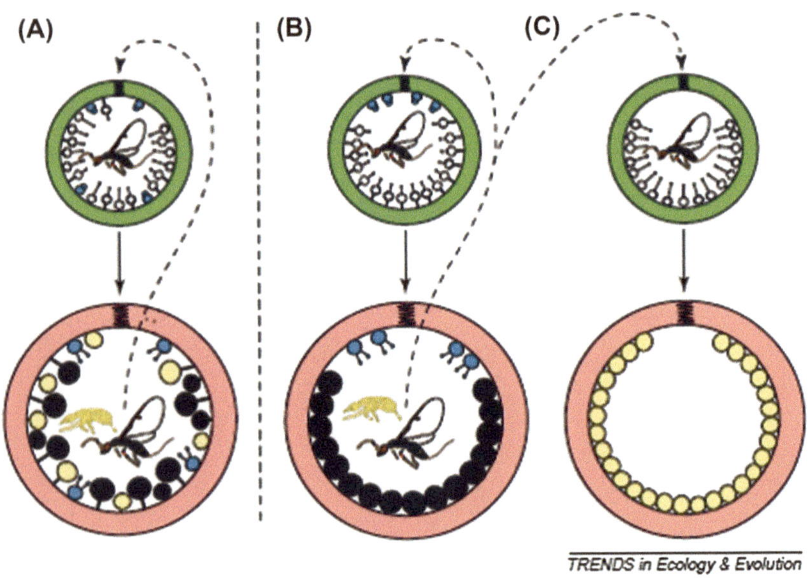

Figure 3 Figs develop from small receptive forms (green) to larger ripe ones (pink). In monoecious species (A) all figs are essentially the same and produce both wasps (black) and seeds (yellow). In dioecious species, male figs (B) give rise only to wasps, whereas female figs (C) nurture only seeds and contain no male flowers (blue) (Cook & Rasplus, 2003).

fate of fig flowers, the plants are selected to produce both seeds and wasps (to carry their pollen). However, short-term natural selection will favour wasps that maximize offspring production without setting aside some flowers for seed production. This conflict must be stabilized for long-term persistence of the mutualism. In the case of dioecious figs, the plants appear to be in control because they lure female wasps into female figs, where they cannot lay eggs. Intersexual mimicry through fig chemical cues contributes to this plant control of the conflict situation (Soler, Proffit, Bessiere, Hossaert-McKey, & Schatz, 2012). In *F. carica*, a particular form of seasonality also ensures that many adult female wasps are emerging at a time when the vast majority of available receptive figs are female.

However, over half of all fig species are monoecious, and every fig tends to produce both seeds and wasp offspring. In some respects this makes things simpler: searching female wasps can enter any receptive fig of the right species to lay eggs. However, it makes it harder to understand how the tree limits seed predation and stabilizes the mutualism in the longer term. In contrast to the situation in dioecious trees, the distribution of style lengths is continuous. Moreover, in many species the wasps appear to have ovipositors long enough to lay eggs in most or all of the flowers that they encounter. Current evidence suggests that multiple mechanisms may have a role. In some species, figs may penalize wasps that do not pollinate actively by aborting fruits (Jandér, Herre, & Simms, 2012). In others, plant control of the closing of the ostiole (the only way into the fig) limits the number of female wasps that can lay egg inside and there are 'not enough eggs' to prevent seed production (Nefdt & Compton, 1996). Finally, different flowers may offer different fitness outcomes for wasp offspring, so that 'optimal foraging' for oviposition sites by female wasps may cause a slowing of the rate of seed destruction (Dunn, Segar, et al., 2008; Dunn, Yu, Ridley, & Cook, 2008; Yu et al., 2004). This is an exciting but complex topic (Cook & Rasplus, 2003; Herre, 1989) and an active area of debate and research (Herre, Jandér, & Machado, 2008).

4.2.3 Coadaptation and Cospeciation

The conflict over the fate of seeds is a key coevolutionary issue, but other impressive examples of coadaptation (i.e., mutual adaptation) can be found in the fig–pollinator mutualism. Kjellberg et al. (2001) showed that fig species with active pollinators invest far less in pollen production than those with passive pollinators. In fact, those with active pollinators have only about 10% male flowers while those with passive pollinators have about 30%.

Moreover, there is also a key difference in pollen liberation—passively pollinated species tend to burst their pollen sacs so that the pollen grains cover wasps, while this is not necessary in actively pollinated figs because the wasps deliberately collect the pollen. Importantly, there have been several changes between active and passive pollination in the evolutionary history of the mutualism but all cases investigated fit the coadaptive pattern of active wasps/low pollen or passive wasps/high pollen (Cook, Bean, Power, & Dixon, 2004; Kjellberg et al., 2001).

Another example of coadaptation is provided by correlated changes in wasp ovipositor length and fig breeding system (monoecy/dioecy) (Weiblen, 2004). In dioecious figs, wasps are faced with two highly divergent types of flowers, those with very short (in male figs) and those with very long (in female figs) styles. The long styles are too long for any wasps to lay eggs as far down as the fig ovule while the short styles require only a short ovipositor, so wasps from dioecious figs have short ovipositors. In contrast, wasps from monoecious figs face a continuous distribution of styles, lengths and a longer ovipositor makes more of these available for egg-laying. In fact, there is another, intriguing, twist to this story. Wasps that enter female dioecious figs are faced by flowers with long styles; however, they are so long that the wasps fail to reproduce. Consequently, all wasps in dioecious figs stem from ancestors that always reproduced in male figs, faced by short styles, so there is no successful natural selection for longer ovipositors. *In Ficus* veritas!

It is clear that figs and pollinators have coevolved extensively, leading to strong patterns of coadaptation of key traits. Consequently, it has long been suspected that figs and fig wasps have cospeciated, but is this the case? Strict cospeciation from a single origin of the mutualism should lead to 1:1 mapping of extant species associations and perfect phylogenetic correspondence. Such patterns are seen in some mutualisms (e.g., insects and obligate bacterial endosymbionts), where the symbiont spends its whole life cycle inside the host, but this is not the case for figs and wasps. Each adult female wasp chooses a new host individual, which is unlikely to be the same one she was born in (so not strict vertical transmission) and could even be a different species (host shift). Despite this, evidence from paired sampling of about 200 fig species and their pollinators support a long-term evolutionary history dominated by cospeciation (Cruaud et al., 2012), although with evidence of some host shifts and duplication of wasp lineages on a single host plant.

The other pattern to consider is present day fig—wasp associations. It was long thought that most fig species had a single (and unshared) pollinator species. However, it has become increasingly obvious that many fig species have more than one pollinator species (Cook & Segar, 2010; Cook & Rasplus, 2003; Machado, Robbins, Gilbert, & Herre, 2005), though the number of pollinator species involved with a given fig is still small (2–5) (Cook & Segar, 2010; Darwell, al-Beidh, & Cook, 2014). Nevertheless, most of these pollinator species are specific to a single fig species, although there are exceptions to this, particularly from studies of the *Ficus* section *Americana* in Panama (Machado et al., 2005). Where one fig species is associated with multiple pollinator species, these may essentially replace each other geographically over a large host plant range (Silvieus, Clement, & Weiblen, 2007), or in other cases co-occur extensively at the same sites (Haine, Martin, & Cook, 2006; Machado et al., 2005). Yang et al. (2015) revealed that the former may be more common in dioecious figs and the latter in monoecious species (Fig. 3).

So, the long-term pattern of fig–pollinator diversification is dominated by cospeciation (Cruaud et al. 2012), but examination of current partner fidelity shows that many figs have multiple pollinator species and a few figs share pollinator species (Cook & Segar, 2010; Machado et al., 2005). These patterns are not necessarily incompatible, but may reflect processes over different timescales (Cook & Segar, 2010). Figs that currently have multiple pollinator species may have sister species of wasps that speciated recently without the host fig speciating, or they have acquired a new wasp species by host shift. These cases may generally be transient so that over longer time span only one wasp lineage persists, usually contributing to the long-term cospeciation signal, but occasionally to a host shift. In the medium term two wasp lineages may occasionally coexist for several millennia, as seen in some groups of African *Galogylychia* figs (McLeish & van Noort, 2012), but this is probably rare. One of the biggest unanswered questions is why the wasps apparently speciate frequently without fig speciation. This might be driven by factors largely unconnected to the mutualism with figs.

4.3 Obligate Pollination Mutualisms Interacting With Other Trophic Levels

Emitting an olfactive signal is a good way to signal to pollinators; however, a signal linked to a resource is also subject to exploitation by other kinds of insects. Nursery pollination mutualisms do not interact in isolation, but are integrated into ecosystems. Indeed, the scent emitted by host plants is also perceived by other insects that use this signal to locate and exploit resources exchanged by mutualists. In nursery pollination mutualisms, both parasites (Cook & Rasplus, 2003; Jaeger et al., 2000) and predators (Bronstein, 1988; Schatz & Hossaert-McKey, 2003; Hossaert-McKey et al., 2010) have been reported to exploit plant seeds or ovaries, pollinators and/or associated insects. Hossaert-McKey et al. (2010) identified that interactions with third actors are known only in four types of nursery pollination mutualisms, but have been mainly investigated in the fig—fig wasp mutualism (Fig. 2). The role of these third actors certainly remains underestimated given that most of these mutualism occur in the tropics where biodiversity and interactions are often high. They are increasingly investigated with approaches from community ecology and network interactions (Schatz, Proffit, Kjellberg, & Hossaert-McKey, 2013; Wang, Geng, Ma, Cook, & Wang, 2014). However, investigations in other nursery pollination mutualisms would be relevant for both fundamental knowledge and comparative analyses.

4.3.1 A Community of Parasites on the Fig—Fig Wasp Mutualism

Several nonpollinating fig wasps (NPFWs) are known to exploit the OPM between figs and their pollinators (e.g., Kerdelhué, Rossi, & Rasplus, 2000; Segar, Dunn, Darwell, & Cook, 2014). Three main feeding strategies of NPFWs have been distinguished: gall makers, which transform fig ovaries into galls (their larvae feeding on these galled ovaries); inquilines, which feed on the gall tissue induced by pollinator larvae or by nonpollinating gall makers; and parasitoids, whose larvae feed on the larvae of other wasps (pollinators or NPFWs). Most NPFWs are assumed to be associated with a single *Ficus* species, but sometimes a single NPFW species exploits multiple fig species (McLeish, van Noort, & Tolley, 2010; Silvieus et al., 2007). A fig species can host up to 30 NPFW species (Cook & Rasplus, 2003), which raises the question of niche partitioning (Kerdelhué et al., 2000; Segar et al., 2014).

NPFWs must detect the exact time window when a particular resource is present at the right developmental stage in order to exploit it efficiently (Kerdelhué et al., 2000; Marr, Brock, & Pellmyr, 2001; Schatz & Hossaert-McKey, 2010) (Fig. 4). Such synchronization is achieved thanks to chemical mediation and NPFWs detect and use the progressive change of scent emitted by figs along the developmental stages (Proffit, Schatz, Borges, & Hossaert-McKey, 2007; Proffit et al., 2009). Extraction of volatile compounds during fig development and olfactometer experiments have been combined to demonstrate that NPFWs are attracted only by the specific scent of their own fig host and not those emitted by other sympatric

Figure 4 Nonpollinating fig wasps laying their eggs, thanks to their extraordinary long ovipositor, from the outside surface of figs. Left: *Sycoscapter australis* on *Ficus macrophylla*; right: *Phylotripesis caricae* on *Ficus carica*. *(Left) photo: J. Cook; (Right) photo: B. Schatz.*

fig species (Proffit et al., 2009). Moreover, each NPFW species is able to detect the appropriate phase of fig development it uses for oviposition (Proffit et al., 2007).

Interestingly, it has also been suggested that NPFWs may help to stabilize the conflict between monoecious figs and pollinators over the fate of fig ovules (Dunn, Segar, et al., 2008). The logic here is that outer fig ovules, close to the fig wall, are more accessible than inner fig ovules to female NPFWs laying eggs through the fig wall (Al Beidh, Dunn, Power, & Cook, 2012). Consequently, pollinator larvae face a higher risk of parasitism in the more numerous outer ovules than the smaller number of inner ovules (Dunn, Segar, et al., 2008). This should result in selection for female pollinators, when laying eggs inside the fig, to focus strongly on finding inner ovules in which to lay eggs, thus reducing the overall rate of seed predation.

The diversity of parasites remains poorly documented in other OPMs, with limited information on only a few other cases, e.g., *Yucca filamentosa* L. and *Trollius europaeus* L. In the latter case, several species of *Chiastochaeta* flies (Diptera, Anthomyiidae) visit the flowers successively. The earliest-arriving fly species acts as the pollinator, while later-arriving species are seed parasites only (Jaeger et al., 2000). However, the proximate mechanism explaining the succession of species is still to be discovered in this system. For *Yucca filamentosa*, two species of nonpollinating yucca moths co-occur but we lack knowledge about the signals used by these parasites to find their host. The moths appear to have evolved from a mutualist species, representing a reversion from mutualism to parasitism (Marr et al., 2001).

4.3.2 A Community of Predators on the Fig—Fig Wasp Mutualism

Insect predators act directly and indirectly in populations of herbivores and parasites involved in various plant—insect interactions. Among the different OPMs, ants are well known to be major and opportunistic predators in the fig—fig wasp mutualism (Bronstein, 1988; Rodriguez-Girones, Gonzalvez, Llandres, Corlett, & Santamaria, 2013; Schatz, Proffit, Rakhi, Borges, & Hossaert-McKey, 2006; Schatz, Kjellberg, Nyawa, & Hossaert—McKey, 2008). OPMs are generally associated with the presence near flowers of both pollinators and other insect visitors, which constitute a potential source of prey. As a result, the presence during the emergence phase of numerous fig wasps on fig surfaces often attracts various arthropod predators such as mantids, spiders and predatory beetles (Bronstein, 1988; Schatz & Hossaert-McKey, 2003; Zachariades, 1994; Zachariades, Compton, & Schatz, 2009). A large number of ant species are often present on figs (exploiting sugar from open fruits, other prey, attending aphids, etc.), but the main effect on the fig mutualism is often from ants specialized in arboreal predation such as *Oecophylla* and *Crematogaster* species (Hymenoptera, Formicidae). Bain, Harrison, and Schatz (2014) recently reviewed the known diversity of 'how to be an ant on figs'.

In fig—wasp systems, ants are the dominant predators of fig wasps (both pollinators and NPFWs) (Bain et al., 2014; Bronstein, 1988; Schatz et al., 2008). Ants associated with *F. carica* prey mostly on pollinators (Schatz & Hossaert-McKey, 2003), while in other cases they also prey on pollinators and NPFWs on tropical fig species (Bain et al., 2014; Schatz et al., 2006; Wang et al., 2014). Unsuccessful predation attempts during fig patrolling greatly disturb the oviposition behaviour of NPFWs on the outer surface of figs. Both predation and disturbance can reduce the negative effects of NPFW and benefit the mutualists. Ants are particularly efficient in the capture of fig wasps, detecting the scent they emit and exhibiting olfactory learning (Ranganathan & Borges, 2009; Schatz, Anstett, Out, & Hossaert-McKey, 2003; Schatz & Hossaert-McKey, 2010). Moreover, as fig scent change with time (Proffit et al., 2008), some ants are also able to learn the specific scent associated with developmental phases when the probability of the presence of fig wasps is higher, and to modulate their presence and their aggressiveness using these olfactory cues (Schatz & Hossaert-McKey, 2010). Such abilities explain why ants may be observed to anticipate the exit of fig wasps and how they exert a strong predation pressure on fig wasps (pollinators and NPFWs) (Bain et al., 2014; Schatz & Hossaert-McKey,

2010) and then be more abundant to male trees where fig wasps are more occurrent (Schatz et al., 2008).

Among other nursery pollination mutualisms, ants may also play a role as indirect mutualistic partners in yucca—yucca moth mutualisms. Indeed, Perry, Mondor, and Addicott (2004) showed that *Formica* species on *Yucca glauca* Nutt. deter oviposition of the nonpollinating moth *Tegeticula corruptrix* (Pellmyr, 1999) but not that of the pollinating moth *Tegeticula yuccasella* (Riley, 1872) (Lepidoptera, Prodoxidae). Unfortunately, no studies have investigated the possible role of chemical mediation in this system.

4.3.3 Towards a Network of Interactions

The OPM between figs and fig wasps is associated with numerous resources (galls, larvae, various insects used as prey, carbohydrates, etc.), which are exploited by many insect species, building a complex network of interactions (Fig. 4) (Bain et al., 2014; Bronstein, 1988; Schatz et al., 2006, 2008; Schatz, Proffit, et al., 2013; Wang et al., 2014; Zachariades, 1994). The majority of insects flying above the tropical forest canopy are linked to figs, and notably fig wasps (Harrison, 1996). Hitherto, NPFWs and ants on figs are the sole species acting as third actors in which the exploitation of the chemical mediation linked to nursery pollination mutualism has been demonstrated (Schatz & Hossaert-McKey, 2010). In this network, interactions exists among third actors; experimental exclusion of ants conduced on three fig species led to higher abundance of NPFWs, showing that ant presence can confer substantial protection upon pollinator larvae (Schatz et al., 2006). Wang et al. (2014) even demonstrated that predation by weaver ant *Oecophylla smaragdina* (Fabricius, 1775) (Hymenoptera, Formicidae) limits the success of the NPFWs *Platyneura mayri* Rasplus (Hymenoptera, Agaonidae) in *F. racemosa* and therefore indirectly benefits the mutualism by increasing the reproductive success of both the pollinator and the plant. Maschwitz, Fiala, Saw, Norma-Rashid, and Azarae (1994) and Bain et al. (2012) discovered two fig species *Ficus obscura* Blume var. *borneensis* (Miq.) and *F. subpisocarpa* Gagnep that have hollow internodes of young branches that ants may inhabit, suggesting that figs are sometimes involved in a plant—ant mutualism.

Taken together, the complexity of interactions and the diversity of involved species suggest that figs represent a keystone genus in numerous tropical habitats, since their presence permits that of numerous species of insects, particularly Hymenoptera (Borges, Bessière, & Hossaert-McKey, 2008; Lambert, & Marshall, 1991). In an era of global change, which tends to simplify interspecific interactions, figs constitute fascinating models to investigate the impact of third actors on mutualisms and make comparative analyses (Bain et al., 2014; Schatz et al., 2008; Wang et al., 2014).

5. CONCLUSION

The diversity of plant utilization strategies by insects offers an array of possible human actions aiming at protecting crops and improving their productivity. The target of these actions must necessarily be the mechanisms that mediate insect–plant interactions. The various ways that phytophagous insects exploit host plant chemicals to defend themselves against their predators create a number of possibilities. Among others, ways to disturb the metabolism by which the herbivore resists toxins produced by host plants merit investigation. A complementary approach would be to modify or combine plant toxins to prevent the adaption of the herbivore. The various mechanisms underlying plant manipulation by insects also could be exploited to improve the nutritional value of crops by mimicking the nutrient sinks induced by plant-manipulating insects. Alternatively, a better understanding of the ecology of plant-manipulating insects and their interactions with their host plant could ultimately result in the selection of resistant plant genotypes and in designing cultural practices which might, through a modulation of the physiological status of the host plant, limit plant manipulation. Finally, the microorganisms that assist herbivores in their manipulation of plants also could be targets of human actions. In the field of plant–insect mutualisms, new possible actions may be considered to improve the efficacy of pollination services, including manipulation of the plant's chemical signals to facilitate its discovery by pollinators.

From a pure scientific aspect, utilization of plants by phytophagous insects represents a fascinating interaction. Phytophagous insects indeed face a very particular challenge. They feed on plants that both are not nutritionally adequate for them and possess efficient defence mechanisms. Under the strong selection pressure resulting from such a situation, insects have adapted to the constraints imposed by the plants through much diversified strategies, which can fit in three levels of complexity. (1) They have exploited and misappropriated the plant secondary metabolites at their own advantage, for protection or communication purposes; (2) they have modified plant tissues and adapted them to their needs using various more or less sophisticated ways, ranging from simple physical to complex molecular interactions;

(3) they have even sometimes developed elaborated mutualism coevolutions resulting in benefits for both the plant and the insect. All these strategies are essential subjects of intense research activities in the field of plant—insect interactions. A particularly interesting aspect which deserves to be clarified is how antagonistic relations have evolved to mutualistic situations.

REFERENCES

Abrahamson, W. G., & Weis, A. E. (1987). Nutritional ecology of arthropod gall makers. In F. Slansky, & J. G. Rodriguez (Eds.), *Nutritional ecology of insects, mites spiders, and related invertebrates* (pp. 235—258). New York: Wiley.

Adams, A. S., Currie, C. R., Cardoza, Y., Klepzig, K. D., & Raffa, K. F. (2009). Effects of symbiotic bacteria and tree chemistry on the growth and reproduction of bark beetle fungal symbionts. *Canadian Journal of Forest Research, 39*, 1133—1147. http://dx.doi.org/10.1139/X09-034.

Aggarwal, R., Subramanyam, S., Zhao, C., Chen, M. S., Harris, M. O., & Stuart, J. J. (2014). Avirulence effector discovery in a plant galling and plant parasitic arthropod, the hessian fly (*Mayetiola destructor*). *PLoS One, 9*, e100958. http://dx.doi.org/10.1371/journal.pone.0100958.

Agrawal, A. A., & Weber, M. G. (2015). On the study of plant defence and herbivory using comparative approaches: How important are secondary plant compounds. *Ecology Letters, 18*, 985—991. http://dx.doi.org/10.1111/ele.12482.

Al Beidh, S., Dunn, D. W., Power, S. A., & Cook, J. M. (2012). Parasites and mutualism function: Measuring enemy free space in a fig—pollinator symbiosis. *Oikos, 121*, 1833—1839. http://dx.doi.org/10.1111/j.1600-0706.2011.20179.x.

Anderson, K. G., & Harris, M. O. (2006). Does *R* gene resistance to Hessian fly allow wheat seedlings to escape larval-induced growth deficits? *Journal of Economic Entomology, 99*, 1842—1853. http://dx.doi.org/10.1093/jee/99.5.1842.

Bain, A., Chantarasuwan, B., Chou, L. S., Hossaert-McKey, M., Schatz, B., & Kjellberg, F. (2012). A new case of ants nesting within branches of a fig tree: The case of *Ficus subpisocarpa* in Taiwan. *Sociobiology, 59*, 415—434.

Bain, A., Harrison, R., & Schatz, B. (2014). How to be an ant on figs? *Acta Oecologica, 57*, 97—108. http://dx.doi.org/10.1016/j.actao.2013.05.006.

Bartlett, L., & Connor, E. F. (2014). Exogenous phytohormones and the induction of plant galls by insects. *Arthropod—Plant Interactions, 8*, 339—348. http://dx.doi.org/10.1007/s11829-014-9309-0.

Bascompte, J., & Jordano, P. (2007). Plant-animal mutualistic networks: The architecture of biodiversity. *Annual Review of Ecology, Evolution and Systematics, 38*, 567—593. http://dx.doi.org/10.1146/annurev.ecolsys.38.091206.095818.

Bergström, G., Groth, I., Pellmyr, O., Endressb, P. K., Thienb, L. B., Hübenerd, A., & Francke, W. (1991). Chemical basis of a highly specific mutualism: Chiral esters attract pollinating beetles in Eupomatiaceae. *Phytochemistry, 30*, 3221—3225. http://dx.doi.org/10.1016/0031-9422(91)83179-O.

Bernays, E. A., & Janzen, D. H. (1988). Saturniid and Sphingid caterpillars — 2 ways to eat leaves. *Ecology, 69*, 1153—1160.

Berryman, A. A. (1972). Resistance of conifers to invasion by bark beetle-fungus associations. *BioScience, 22*, 298—602. http://dx.doi.org/10.2307/1296206.

Berryman, A. A. (1976). Theoretical explanation of mountain pine beetle dynamics in lodgepole pine forests. *Environmental Entomology, 5*, 1225—1233. http://dx.doi.org/10.1093/ee/5.6.1225.

Betz, O., & Kölsch, G. (2004). The role of adhesion in prey capture and predator defence in arthropods. *Arthropod Structure & Development, 33*, 3–30.
Björkman, C. (1998). Opposite, linear and non-linear effects of plant stress on a galling aphid. *Scandinavian Journal of Forest Research, 13*, 177–183. http://dx.doi.org/10.1016/j.asd.2003.10.002.
Body, M., Kaiser, W., Dubreuil, G., Casas, J., & Giron, D. (2013). Leaf-miners co-opt microorganisms to enhance their nutritional environment. *Journal of Chemical Ecology, 39*, 969–977. http://dx.doi.org/10.1007/s10886-013-0307-y.
Borges, R., Bessière, J. M., & Hossaert-McKey, M. (2008). The chemical ecology of seed dispersal in monoecious and dioecious figs. *Functional Ecology, 22*, 484–493. http://dx.doi.org/10.1111/j.1365-2435.2008.01383.x.
Bowers, D. M. (1990). Recycling plant natural products for insect defence. In D. L. De Evans, & J. O. Schmidt (Eds.), *Insect Defences: Adaptive mechanisms and strategies of prey and predators* (pp. 353–383). New York: SUNY Press.
Bridges, J. R., Nettleton, W. A., & Conner, M. D. (1985). Southern pine beetle (Coleoptera: Scolytidae) infestations without the blue-stain fungus, Ceratocystis minor. *Journal of Economic Entomology, 78*, 325–327. http://dx.doi.org/10.1093/jee/78.2.325.
Bronner, R. (1992). The role of nutritive cells in the nutrition of cynipids and cecidomyiids. In J. D. Shorthouse, & O. Rohfritsch (Eds.), *Biology of insect-induced galls* (pp. 118–140). Oxford: Oxford University Press.
Bronstein, J. L. (1988). Predators of fig wasps. *Biotropica, 20*, 215–219. http://dx.doi.org/10.2307/2388236.
Brower, L. P. (1995). Understanding and misunderstanding the migration of the monarch butterfly (Nymphalidae) in North America: 1857–1995. *Journal of the Lepidopterist's Society, 49*, 304–385.
Brown, M. W., Glenn, D. M., & Wisniewski, M. E. (1991). Functional and anatomical disruption of apple roots by the woolly apple aphid (Homoptera: Aphididae). *Journal of Economic Entomology, 84*, 1823–1826. http://dx.doi.org/10.1093/jee/84.6.1823.
Carneiro, R. G. S., Oliveira, D. C., & Isaias, R. M. S. (2014). Developmental anatomy and immunocytochemistry reveal the neo-ontogenesis of the leaf tissues of *Psidium myrtoides* (Myrtaceae) towards the globoid galls of *Nothotrioza myrtoidis* (Triozidae). *Plant Cell Reports, 33*, 2093–2106. http://dx.doi.org/10.1007/s00299-014-1683-7.
Chen, J., Henderson, G., & Laine, R. A. (1999). Lignoceric acid and hexacosanoic acid: Major components of soldier frontal gland secretions of the Formosan subterranean termite (*Coptotermes formosanus*). *Journal of Chemical Ecology, 25*, 817–824. http://dx.doi.org/10.1023/A:1020844817492.
Chen, C., & Song, Q. (2008). Responses of the pollinating wasp *Ceratosolen solmsi marchali* to odour variation between two floral stages of *Ficus hispida*. *Journal of Chemical Ecology, 34*, 1536–1544. http://dx.doi.org/10.1007/s10886-008-9558-4.
Chen, C., Song, Q. S., Proffit, M., Bessiere, J. M., Li, Z. B., & Hossaert-McKey, M. (2009). Private channel: A single unusual compound assures specific pollinator attraction in *Ficus semicordata*. *Functional Ecology, 23*, 941–950.
Chen, M. S., Zhao, H. X., Zhu, Y. C., Scheffler, B., Liu, X., Liu, X., … Stuart, J. J. (2008). Analysis of transcripts and proteins expressed in the salivary glands of Hessian fly (*Mayetiola destructor*) larvae. *Journal of Insect Physiology, 54*, 1–16. http://dx.doi.org/10.1016/j.jinsphys.2007.07.007.
Christiansen, E., Waring, R. H., & Berryman, A. A. (1987). Resistance of conifers to bark beetle attacks: Searching for general relationships. *Forest Ecology and Management, 22*, 89–106. http://dx.doi.org/10.1016/0378-1127(87)90098-3.
Chung, S. H., Rosa, C., Scully, E. D., Peiffer, M., Tooker, J. F., Hoover, K., … Felton, G. W. (2013). Herbivore exploits orally secreted bacteria to

suppress plant defenses. *Proceedings of the National Academy of Sciences of the United States of America, 110*, 15728–15733. http://dx.doi.org/10.1073/pnas.1308867110.

Clark, B. C., & Faeth, S. H. (1997). The consequences of larval aggregation in the butterfly *Chlosyne lacinia*. *Ecological Entomology, 22*, 408–415. http://dx.doi.org/10.1046/j.1365-2311.1997.00091.x.

Consales, F., Schweizer, F., Erb, M., Gouhier-Darimont, C., Bodenhausen, N., Bruessow, F., … Reymond, P. (2012). Insect oral secretions suppress wound-induced responses in Arabidopsis. *Journal of Experimental Botany, 63*, 727–737. http://dx.doi.org/10.1093/jxb/err308.

Cook, J. M., Bean, D., Power, S. A., & Dixon, D. J. (2004). Evolution of a complex coevolved trait: Active pollination in a genus of fig wasps. *Journal of Evolutionary Biology, 17*, 238–246. http://dx.doi.org/10.1111/j.1420-9101.2003.00683.x.

Cook, J. M., & Rasplus, J. Y. (2003). Mutualists with attitude: Coevolving fig wasps and figs. *Trends in Ecology & Evolution, 18*, 241–248. http://dx.doi.org/10.1016/S0169-5347(03)00062-4.

Cook, J. M., & Segar, S. T. (2010). Speciation in fig wasps. *Ecological Entomology, 35*, 54–66. http://dx.doi.org/10.1111/j.1365-2311.2009.01148.x.

Cosby, L. A., Pardini, R. S., Biagini, R. E., Lambert, T. L., Lin, A. J., Huang, V. M., … Sartorelli, A. C. (1976). Mode of action of the bioreductive alkylating agent, 2,3-bis(chloromethyl)-1,4-naphthoquinone. *Cancer Research, 36*, 4023–4031.

Cruaud, A., Rønsted, N., Chantarasuwan, B., Chou, L. S., Clement, W. L., Couloux, A., … Hossaert-McKey, M. (2012). An extreme case of plant–insect codiversification: Figs and fig-pollinating wasps. *Systematic Biology, 61*, 1029–1047. http://dx.doi.org/10.1093/sysbio/sys068.

Dardeau, F., Body, M., Berthier, A., Miard, F., Christidès, J. P., Feinard-Duranceau, M., … Sallé, A. (2015). Effects of fertilisation on amino acid mobilisation by a plant-manipulating insect. *Ecological Entomology, 40*, 814–822. http://dx.doi.org/10.1111/een.12274.

Dardeau, F., Deprost, E., Laurans, F., Lainé, V., Lieutier, F., & Sallé, A. (2014). Resistant poplar genotypes inhibit pseudogall formation by the wooly poplar aphid, *Phloeomyzus passerinii* Sign. *Trees, 28*, 1007–1019. http://dx.doi.org/10.1007/s00468-014-1014-1.

Dardeau, F., Pointeau, S., Ameline, A., Laurans, F., Cherqui, A., Lieutier, F., & Sallé, A. (2014). Host manipulation by a herbivore optimizes its feeding behaviour. *Animal Behaviour, 95*, 49–56. http://dx.doi.org/10.1016/j.anbehav.2014.06.002.

Darwell, C. T., al-Beidh, S., & Cook, J. M. (2014). Molecular species delimitation of a symbiotic fig-pollinating wasp species complex reveals extreme deviation from reciprocal partner specificity. *BMC Evolutionary Biology, 14*, 189–199. http://dx.doi.org/10.1186/s12862-014-0189-9.

Dettner, K., Scheuerlein, A., Fabian, P., Schulz, S., & Francke, W. (1996). Chemical defence of the giant springtail *Tetrodontophora bielanensis* (Waga) (Insecta: Collembola). *Journal of Chemical Ecology, 22*, 1051–1074. http://dx.doi.org/10.1007/BF02029954.

Dobler, S. (2001). Evolutionary aspects of defence by recycled plant compounds in herbivorous insects. *Basic in Applied Ecology, 2*, 15–26. http://dx.doi.org/10.1078/1439-1791-00032.

Dor, A., Machkour-M'Rabet, S., Legal, L., Williams, T., & Hénaut, Y. (2008). Chemically-mediated intraspecific recognition in the Mexican tarantula *Brachypelma vagans*. *Naturwissenschaften, 95*, 1189–1193. http://dx.doi.org/10.1007/s00114-008-0441-5.

Dötterl, S., Burkhardt, D., Jürgens, A., & Mosandl, A. (2007). Stereoisomeric pattern of lilac aldehyde in *Silene latifolia*, a plant involved in a nursery pollination system. *Phytochemistry, 68*, 499–504.

Dötterl, S., Jürgens, A., Seifert, K., Laube, T., Weissbecker, B., & Schutz, S. (2006). Nursery pollination by a moth in *Silene latifolia*: The role of odours in eliciting antennal and behavioural responses. *New Phytologist, 169*, 707–718. http://dx.doi.org/10.1111/j.1469-8137.2005.01509.x.

Dufaÿ, M., & Anstett, M. C. (2003). Conflicts between plants and pollinators that reproduce within inflorescences: Evolutionary variations on a theme. *Oikos, 100*, 3–14. http://dx.doi.org/10.1034/j.1600-0706.2003.12053.x.

Dufaÿ, M., Hossaert-McKey, M., & Anstett, M. C. (2003). When leaves act like flowers: Howdwarf palms attract their pollinators. *Ecology Letters, 6*, 28–34. http://dx.doi.org/10.1046/j.1461-0248.2003.00382.x.

Dufaÿ, M., Hossaert-McKey, M., & Anstett, M. C. (2004). Temporal and sexual variation of leaf produced pollinator-attracting odours in the dwarf palm. *Oecologia, 139*, 392–398. http://dx.doi.org/10.1007/s00442-004-1538-y.

Duffey, S. S. (1980). Sequestration of plant natural products by insects. *Annual Review of Entomology, 25*, 447–477. http://dx.doi.org/10.1146/annurev.en.25.010180.002311.

Dunn, D. W., Segar, S. T., Ridley, J., Chan, R., Crozier, R. H., Douglas, W. Y., & Cook, J. M. (2008). A role for parasites in stabilizing the fig-pollinator mutualism. *PLoS Biology, 6*, 59–70. http://dx.doi.org/10.1371/journal.pbio.0060059.

Dunn, D. W., Yu, D. W., Ridley, J., & Cook, J. M. (2008). Longevity, early emergence and body size in a pollinating fig wasp—implications for stability in a fig—pollinator mutualism. *Journal of Animal Ecology, 77*, 927–935. http://dx.doi.org/10.1111/j.1365-2656.2008.01416.x.

Dyer, L. A. (1995). Tasty generalists and nasty specialists? Antipredator mechanisms in tropical lepidopteran larvae. *Ecology, 76*, 1483–1496. http://dx.doi.org/10.2307/1938150.

Eisner, T., Aneshansley, D. J., Eisner, M., Attygalle, A. B., Alsop, D. W., & Meinwald, J. (2000). Spray mechanism of the most primitive bombardier beetle (*Metrius contractus*). *The Journal of Experimental Biology, 203*, 1265–1275.

Eisner, T., Goetz, M. A., Hil, D. E., Smedley, S. R., & Meinwald, J. (1997). Firefly "femmes fatales" acquire defensive steroids (lucibufagins) from their firefly prey. *Proceedings of the National Academy of Sciences of the United States of America, 94*, 9723–9728.

Erb, M., Meldau, S., & Howe, G. A. (2012). Role of phytohormones in insect-specific plant reactions. *Trends in Plant Science, 17*, 250–259. http://dx.doi.org/10.1016/j.tplants.2012.01.003.

Fink, L. S., & Brower, L. P. (1981). Birds can overcome the cardenolide defence of monarch butterflies in Mexico. *Nature, 291*, 67–70. http://dx.doi.org/10.1038/291067a0.

Frick, C., & Wink, M. (1995). Uptake and sequestration of ouabain and other cardiac glycosides in *Danaus plexippus* (Lepidoptera: Danaidae): Evidence for a carrier-mediated process. *Journal of Chemical Ecology, 21*, 557–575. http://dx.doi.org/10.1007/BF02033701.

Fürstenberg-Hägg, J., Zagrobelny, M., Jørgensen, K., Vogel, H., Møller, B. L., & Bak, S. (2014). Chemical defence balanced by sequestration and de novo biosynthesis in a Lepidopteran specialist. *PLoS One, 9*(10), e108745. http://dx.doi.org/10.1371/journal.pone.0108745.

Gimenez-Benavides, L., Dötterl, S., Jürgens, A., Escudero, A., & Iriondo, J. M. (2007). Generalist diurnal pollination provides greater fitness in a plant with nocturnal pollination syndrome: Assessing the effects of a *Silene—Hadena* interaction. *Oikos, 116*, 1461–1472. http://dx.doi.org/10.1007/BF02033701.

Giron, D., Dedeine, F., Dubreuil, G., Huguet, E., Mouton, L., Outreman, Y., & Simon, J.-C. (2017). Influence of microbial symbionts on plant-insect interactions. In N. Sauvion, D. Thiery, & P. A. Calatayud (Eds.), *Insect-Plant Interactions in a Crop Protection Perspective* (Vol. 81, pp. 225–257).

Giron, D., Frago, E., Glevarec, G., Pieterse, C. M. J., & Dicke, M. (2013). Cytokinins as key regulators in plant—microbe—insect interactions: Connecting plant growth and defence. *Functional Ecology, 27*, 599–609. http://dx.doi.org/10.1111/1365-2435.12042.

Giron, D., Huguet, E., Stone, G. N., & Body, M. (2016). Insect-induced effects on plants and possible effectors used by galling and leaf-mining insects to manipulate their host-plant. *Journal of Insect Physiology, 84*, 70–89. http://dx.doi.org/10.1016/j.jinsphys.2015.12.009.

Giron, D., Kaiser, W., Imbault, N., & Casas, J. (2007). Cytokinin-mediated leaf manipulation by a leafminer caterpillar. *Biology Letters, 3*, 340–343. http://dx.doi.org/10.1098/rsbl.2007.0051.

Glendinning, J. (2007). How do predators cope with chemically defended foods? *The Biological Bulletin, 213*, 252–266.

Goel, G., Makkar, P. S., Francis, G., & Becker, K. (2007). Phorbol esters: Structure, biological activity and toxicity in animals. *International Journal of Toxicology, 26*, 279–288. http://dx.doi.org/10.1080/10915810701464641.

Gols, R. (2014). Direct and indirect chemical defences against insects in a multitrophic framework. *Plant, Cell & Environment, 37*, 1741–1752. http://dx.doi.org/10.1111/pce.12318.

González, A., Hare, J. H., & Eisner, T. (1999). Chemical egg defence in *Photuris* firefly "femmes fatales". *Chemoecology, 9*, 177–185. http://dx.doi.org/10.1007/s000490050051.

González, A., Schroeder, F., Meinwald, J., & Eisner, T. (1999). N-methylquinolinium-2-carboxylate, a defensive betaine from *Photuris versicolor* fireflies. *Journal of Natural Products, 62*, 378–380. http://dx.doi.org/10.1021/np980400o.

Granett, J., Walker, M. A., Kocsis, L., & Omer, A. D. (2001). Biology and management of grape phylloxera. *Annual Review of Entomology, 46*, 387–412. http://dx.doi.org/10.1146/annurev.ento.46.1.387.

Guiguet, A., Dubreuil, G., Harris, M. O., Appel, H., Schultz, J. C., Pereira, M. H., & Giron, D. (2016). Shared weapons of blood- and plant-feeding insects: Surprising commonalities for manipulating hosts. *Journal of Insect Physiology, 84*, 4–21. http://dx.doi.org/10.1016/j.jinsphys.2015.12.006.

Gündüz, E. A., & Douglas, A. E. (2009). Symbiotic bacteria enable insect to use a nutritionally inadequate diet. *Proceedings of the Royal Society B – Biological Sciences, 279*, 987–991. http://dx.doi.org/10.1098/rspb.2008.1476.

Guo, W., Reigan, P., Siegel, D., & Ross, D. (2008). Enzymatic reduction and glutathione conjugation of benzoquinone ansamycin heat shock protein 90 inhibitors: Relevance for toxicity and mechanism of action. *Drug Metabolism and Disposition, 36*, 2050–2057. http://dx.doi.org/10.1124/dmd.108.022004.

Gutzwiller, F., Dedeine, F., Kaiser, W., Giron, D., & Lopez-Vaamonde, C. (2015). Correlation between the green-island phenotype and wolbachia infections during the evolutionary diversification of Gracillariidae leaf-mining moths. *Ecology and Evolution, 5*, 4049–4062. http://dx.doi.org/10.1002/ece3.1580.

Haine, E. R., Martin, J., & Cook, J. M. (2006). Deep mtDNA divergences indicate cryptic species in a fig-pollinating wasp. *BMC Evolutionary Biology, 6*, 83–94. http://dx.doi.org/10.1186/1471-2148-6-83.

Harris, M. O., Freeman, T. P., Rohfritsch, O., Anderson, K. G., Payne, S. A., & Moore, J. A. (2006). Virulent Hessian fly (Diptera: Cecidomyiidae) larvae induce a nutritive tissue during compatible interactions with wheat. *Annals of the Entomological Society of America, 99*, 305–316. http://dx.doi.org/10.1603/0013-8746(2006)099[0305:VHFDCL]2.0.CO;2.

Harris, M. O., Friessen, T. L., Xu, S. S., Chen, M. S., Giron, D., & Stuart, J. J. (2015). Pivoting from *Arabidopsis* to wheat to understand how agricultural plants integrate response to biotic stress. *Journal of Experimental Botany, 66*, 513–531. http://dx.doi.org/10.1093/jxb/eru465.

Harris, M. O., Stuart, J. J., Mohan, M., Nair, S., Lamb, R. J., & Rohfritsch, O. (2003). Grasses and gall midges: Plant defense and insect adaptation. *Annual Review of Entomology, 48*, 549–577. http://dx.doi.org/10.1146/annurev.ento.48.091801.112559.

Harrison, R. (1996). *The ecology of the fig-fig wasp mutualism in a lowland tropical forest in Sarawak, Malaysia (PhD thesis) Kyoto*. Japan: Kyoto University.
Harrison, R. D. (2005). Figs and the diversity of tropical rainforests. *BioScience, 55*, 1053–1064. http://dx.doi.org/10.1641/0006-3568(2005)055[1053.
Hartley, S. E. (1998). The chemical composition of plant galls: Are levels of nutrients and secondary compounds controlled by the gall-former? *Oecologia, 113*, 492–501. http://dx.doi.org/10.1007/s004420050401.
Hartley, S. E., & Lawton, J. H. (1992). Host-plant manipulation by gall-insects: A test of the nutrition hypothesis. *Journal of Animal Ecology, 61*, 113–119. http://www.jstor.org/stable/5514.
Heckel, D. G. (2014). Insect detoxification and sequestration strategies. *Annual Plant Review, 47*, 77–114. http://dx.doi.org/10.1002/9781118829783.ch3.
Hénaut, Y., Delmé, J., Legal, L., & Williams, T. (2005). Host selection by a kleptobiotic spider. *Naturwissenschaften, 92*, 95–99. http://dx.doi.org/10.1007/s00114-004-0597-6.
Herms, D. A., & Mattson, W. J. (1992). The dilemma of plants: To grow or defend. *Quarterly Review of Biology, 67*, 283–335. www.jstor.org/stable/2830650.
Herre, E. A. (1989). Coevolution of reproductive characteristics in 12 species of New World figs and their pollinator wasps. *Experientia, 45*, 637–647. http://dx.doi.org/10.1007/BF01975680.
Herre, E. A., Jandér, K. C., & Machado, C. A. (2008). Evolutionary ecology of figs and their associates: Recent progress and outstanding puzzles. *Annual Review of Ecology, Evolution and Systematics, 39*, 439–458. www.jstor.org/stable/30245171.
Hofstetter, R. W., Dinkins-Bookwalter, J., Davis, T. S., & Klepzig, K. D. (2015). Symbiotic associations of bark beetles. In F. E. Vega, & R. W. Hostetter (Eds.), *Bark beetles, Biology and ecology of native and invasive species* (pp. 209–245). Oxford: Elsevier.
Hogenhout, S. A., & Bos, J. I. (2011). Effector proteins that modulate plant-insect interactions. *Current Opinion in Plant Biology, 14*, 422–428. http://dx.doi.org/10.1016/j.pbi.2011.05.003.
Holzinger, F., & Wink, M. (1996). Mediation of cardiac glycoside insensitivity in the monarch butterfly (*Danaus plexippus*) – Role of an amino-acid substitution in the ouabain binding-site of Na+, K+ ATPase. *Journal of Chemical Ecology, 22*, 1921–1937. http://dx.doi.org/10.1007/BF02028512.
Hossaert-McKey, M., Soler, C., Schatz, B., & Proffit, M. (2010). Floral scents: Their role in nursery pollination mutualism. *Chemoecology, 20*, 75–88. http://dx.doi.org/10.1007/s00049-010-0043-5.
Hundsdörfer, A., Tshibangu, J. N., Wetterauer, B., & Wink, M. (2005). Sequestration of phorbol esters by aposematic larvae of *Hyles euphorbiae* (Lepidoptera: Sphingidae)? *Chemoecology, 15*, 261–267. http://dx.doi.org/10.1007/s00049-005-0321-9.
Ibanez, S., Dötterl, S., Anstett, M. C., Baudino, S., Caissard, J. C., Gallet, C., & Desprès, L. (2010). The role of volatile organic compounds, morphology and pigments of globeflowers in the attraction of their specific pollinating flies. *New Phytologist, 188*, 451–463. http://dx.doi.org/10.1111/j.1469-8137.2010.03317.x.
Ikai, N., & Hijii, N. (2007). Manipulation of tannins in oaks by galling cynipids. *Journal of Forest Research, 12*, 316–319. http://dx.doi.org/10.1007/s10310-007-0016-x.
IPBES. (2016). In S. G. Potts, V. L. Imperatriz-Fonseca, H. T. Ngo, J. C. Biesmeijer, T. D. Breeze, L. V. Dicks, et al. (Eds.), *Summary for policymakers of the assessment report of the intergovernmental science-policy platform on biodiversity and ecosystem services on pollinators, pollination and food production* (pp. 1–30).
Jaeger, N., Till-Bottraud, I., & Desprès, L. (2000). Evolutionary conflict between *Trollius europaeus* and its seed-parasite pollinators *Chiastocheta* flies. *Evolutionary & Ecological Research, 2*, 885–896.

Jandér, C. K., Herre, E. A., & Simms, E. L. (2012). Precision of host sanctions in the fig tree—fig wasp mutualism: Consequences for uncooperative symbionts. *Ecology Letters, 15*, 1362—1369. http://dx.doi.org/10.1111/j.1461-0248.2012.01857.

Jones, D. G., & Dangl, J. L. (2006). The plant immune system. *Nature, 444*, 323—329. http://dx.doi.org/10.1038/nature05286.

Jürgens, A., Witt, T., & Gottsberger, G. (2002). Flower scent composition in night-flowering *Silene* species (Caryophyllaceae). *Biochemistry Systematics Ecology, 30*, 383—397. http://dx.doi.org/10.1016/S0305-1978(01)00106-5.

Jürgens, A., Witt, T., & Gottsberger, G. (2003). Flower scent composition in *Dianthus* and *Saponaria* species (Caryophyllaceae) and its relevance for pollination biology and taxonomy. *Biochemistry Systematics Ecology, 31*, 345—357. http://dx.doi.org/10.1016/S0305-1978(02)00173-4.

Kaiser, W., Huguet, E., Casas, J., Commin, C., & Giron, D. (2010). Plant green-island phenotype induced by leaf-miners is mediated by bacterial symbionts. *Proceedings of the Royal Society B, 277*, 2311—2319. http://dx.doi.org/10.1098/rspb.2010.0214.

Kato, M., Takimura, A., & Kawakita, A. (2003). An obligate pollination mutualism and reciprocal diversification in the tree genus *Glochidion* (Euphorbiaceae). *Proceedings of the National Academy of Sciences of the United States of America, 100*, 5264—5267.

Kerdelhué, C., Rossi, J.-P., & Rasplus, J.-Y. (2000). Comparative community ecology studies on old world figs and fig wasps. *Ecology, 81*, 2832—2849. http://dx.doi.org/10.1890/0012-9658(2000)081[2832:CCESOO]2.0.CO;2.

Kirisits, T. (2004). Fungal associates of European bark beetles with special emphasis on the Ophiostomatoid fungi. In F. Lieutier, K. Day, A. Battisti, J. C. Grégoire, & H. Evans (Eds.), *Bark and wood boring insects in living trees in Europe, a synthesis* (pp. 181—235). Dordrecht: Kluwer.

Kjellberg, F., Jousselin, E., Bronstein, J. L., Patel, A., Yokoyama, J., & Rasplus, J.-Y. (2001). Pollination mode in fig wasps: The predictive power of correlated traits. *Proceedings of the Royal Society, London B, 268*, 1113—1121.

Klepzig, K. D., & Six, D. L. (2004). Bark beetle-fungal symbiosis: Context dependency in complex associations. *Symbiosis, 37*, 189—205.

Koricheva, J., Larsson, S., & Haukioja, E. (1998). Insect performance on experimentally stressed woody plants: A meta-analysis. *Annual Review of Entomology, 43*, 195—216. http://dx.doi.org/10.1146/annurev.ento.43.1.195.

Krall, B. S., Bartelt, R. J., Lewis, C. J., & Whitman, D. W. (1999). Chemical defence in the stink bug *Cosmopepla bimaculata*. *Journal of Chemical Ecology, 25*, 2477—2494. http://dx.doi.org/10.1023/A:1020822107806.

Kurzfeld-Zexer, L., Wool, D., & Inbar, M. (2010). Modification of tree architecture by a gall-forming aphid. *Trees, 24*, 13—18. http://dx.doi.org/10.1007/s00468-009-0374-4.

Lambert, F. R., & Marshall, A. G. (1991). Keystone characteristics of bird dispersed *Ficus* in a Malaysian lowland rain forest. *Journal of Ecology, 79*, 793—809. www.jstor.org/stable/2260668.

Lampert, E. C., & Bowers, M. D. (2010). Host plant influences on iridoid glycoside sequestration of generalist and specialist caterpillars. *Journal of Chemical Ecology, 36*, 1101—1104. http://dx.doi.org/10.1007/s10886-010-9849-4.

Lampert, E. C., Dyer, L. A., & Bowers, M. D. (2014). Dietary specialization and the effects of plant species on potential multitrophic interactions of three species of nymphaline caterpillars. *Entomologia Experimentalis et Applicata, 153*, 207—216.

Larson, K. C., & Whitham, T. G. (1991). Manipulation of food resources by a gall-forming aphid: The physiology of sink-source interactions. *Oecologia, 88*, 15—21. http://dx.doi.org/10.1007/BF00328398.

Larson, K. C., & Whitham, T. G. (1997). Competition between gall aphids and natural plant sinks: Plant architecture affects resistance to galling. *Oecologia, 109*, 575–582. http://dx.doi.org/10.1007/s004420050119.

Laurent, P., Braekman, J. C., Daloze, D., & Pasteels, J. (2003). Biosynthesis of defensive compounds from beetles and ants. *European Journal of Organic Chemistry, 15*, 2733–2743. http://dx.doi.org/10.1002/ejoc.200300008.

Laurent, P., Dooms, C., Braekman, J. C., Daloze, D., Habib-Jiwan, J. L., Rozenberg, A., … Pasteels, J. M. (2003). Oleanane glycosides in defensive glands of chrysomelid beetles are produced from dietary β-amyrin. *Naturwissenschaften, 90*, 524–527.

Lawson, S. P., Christian, N., & Abbot, P. (2014). Comparative analysis of the biodiversity of fungal endophytes in insect-induced galls and surrounding foliar tissue. *Fungal Diversity, 66*, 89–97. http://dx.doi.org/10.1007/s13225-013-0268-z.

Lees, D., Lopez-Vaamonde, C., & Augustin, S. (2011). Taxon page for *Cameraria ohridella* Deschka & Dimic 1986. In *Encyclopedia of life*. http://eolspecies.lifedesks.org/pages/8675.

Lemke, T., & Porembski, S. (2013). Variation in the reproductive performance of the *Trollius–Chiastocheta* mutualism at the edge of its range in north-east Germany. *Oecologia, 172*, 437–447.

Lieutier, F. (2004). Host resistance to bark beetles and its variations. In F. Lieutier, K. Day, A. Battisti, J. C. Grégoire, & H. Evans (Eds.), *Bark and wood boring insects in living trees in Europe, A synthesis* (pp. 135–180). Dordrecht: Kluwer.

Lieutier, F., Garcia, J., Yart, A., & Romary, P. (1995). Wound reaction of scots pine (*Pinus sylvestris* L.) to attacks by *Tomicus piniperda* L. and *Ips sexdentatus* Boern. (Coleoptera: Scolytidae). *Journal of Applied Entomology, 119*, 591–600. http://dx.doi.org/10.1111/j.1439-0418.1995.tb01341.x.

Lieutier, F., Yart, A., & Sallé, A. (2009). Stimulation of tree defenses by Ophiostomatoid fungi can explain attack success of bark beetles on conifers. *Annals of Forest Science, 66*, 1–22. http://dx.doi.org/10.1051/forest/2009066.

Lieutier, F., Yart, A., Ye, H., Sauvard, D., & Gallois, V. (2004). Between-isolate variations in the performances of *Leptographium wingfieldii* Morelet, a fungus associated with the bark beetle *Tomicus piniperda* L. *Annals of Forest Science, 61*, 45–53. http://dx.doi.org/10.1051/forest:2003083.

Lill, J. T., & Marquis, R. J. (2007). Microhabitat manipulation: Ecosystem engineering by shelter-building insects. *Theoretical Ecology Series, 4*, 107–138.

Liu, X., Bai, J., Huang, L., Zhu, L., Liu, X., Weng, N., … Chen, M. S. (2007). Gene expression of different wheat genotypes during attack by virulent and avirulent Hessian fly (*Mayetiola destructor*) larvae. *Journal of Chemical Ecology, 33*, 2171–2194. http://dx.doi.org/10.1007/s10886-007-9382-2.

Lucas, E., Coderre, D., & Brodeur, J. (1997). Instar-specific defense of *Coleomegilla maculata lengi* (Col.: Coccinellidae): Influence on attack success of the intraguild predator *Chrysoperla rufilabris* (Neur.: Chrysopidae). *BioControl, 42*, 3–12. http://dx.doi.org/10.1007/BF02769874.

Lucas-Silva, K., & Trigo, J. R. (2002). Structure-activity relationshipis of pyrrolizidine alkaloids in insect chemical defence against the orb-weaving spider *Nephila clavipes*. *Journal of Chemical Ecology, 28*, 657–668. http://dx.doi.org/10.1023/A:1015214422971.

Machado, C. A., Robbins, N., Gilbert, M. T. P., & Herre, E. A. (2005). Critical review of host specificity and its coevolutionary implications in the fig/fig-wasp mutualism. *Proceedings of the National Academy of Sciences of the United States of America, 102*, 6558–6565. http://dx.doi.org/10.1073/pnas.0501840102.

Malcolm, S. B., & Brower, L. P. (1989). Evolutionary and ecological implications of cardenolide sequestration in the monarch butterfly. *Cellular and Molecular Life Sciences, 45*, 284–295. http://dx.doi.org/10.1007/BF01951814.

Mani, M. S. (1964). In W. Junk (Ed.), *Ecology of plant galls*. The Hague, 156 pages.

Marini-Filho, O. J., & Fernandes, G. W. (2012). Stem galls drain nutrients and decrease shoot performance in *Diplusodon orbicularis* (Lythraceae). *Arthropod–Plant Interactions, 6*, 121–128. http://dx.doi.org/10.1007/s11829-011-9147-2.

Marr, D. L., Brock, M. T., & Pellmyr, O. (2001). Coexistence of mutualists and antagonists: Exploring the impact of cheaters on the yucca – yucca moth mutualism. *Oecologia, 128*, 454–463. http://dx.doi.org/10.1007/s004420100669.

Maschwitz, U., Fiala, B., Saw, L. G., Norma-Rashid, Y., & Azarae, H. I. (1994). *Ficus obscura* var. *borneensis* (Moraceae), a new non-specific ant-plant from Malaysia. *The Malayan Nature Journal, 47*, 409–416.

McKinnon, M. L., Quiring, D. T., & Bauce, E. (1999). Influence of tree growth rate, shoot size and foliar chemistry on the abundance and performance of a galling adelgid. *Functional Ecology, 13*, 859–867. http://dx.doi.org/10.1046/j.1365-2435.1999.00376.x.

McLeish, M. J., & van Noort, S. (2012). Codivergence and multiple host species use by fig wasp populations of the Ficus pollination mutualism. *BMC Evolutionary Biology, 12*, 1–12. http://dx.doi.org/10.1186/1471-2148-12-1.

McLeish, M. J., van Noort, S., & Tolley, K. A. (2010). African parasitoid fig wasp diversification is a function of *Ficus* species ranges. *Molecular Phylogenetics and Evolution, 57*, 122–134.

Mithöfer, A., & Boland, W. (2012). Plant defense against herbivores: Chemical aspects. *Annual Review of Plant Biology, 63*, 431–450. http://dx.doi.org/10.1146/annurev-arplant-042110-103854.

Müller, C., Agerbirk, N., Olsen, C. E., Boevé, J. L., Schaffner, U., & Brakefield, P. M. (2001). Sequestration of host-plant glucosinolates in the defensive hemolymph of the sawfly *Athalia rosae*. *Journal of Chemical Ecology, 27*, 2505–2516. http://dx.doi.org/10.1023/A:1013631616141.

Nabity, P. D., Haus, M. J., Berenbaum, M. R., & DeLucia, E. H. (2013). Leaf-galling phylloxera on grapes reprograms host metabolism and morphology. *Proceedings of the National Academy of Sciences of the United States of America, 110*, 16663–16668. http://dx.doi.org/10.1073/pnas.1220219110.

Nefdt, R. J. C., & Compton, S. G. (1996). Regulation of seed and pollinator production in the fig-fig wasp mutualism. *Journal of Animal Ecology, 65*, 170–182. www.jstor.org/stable/5720.

Neiland, M. R. M., & Wilcock, C. C. (1999). The presence of heterospecific pollen on stigmas of nectariferous and nectarless orchids and its consequences for their reproductive success. *Protoplasma, 1*, 65–75. http://dx.doi.org/10.1007/BF01279076.

Nishida, R., & Fukami, H. (1990). Sequestration of distasteful compounds by some pharmacophagous insects. *Journal of Chemical Ecology, 16*, 151–164. http://dx.doi.org/10.1007/BF01021276.

Nyman, T., & Julkunen-Tiitto, R. (2000). Manipulation of the phenolic chemistry of willows by gall-inducing sawflies. *Proceedings of the National Academy of Sciences of the United States of America, 97*, 13184–13187. http://dx.doi.org/10.1073/pnas.230294097.

Okamoto, T., Kawakita, A., & Kato, M. (2007). Interspecific variation of floral scent composition in *Glochidion* and its association with host-specific pollinating seed parasite (*Epicephala*). *Journal of Chemical Ecology, 33*, 1065–1081. http://dx.doi.org/10.1007/s10886-007-9287-0.

Oyeyele, S. O., & Zalucki, M. P. (2008). Cardiac glycosides and oviposition by *Danaus plexippus* on *Asclepias fruticosa* in south-east Queensland (Australia), with notes on the effect of plant nitrogen content. *Ecological Entomology, 15*, 177–185. http://dx.doi.org/10.1111/j.1365-2311.1990.tb00799.x.

Paine, T. D., Raffa, K. F., & Harrington, T. C. (1997). Interactions among Scolytid bark beetles, their associated fungi, and live host conifers. *Annual Review of Entomology, 42*, 179–206. http://dx.doi.org/10.1146/annurev.ento.42.1.179.

Palmeira, C. M., & Wallace, K. B. (1996). Benzoquinone inhibits the voltage-dependent induction of the mitochondrial permeability transition caused by redox-cycling naphthoquinones. *Toxicology and Applied Pharmacology, 143*, 338–347. http://dx.doi.org/10.1006/taap.1996.8099.

Pasteels, J., Duffey, S., & Rowell-Rahier, M. (1990). Toxins in chrysomelid beetles. Possible evolutionary sequence from de novo synthesis to derivation from food-plant chemicals. *Journal of Chemical Ecology, 16*, 211–222. http://dx.doi.org/10.1007/BF01021280.

Pasteels, J., Theuring, C., Witte, L., & Hartmann, T. (2003). Sequestration and metabolism of protoxic pyrrolizidine alkaloids by larvae of the leaf beetle *Platyphora boucardi* and their transfer via pupae into defensive secretions of adults. *Journal of Chemical Ecology, 29*, 337–355. http://dx.doi.org/10.1023/A:1022629911304.

Patt, J. M., Hartman, T. G., Creekmore, R. W., Elliott, J. J., Schal, C., Lech, J., & Rosen, R. T. (1992). The floral odor of *Peltandra virginica* contains novel trimethyl-2, 5-dioxa-bicyclo[3.2.l]nonanes. *Phytochemistry, 31*, 437–491. http://dx.doi.org/10.1016/0031-9422(92)90022-I.

Pellmyr, O., & Huth, C. J. (1994). Evolutionary stability of mutualism between yuccas and yuccas moths. *Nature, 372*, 257–260.

Perry, J. C., Mondor, E. B., & Addicott, J. F. (2004). An indirect mutualism: Ants deter seed predators from ovipositing in yucca fruit. *Canadian Journal of Zoology, 82*, 823–827. http://dx.doi.org/10.1139/z04-039.

Peschke, K., & Eisner, T. (1987). Defensive secretion of the tenebrionid beetle, *Blaps mucronata*: Physical and chemical determinants of effectiveness. *Journal of Comparative Physiology A Sensory Neural and Behavioral Physiology, 161*, 377–388. http://dx.doi.org/10.1007/BF00603963.

Petschenka, G., & Anurag, A. A. (2016). How herbivores co-opt plant defences: Natural selection, specialization, and sequestration. *Current Opinion in Insect Science, 14*, 17–24. http://dx.doi.org/10.1016/j.cois.2015.12.004.

Pincebourde, S., & Casas, J. (2006). Multitrophic biophysical budgets: Thermal ecology of an intimate herbivore insect-plant interaction. *Ecological Monographs, 76*, 175–194. http://dx.doi.org/10.1890/0012-9615(2006)076[0175:MBBTEO]2.0.CO;2.

Post, J. A., Kleinjan, C. A., Hoffmann, J. H., & Impson, F. A. C. (2010). Biological control of *Acacia cyclops* in South Africa: The fundamental and realized host range of *Dasineura dielsi* (Diptera: Cecidomyiidae). *Biological Control, 53*, 68–75. http://dx.doi.org/10.1016/j.biocontrol.2009.10.014.

Price, P. W. (1991). The plant vigor hypothesis and herbivore attack. *Oikos, 62*, 244–251. www.jstor.org/stable/3545270.

Price, P. W., Fernandes, G. W., & Waring, G. L. (1987). Adaptive nature of insect galls. *Environmental Entomology, 16*, 15–24. http://dx.doi.org/10.1093/ee/16.1.15.

Price, P. W., Waring, G. L., & Fernandes, G. W. (1986). Hypotheses on the adaptive nature of galls. *Proceedings of the Entomological Society, 88*, 361–363.

Proffit, M., Chen, C., Soler, C., Bessière, J. M., Schatz, B., & Hossaert-McKey, M. (2009). Can chemical signals responsible for mutualistic partner encounter promote the specific exploitation of nursery pollination mutualisms? The case of figs and fig wasps. *Entomologia Experimentalis et Applicata, 131*, 46–57. http://dx.doi.org/10.1111/j.1570-7458.2009.00823.x.

Proffit, M., Schatz, B., Bessière, J. M., Chen, C., Soler, C., & Hossaert-McKey, M. (2008). Signalling receptivity: Comparison of the emission of volatile compounds by figs of *Ficus hispida* before, during and after phase of receptivity to pollinators. *Symbiosis, 45*, 15–24.

Proffit, M., Schatz, B., Borges, R., & Hossaert-McKey, M. (2007). Chemical mediation and niche partitioning in non-pollinating fig-wasp communities. *Journal of Animal Ecology, 76*, 296–303. http://dx.doi.org/10.1111/j.1365-2656.2007.01213.x.

Quintana, A., Reinhard, J., Faure, R., Uva, P., Bagnères, A. G., Massiot, G., & Clément, J. L. (2003). Interspecific variation in terpenoid composition of defensive secretions of European *Reticulitermes* termites. *Journal of Chemical Ecology, 29*, 639–652. http://dx.doi.org/10.1023/A:1022868603108.

Quiring, D., Flaherty, L., Johns, R., & Morrison, A. (2006). Variable effects of plant module size on abundance and performance of galling insects. In *Galling arthropods and their associates* (pp. 189–197). Japan: Springer.

Raffa, K. F., & Berryman, A. A. (1983). The role of host plant resistance in the colonization behavior and ecology of bark beetles (Coleoptera: Scolytidae). *Ecological Monographs, 53*, 27–49. http://dx.doi.org/10.2307/1942586.

Raguso, R. A. (2008). Wake up and smell the roses: The ecology and evolution of floral scent. *Annual Review of Ecology, Evolution and Systematics, 39*, 549–569. www.jstor.org/stable/30245176.

Raman, A. (2012). Gall induction by hemipteroid insects. *Journal of Plant Interactions, 7*, 29–44. http://dx.doi.org/10.1080/17429145.2011.630847.

Ranganathan, Y., & Borges, R. M. (2009). Predatory and trophobiont-tending ants respond differently to fig and fig wasp volatiles. *Animal Behaviour, 77*, 1539–1545. http://dx.doi.org/10.1016/j.anbehav.2009.03.010.

Rehill, B. J., & Schultz, J. C. (2003). Enhanced invertase activities in the galls of Hormaphis hamamelidis. *Journal of Chemical Ecology, 29*, 2703–2720. http://dx.doi.org/10.1023/B:JOEC.0000008014.12309.04.

Reid, R. W., Whitney, H. S., & Watson, J. A. (1967). Reactions of lodgepole pine to attack by *Dendroctonus ponderosae* Hopkins and blue stain fungi. *Canadian Journal of Botany, 45*, 115–126.

Rocha, S., Branco, M., Boas, L. V., Almeida, M. H., Protasov, A., & Mendel, Z. (2013). Gall induction may benefit host plant: A case of a gall wasp and eucalyptus tree. *Tree Physiology, 33*, 388–397. http://dx.doi.org/10.1093/treephys/tpt009.

Rodriguez-Girones, M. A., Gonzalvez, F. G., Llandres, A. L., Corlett, R. T., & Santamaria, L. (2013). Possible role of weaver ants, *Oecophylla smaragdina*, in shaping plant–pollinator interactions in South-East Asia. *Journal of Ecology, 101*, 1000–1006. http://dx.doi.org/10.1111/1365-2745.12100.

Rohfritsch, O. (1992). Patterns in gall development. In J. D. Shorthouse, & O. Rohfritsch (Eds.), *Biology of insect-induced galls* (pp. 60–86). New York: Oxford University Press.

Sallé, A., & Raffa, K. F. (2007). Interactions among intraspecific competition, emergence patterns, and host selection behaviour in *Ips pini* (Coleoptera: Scolytinae). *Ecological Entomology, 32*, 162–171. http://dx.doi.org/10.1111/j.1365-2311.2006.00833.x.

Sallé, A., Ye, H., Yart, A., & Lieutier, F. (2008). Seasonal water stress and the resistance of *Pinus yunnanensis* to a bark-beetle-associated fungus. *Tree Physiology, 28*, 679–687. http://dx.doi.org/10.1093/treephys/28.5.679.

Sandström, J., Telang, A., & Moran, N. A. (2000). Nutritional enhancement of host plants by aphids—a comparison of three aphid species on grasses. *Journal of Insect Physiology, 46*, 33–40. http://dx.doi.org/10.1016/S0022-1910(99)00098-0.

Schatz, B., Anstett, M. C., Out, W., & Hossaert-McKey, M. (2003). Olfactory prey detection of fig wasps by the ant *Crematogaster scutellaris* (Formicidae: Myrmicinae). *Naturwissenschaften, 90*, 456–459. http://dx.doi.org/10.1007/s00114-003-0457-9.

Schatz, B., & Hossaert-McKey, M. (2003). Interactions of the ant *Crematogaster scutellaris* with the fig/fig wasp mutualism. *Ecological Entomology, 28*, 359–368. http://dx.doi.org/10.1046/j.1365-2311.2003.00519.x.

Schatz, B., & Hossaert-McKey, M. (2010). Ants use odour cues to exploit fig-fig wasp interactions. *Acta Oecologica, 36*, 107–113. http://dx.doi.org/10.1016/j.actao.2009.10.008.

Schatz, B., Hossaert-McKey, M., & Kjellberg, F. (2013). Histoire évolutive et diversité des interactions entre insectes pollinisateurs et plantes. In N. Sauvion, P. A. Calatayud, F. Marion-Poll, & D. Thiéry (Eds.), *Interactions insectes-plantes* (pp. 565–576). Editions Quæ.

Schatz, B., Kjellberg, F., Nyawa, S., & Hossaert–McKey, M. (2008). Fig wasps: A staple food for ants on Ficus. *Biotropica, 40*, 190–195. http://dx.doi.org/10.1111/j.1744-7429.2007.00352.x.

Schatz, B., Proffit, M., Kjellberg, F., & Hossaert-McKey, M. (2013). Un réseau trophique complexe: le cas des figuiers associés à différentes communautés d'insectes. In N. Sauvion, P. A. Calatayud, F. Marion-Poll, & D. Thiéry (Eds.), *Interactions insectes-plantes* (pp. 483–495). Editions Quæ.

Schatz, B., Proffit, M., Rakhi, B. V., Borges, R. M., & Hossaert-McKey, M. (2006). Complex interactions on fig trees: Ants capturing parasitic wasps as possible indirect mutualists of the fig-fig wasp interaction. *Oikos, 113*, 344–352. http://dx.doi.org/10.1111/j.2006.0030-1299.13986.x.

Schoonhoven, L. M., Van Loon, J. J. A., & Dicke, M. (2005). *Insect–plant biology*. Oxford University Press, 421 pages.

Schuler, M. A. (1996). The role of cytochrome P450 monooxygenases in plant-insect interactions. *Plant Physiology, 112*, 1411–1419.

Schuman, M. C., & Baldwin, I. T. (2016). The layers of plant responses to insect herbivores. *Annual Review of Entomology, 61*, 373–394. http://dx.doi.org/10.1146/annurev-ento-010715-023851.

Schwachtje, J., & Baldwin, I. T. (2008). Why does herbivore attack reconfigure primary metabolism? *Plant Physiology, 146*, 845–851. http://dx.doi.org/10.1104/pp.107.112490.

Scott, J. G., Liu, N., & Wen, Z. (1998). Insect cytochromes P450: Diversity, insecticide resistance and tolerance to plant toxins. *Comparative Biochemistry and Physiology Part C: Comparative Pharmacology, 121*, 147–155. http://dx.doi.org/10.1016/S0742-8413(98)10035-X.

Scott, J. J., Oh, D. C., Yuceer, M. C., Klepzig, K. D., Clardy, J., & Currie, C. R. (2008). Southern pine beetles use antibiotic-producing bacteria to regulate fungal symbionts. *Science, 320*, 63.

Scott, J. G., & Wen, Z. (2001). Cytochromes P450 of insects: The tip of the iceberg. *Pest Management Science, 57*, 958–967. http://dx.doi.org/10.1002/ps.354.

Segar, S. T., Dunn, D. W., Darwell, C. T., & Cook, J. M. (2014). How to be a fig wasp down under: The diversity and structure of an Australian fig wasp community. *Acta Oecologica, 57*, 17–27. http://dx.doi.org/10.1016/j.actao.2013.03.014.

Shorthouse, J. D., Wool, D., & Raman, A. (2005). Gall-inducing insects – nature's most sophisticated herbivores. *Basic in Applied Ecology, 6*, 407–411. http://dx.doi.org/10.1016/j.baae.2005.07.001.

Silvieus, S. I., Clement, W. L., & Weiblen, G. D. (2007). Cophylogeny of figs, pollinators, gallers and parasitoids. In *Specialization, speciation, and radiation: The evolutionary biology of herbivorous insects* (pp. 225–239).

Six, D. L. (2012). Ecological and evolutionary determinants of bark beetle – Fungus symbiosis. *Insects, 3*, 339–366. http://dx.doi.org/10.3390/insects3010339.

Six, D. L., & Paine, T. D. (1998). Effects of mycangial fungi and host tree species on progeny survival and emergence of *Dendroctonus ponderosae* (Coleoptera: Scolytidae). *Environmental Entomology, 27*, 1393–1401. http://dx.doi.org/10.1093/ee/27.6.1393.

Skow, C. D., & Jakob, E. M. (2006). Jumping spiders attend to context during learned avoidance of aposematic prey. *Behavioral Ecology, 17*, 34–40. http://dx.doi.org/10.1093/beheco/ari094.

Soler, C., Proffit, M., Bessiere, J. M., Hossaert-McKey, M., & Schatz, B. (2012). Evidence for intersexual chemical mimicry in a dioecious plant. *Ecology Letters, 9*, 978–985. http://dx.doi.org/10.1111/j.1461-0248.2012.01818.x.

Stam, J. M., Kroes, A., Li, Y., Gols, R., van Loon, J. J., Poelman, E. H., & Dicke, M. (2014). Plant interactions with multiple insect herbivores: From community to genes. *Plant Biology, 65*, 689. http://dx.doi.org/10.1146/annurev-arplant-050213-035937.

Stone, G. N., & Schönrogge, K. (2003). The adaptive significance of insect gall morphology. *Trends in Ecology & Evolution, 18*, 512–522. http://dx.doi.org/10.1016/S0169-5347(03)00247-7.

Stone, G. N., Schönrogge, K., Atkinson, R. J., Bellido, D., & Pujade-Villar, J. (2002). The population biology of oak gall wasps (Hymenoptera: Cynipidae). *Annual Review of Entomology, 47*, 633–668. http://dx.doi.org/10.1146/annurev.ento.47.091201.145247.

Stuart, J. J., Chen, M. S., Shukle, R., & Harris, M. O. (2012). Gall midges (Hessian flies) as plant pathogens. *Annual Review of Phytopathology, 50*, 339–357. http://dx.doi.org/10.1146/annurev-phyto-072910-095255.

Sugio, A., Dubreuil, G., Giron, D., & Simon, J.-C. (2015). Plant–insect interactions under bacterial influence: Ecological implications and underlying mechanisms. *Journal of Experimental Botany, 66*, 467–478. http://dx.doi.org/10.1093/jxb/eru435.

Suzuki, A. Y. M., Bedetti, C. S., & Isaias, R. M. S. (2015). Detection and distribution of cell growth regulators and cellulose microfibrils during the development of *Lopesia* sp. *Galls on Lonchocarpus cultratus (Fabaceae). Botany, 93*, 435–444. http://dx.doi.org/10.1139/cjb-2015-0012.

Svensson, G. P., Hickman, M. O., Bartram, S., Boland, W., Pellmyr, O., & Raguso, R. A. (2005). Chemistry and geographic variation of floral scent in Yucca filamentosa (Agavaceae). *American Journal of Botany, 92*, 1624–1631.

Svensson, G. P., Pellmyr, O., & Raguso, R. A. (2006). Strong conservation of floral scent composition in two allopatric yuccas. *Journal of Chemical Ecology, 32*, 2657–2665.

Terry, I., Moore, C. J., Walter, G. H., Forster, P. I., Roemer, R. B., Donaldson, J., & Machin, P. (2004). Association of cone thermogenesis and volatiles with pollinator specificity in *Macrozamia* cycads. *Plant Systematics & Evolution, 243*, 233–247.

Terry, I., Walter, G. H., Moore, C., Roemer, R., & Hull, C. (2007). Odor-mediated push-pull pollination in cycads. *Science, 318*, 70–73.

Tooker, J. F., & Helms, A. M. (2014). Phytohormone dynamics associated with gall insects, and their potential role in the evolution of the gall-inducing habit. *Journal of Chemical Ecology, 40*, 742–753. http://dx.doi.org/10.1007/s10886-014-0457-6.

Tsoupras, G., Luu, B., Hetru, C., Muller, J. F., & Hoffmann, J. A. (1983). Conversion *in vitro* de 20-hydroxyecdysone en métabolites phosphorylés et acétylés par des complexes tube digestif-tubes de Malpighi de larves de *Locusta migratoria. Compte Rendus de l'Académie des Sciences Paris, 296*, 77–80.

Von Nickisch-Rosenegk, E., & Wink, M. (1993). Sequestration of pyrrolizidine alkaloids in several arctiid moths (Lepidoptera: Arctiidae). *Journal of Chemical Ecology, 19*, 1889–1903. http://dx.doi.org/10.1007/BF00983794.

Vulinec, K. (1990). Collective security: Aggregation by insects as a defence. In D. L. De Evans, & J. O. Schmidt (Eds.), *Insect defences: Adaptive mechanisms and strategies of prey and predators* (pp. 251–288). New York: SUNY Press.

Wang, B., Geng, X. Z., Ma, L. H., Cook, J. M., & Wang, R. W. (2014). A trophic cascade induced by predatory ants in a fig–fig wasp mutualism. *Journal of Animal Ecology, 83*, 1149–1167. http://dx.doi.org/10.1111/1365-2656.12219.

Weiblen, G. D. (2004). Correlated evolution in fig pollination. *Systematic Biology, 53*, 128–139. http://dx.doi.org/10.1080/10635150490265012.

Wink, M. (2016). Evolution of secondary plant metabolism. In *eLS*. Chichester: John Wiley & Sons. http://dx.doi.org/10.1002/9780470015902.a0001922.

Wink, M., & Legal, L. (2001). Evidence for two genetically and chemically defined host races of *Tyria jacobaeae* (Arctiidae, Lepidoptera). *Chemoecology, 11*, 199–207. http://dx.doi.org/10.1007/PL00001852.

Wink, M., Legal, L., & von Nickisch-Rosenegk, E. (1998). How not to use systematics to test adaptative hypothesis, Response to Andrew V.Z. Brower. *Journal of Chemical Ecology, 24*, 1285–1291. http://dx.doi.org/10.1023/A:1021214430174.

Wink, M., & Roberts, M. F. (1998). Compartmentation of alkaloids synthesis, and storage. In M. F. Roberts, & M. Wink (Eds.), *Alkaloids: Biochemistry, ecology and medicinal applications* (pp. 239–262). New York: Plenum.

Wood, D. L. (1982). The role of pheromones, kairomones, and allomones in the host selection and behavior of bark beetles. *Annual Review of Entomology, 27*, 411–446. http://dx.doi.org/10.1146/annurev.en.27.010182.002211.

Wool, D., Aloni, R., Ben-Zvi, O., & Wollberg, M. (1999). A galling aphid furnishes its home with a built-in pipeline to the host food supply. *Entomologia Experimentalis et Applicata, 91*, 183–186. http://dx.doi.org/10.1046/j.1570-7458.1999.00482.x.

Wu, S., Peiffer, M., Luthe, D. S., & Felton, G. W. (2012). ATP hydrolyzing salivary enzymes of caterpillars suppress plant defenses. *PLoS One, 7*, e41947. http://dx.doi.org/10.1371/journal.pone.0041947.

Yang, L. Y., Machado, C. A., Dang, X. D., Peng, Y. Q., Yang, D. R., Zhang, D. Y., & Liao, W. J. (2015). The incidence and pattern of copollinator diversification in dioecious and monoecious figs. *Evolution, 69*, 294–304. http://dx.doi.org/10.1111/evo.12584.

Yu, D. W., Ridley, J., Jousselin, E., Herre, E. A., Compton, S. G., Cook, J. M., … Weiblen, G. D. (2004). Oviposition strategies, host coercion and the stable exploitation of figs by wasps. *Proceedings of the Royal Society of London, Series B, 271*, 1185–1195. http://dx.doi.org/10.1098/rspb.2003.2630.

Zachariades, C. (1994). *Complex interactions involving the cape fig, ficus sur forsskål, and its associated insects* (thesis). South Africa: Rhodes University, 243 p.

Zachariades, C., Compton, S. G., & Schatz, B. (2009). Honey-dew as Danegeld? Ants (Hymenoptera; Formicidae) tending a honeydew-producing homopteran don't not offer protection from its main natural enemies. *Sociobiology, 54*, 471–488.

Zhang, H., de Bernonville, T. D., Body, M., Glevarec, G., Reichelt, M., Unsicker, S., … Giron, D. (2016). Leaf-mining by *Phyllonorycter blancardella* reprograms the host-leaf transcriptome to modulate phytohormones associated with nutrient mobilization and plant defense. *Journal of Insect Physiology, 84*, 114–127. http://dx.doi.org/10.1016/j.jinsphys.2015.06.003.

Zhao, C., Escalante, L. N., Chen, H., Benatti, T. R., Qu, J., Chellapilla, S., … Batterton, M. (2015). A massive expansion of effector genes underlies gall-formation in the wheat pest *Mayetiola destructor*. *Current Biology, 25*, 613–620. http://dx.doi.org/10.1016/j.cub.2014.12.057.

Zhao, C., Shukle, R., Navarro Escalante, L., Chen, M., Richards, S., & Stuart, J. J. (2016). Avirulence gene mapping in the Hessian fly (*Mayetiola destructor*) reveals a protein phosphatase 2C effector gene family. *Journal of Insect Physiology, 84*, 22–31. http://dx.doi.org/10.1016/j.jinsphys.2015.10.001.

Zhu, F., Poelman, E. H., & Dicke, M. (2014). Insect herbivore-associated organisms affect plant responses to herbivory. *New Phytologist, 204*, 315–321. http://dx.doi.org/10.1111/nph.12886.

Zhu, L., Liu, X., Jeanotte, R., Reese, J., Harris, M. O., Stuart, J., & Chen, M.-S. (2008). Hessian fly attack causes dramatic shift in carbon/nitrogen metabolism in wheat. *Molecular Plant-Microbe Interactions, 21*, 70–78. http://dx.doi.org/10.1094/MPMI-21-1-0070.

Zhu, H., Sauman, I., Yuan, Q., Casselman, A., Emery-Le, M., Emery, P., & Reppert, S. M. (2008). Cryptochromes define a novel circadian clock mechanism in monarch butterflies that may underlie sun compass navigation. *PLoS Biology, 6*, e4. http://dx.doi.org/10.1371/journal.pbio.0060004.

CHAPTER FOUR

Food Webs and Multiple Biotic Interactions in Plant–Herbivore Models

E. Corcket*[,1], B. Giffard[§], R.F.H. Sforza[¶]
*Université de Bordeaux, INRA, UMR1202 BioGeCo, Pessac, France
[§]Bordeaux Sciences Agro, Université de Bordeaux, Gradignan, France
[¶]USDA-ARS-European Biological Control Laboratory, Montferrier-sur-Lez, France
[1]Corresponding author: E-mail: emmanuel.corcket@u-bordeaux.fr

Contents

Advances in Botanical Research, Volume 81
ISSN 0065-2296
http://dx.doi.org/10.1016/bs.abr.2016.10.002

Abstract

Trophic relationships between plants and insects are not confined to biological interactions such as herbivory (i.e., direct consumption of one primary producer by a predator). In an ecological approach, many other interactions, trophic or even nontrophic, may influence plant herbivory by insects. These interactions are related to the complexity and the diversity of the fauna and flora.

Plants and insects own to food webs, characterized by properties emerging from the identity of organisms and from the organisation of the links they develop with each other. These organisms can be considered as both taxonomic and functional units of biodiversity highlighting the complexity of food webs. Insects can be predated (e.g., by other arthropods or birds), parasitized and competed with for resources. Plant competition and availability of resources for primary producers may also influence the relationship between plants and their consumers.

The inclusion of predation mechanisms in plant–insect interactions leads to the theory of trophic cascades that advanced our understanding of the process of herbivory by insects. Ecological theories give a framework to assess the way by which these biotic interactions may be controlled in natural systems. Trophic cascades provide also biological tools for the management of agroecosystems, especially for biological control of herbivores and the promotion of biodiversity.

Biological interactions within food webs may also be nontrophic and alter plant–insect herbivory. Intraguild competition, interference, abiotic resources, microclimate and changes in animal behaviour are some of the drivers which may influence plant and insects, and thus their trophic relationships.

1. INTRODUCTION

To fully understand plant–insect interactions, one should consider the entire biological system in which such organisms are living. It includes of course all the biotic relationships that species may develop with other organisms, as well as the abiotic components of the environment. This includes the effects that these species themselves have on their environment. In this context, we must view plant–insect interactions as basic components of an entire ecological system, controlled by particular processes whether we consider natural ecosystems or highly managed agrosystems. Whatever the eco- or agrosystem, the nature of the processes embedding the insect–plant interactions is similar. The difference lies in the number of organisms involved in the system and the complexity of those interactions.

In this chapter, we aim to describe the many trophic interactions experienced by insect and plants. Concepts and examples are taken primarily from ecological studies, i.e., systems much more complex than croplands or intensively managed agrosystems. We hope that this approach will allow us to understand how an agronomic system may function in a context of high biological diversity.

2. FOOD WEBS

Agro- or ecosystems include many organisms of numerous species and characterized by a wide range of biological features. Such extraordinarily rich biotic assemblages lead to a multiplicity of trophic interactions, which can be modelled as food webs. These networks are defined by structural attributes, shaping different topologies and giving to the system intrinsic properties.

A food web is characterized by the number of different organism types (i.e., its number of species 'S') as well as by the linkage between these species. Connectivity expresses the proportion of links established between species and is calculated as the number of realized trophic interactions compared to the total potential trophic interactions in the web (Pimm, Lawton, & Cohen, 1991). The length of a food web is the maximal number of successive links between the organisms that can exist in the food web. A food web describes either the trophic interactions existing around a species or group of species, either the trophic interactions at the scale of a habitat (Pimm et al., 1991). 'Source' web includes target organisms and all the organisms consuming these target species, whereas 'sink' web represents target organisms as well as all the organisms eaten by target species. To keep a more integrative view of agroecological processes, food webs should be considered at the scale of habitat and should not be reduced to the organisms scale. This allows to consider every potential interaction existing within the assemblage of species. Food webs are usually graphically represented as 'motifs,' with 'nodes' as species and arrows symbolizing the energy fluxes by predation (Fig. 1A; e.g., Baiser, Elhesha, & Kahveci, 2016). They also can be represented using a matrix indicating the trophic interactions occurring between all the species of two successive trophic levels (Fig. 1B; Allesina, Alonso, & Pascual, 2008). Topologies of wood webs are numerous and reflect different types of interactions processes shaping organisms assemblages (Baiser et al., 2016). Attempts to model such complex networks show that niche basis models better reflect the structure of empirical food webs, suggesting that the consumption of one organism by another is mainly mediated by the ecological context in which species are coexisting (Allesina et al., 2008).

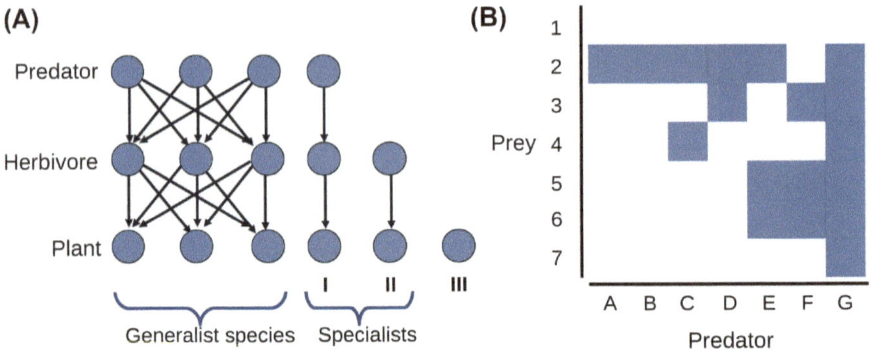

Figure 1 Graphical representations of different types of trophic interactions. (A) Trophic webs for three successive trophic levels; circles = nodes, that is species of a given trophic level; II and III: incomplete trophic webs, respectively, lacking specialist predator and specialist predator + herbivore. (B) Matrix representing food connections between two successive trophic levels; in dark the occurrence of a trophic relation, in white no trophic relationship between both trophic levels. Prey #1 is unpalatable, predator A is a specialist and predator G is a generalist.

The intensity of trophic interactions varies according to the organisms involved, i.e., the probability for an organism type to be eaten is not the same according to the biological properties of the predator and the prey. Some species show a very low number of interactions with the other, despite the fact that these interactions are strong (i.e., the probability of consumption of this species by the other is high). These 'specialist species' will then have important consequences on the complete food web. Some other species develop many interactions but with a weaker intensity, i.e., with a lower probability to realize the trophic interaction. These 'generalist species' are then relatively independent from the rest of the trophic chain (Paine, 1980). The strength of the linkages is not constant since it is mainly due to population density and foraging behaviours of species; it may change over time due to phenology of organisms or environmental variations (Paine, 1980). The proportion of these two kinds of organisms is usually unbalanced, with a greater number of generalists compared to specialists. This imbalance suggests that food webs are marked by a relatively greater amount of weak interactions. However, to assess the relative importance of generalists and specialists on food webs, the question is to disentangle the respective influence of the number of interactions versus the strength of interactions. A large number of weak interactions (by generalists) may be able to balance a few strong interactions (i.e., by specialists; Wootton & Stouffer, 2016). Food webs may be split into two groups: those characterized by weak interactions and those characterized by strong interactions (Prado & Lewinsohn, 2004). Within food webs, functional units (or 'modules') group species showing strong associations, or linked with specific types of resources (McMeans et al., 2016). Strong relationships between organisms create a synchrony of appearance and disappearance of species, and favour coevolution within a module (Paine, 1980).

Food webs show emergent properties rising from their structural parameters (length, connectivity, strength of interactions; Pimm et al., 1991). Among these structural parameters, transitivity is the repercussion of a trophic effect along a food chain. If the effect of a predation by insectivorous birds does not affect plant through a release of insect herbivory, then the transitivity in food webs is low. Transitivity depends on the length of the web, as there tends to be an attenuation of trophic effects with increasing chain length. Transitivity decreases with connectivity but increases with the strength of interactions: a numerical change of specialists, i.e., of organisms weakly connected but strongly related with the species they are connected with, will have great consequences along the chain of organisms. Another property of a food web is its stability over time and resilience to disturbance. If we consider the asymmetric arrangement of interactions, weak interactions are also important in stabilizing food webs because weak interactions related to generalist species have a greater number of interactions within the network. The type of trophic interaction that experience species, positive (predation by the focal species) or negative (the focal species is preyed upon), is also a parameter that may affect the dynamics of food web (Wootton & Stouffer, 2016).

Food web topology can affect ecosystem functioning. This topology is mainly shaped by the number of links between species, the strength and connectivity of these links (i.e., the food 'status' of the species). The presence of specialist versus generalist species, or edible versus inedible ones for instance, may alter herbivore biomass, plant biomass and therefore soil resource availability. These relationships are dependent on the diversity of plant communities, indicating that biodiversity loss may alter ecosystems through food web modifications (Thébault & Loreau, 2003). Structure and topology of food webs, as in many interactions networks, may vary over space and time due to population processes involving species traits (Poisot, Stouffer, & Gravel, 2015). McMeans et al. (2016) argue that the biological properties of food webs and not species diversity per se are important to adapt to environmental changes and underline links between biodiversity and ecosystem functioning.

Food web dynamics have been related to 'spatiality', i.e., what happens at the regional scale in 'meta-food webs' (Mougi & Kondoh, 2016). In large areas, the number of local food webs and their connectedness regulates the dynamics of meta-communities, suggesting a positive complexity–stability relationship for food webs controlled by landscape ecological factors.

3. FUNCTIONAL TYPES OF ORGANISMS/ CLASSIFICATION OF SPECIES WITHIN FOOD WEBS

3.1 Trophic Levels

Living organisms can be grouped according to their similar position in the food web. Trophic levels permit to aggregate species that are very different from a taxonomical point of view (Pimm et al., 1991). The trophic level of a particular organism can be determined as the total number of links below this organism plus one. The levels most often described are decomposers, primary producers, primary consumers (named as herbivores or phytophagous species) and consumers of higher trophic levels. Within this latter level, four classes are classically distinguished: parasites (i.e., organisms that live in or on another organism – their hosts – and benefits by deriving nutrients at the host's expense), parasitoids (i.e., insects whose larvae live as parasites that kill their hosts, typically other insects), pathogens (i.e., bacterium, virus, or other microorganisms that can cause disease) and predators (i.e., animals that naturally prey on others) (Polis, 1999). Trophic relationships occur between two successive levels in the food chain, except for omnivorous species that can feed on more than one trophic level and top predators that can, for example, predate herbivores as well as predator species (food chains with different lengths). Top predator species may then influence the entire food web, preying at different levels below (Paine, 1980): this preying activity is often guided by a relative preying opportunity (i.e., the availability of different preys).

3.2 Functional Groups

Within a trophic level, organisms can also be grouped according to their biological characteristics or functional traits. Species having similar traits may share a similar function in an ecosystem. This is the basis of the question of species or ecological redundancy (Gitay, Wilson, & Lee, 1996), and address the question of the possibility for a trophic web to keep a similar functionality or structure despite a decrease in species number. A 'functional group' of animals that gain and use food resources in a similar way can be classified into 'guilds'. Rather than systematic approaches, researchers currently focus on the classification of organisms according to their functional traits. Interestingly, the main part of functional diversity of organisms such as plants may be accurately captured in a two-dimensional global spectrum based on the size of organs (whole plant, seeds, stem) and the optimization of photosynthesis (leaf area, nitrogen content/leaf mass unit; Díaz et al., 2016). Functional approaches may be useful to assess global mechanisms, such as the effects of global change (climate and land-use change) on ecosystem functioning (Díaz et al., 2007; Lavorel & Garnier, 2002). Most studies assessing global patterns and responses of groups to environmental factors have been conducted using plant traits only. Animal functional groups, except for aquatic systems, have been conducted using well-known taxonomic groups (birds, arthropods, see examples hereafter).

For primary producers, different functional groups have been described based on their response to ecosystem processes. Plant species occurring in ecosystems strongly impacted by herbivory are mostly annual plants of small size and with high seed dispersal ability (Lavorel, McIntyre, Landsberg, & Forbes, 1997). In some ecosystems, resistance to herbivory and adaptation to stress tolerance share similar functional traits such as low growth rate, low size and tough leaves with high concentration of secondary compounds (Oksanen & Ranta, 1992). Most of these functional traits have been originally described for plant groups in with mammal herbivory rather than insect herbivory. Studies focusing on host plant recognition by insects (e.g., searching for egg laying sites or food) have highlighted the plant traits that favour the colonization of plants by insect herbivorous species. They have found that leaf traits for example such as toughness, lignin content or nutritional measures, water and/or nitrogen content can influence host colonization and detection by insects (Coley, 1983).

For animals and upper trophic levels (primary and secondary consumers), functional groups are often linked to the particular use of a niche characteristic among different species or within a taxonomic group. The number of traits collected in these groups is often limited and strongly depends on the degree of knowledge of the studied taxonomic group. Species and functional groups could differ based upon their foraging strategies (like spiders – see Buddle, Spence, & Langor, 2000), dispersal ability (like carabids – Ribera, Dolédec, Downie, & Foster, 2001; like saproxylic beetles – Bouget, Brin, Tellez, & Archaux, 2015), or overwintering sites (hoverflies – Raymond et al., 2014).

The concept of a guild is relatively similar even more developed and strongly linked to the diet of herbivorous species. A guild definition allows one to simplify the study of a particular group of species, across taxonomic position, but more detailed than a simple trophic level. Indeed, one guild will be represented by several habitats and/or communities and then will facilitate the comparison of the diversity between different landscape configuration or management intensities (Root, 1967). The guild classification can also be used to gain a greater understanding of spatial patterns of species richness, and assemblage structure and allow a better understanding of the animal traits that are particularly sensitive to disturbance for example (Williams & Hero, 1998).

3.3 Herbivorous or Phytophagous Guilds

Guilds of phytophagous animals have been mostly described for arthropods and are associated with the type of damage these arthropods inflict on the plant that they feed on. Many articles describe the feeding mode used by the herbivorous species and their abundance and relationship to their host across seasons, between regions or function of the host specialization (Müller, 1989; Novotny et al., 2010; Southwood, Wint, Kennedy, & Greenwood, 2004). All these studies described groups of species that exploit similar resources in a similar way based on the definition proposed by Simberloff and Dayan (1991) and that includes the feeding mode (chewers, skeletonizers, miners, leaf-rollers, sap-suckers, gall-makers, etc.) and the part of the plant that is consumed (roots, leaves, stems, etc.). Other characteristics could be added function of the studies (insect dispersal or life stage) or simplified such as they only focus on internally and externally feeding species.

The typology proposed by Mattson, Lawrance, Haack, Herms, and Charles (1988) also includes the effect on the host plant. Gall-makers are often considered as negligible for their impact on plants, whereas chewing species can be responsible during outbreaks of complete defoliation and important growth lack or mortality. Sap-sucker insects may also induce severe indirect damage to plant fitness as they may vector plant pathogens such as viruses, bacteria, or nematodes (see for example, Gray & Banerjee, 1999).

3.4 Bird Guilds

The same idea was applied for several groups of animals in order to characterize their function in ecosystems (and now applied to the concept of ecosystem services described some species and/or communities as particular service providers). From the 1960s onward, birds were classified based on their feeding mode and particularly their behaviour to catch a prey (Root, 1967). For example, MacNally (1994) then described several bird guilds function of the dominant prey searching activities of birds observed in different forest ecosystems in Australia. This information is particularly interesting in the context of an increasing knowledge in the biological control effect provided by birds in agro- or forest ecosystems (Whelan, Wenny, & Marquis, 2008). Avian predation maintains low herbivorous insects or pest populations that may indirectly favour plant growth (Mäntylä, Klemola, & Laaksnonen, 2011; Whelan et al., 2008). Knowing how these important predator species will forage for prey is fundamental: presence and/or abundance of landscape elements, life cycle of the pest species can be then linked to predator behaviour (Barbaro, Giffard, Charbonnier, van Halder, & Brockerhoff, 2014; Ceia & Ramos, 2016; Tscharntke et al., 2007) and ecosystem function of pest regulation.

The concept of functional group was originally associated with plants, invertebrates or microorganisms, but has been progressively extended to vertebrates and birds. As defined by Blondel (2003), '*in the guild approach, the subject is the species which exploits resources for itself whereas in the functional group approach the subject is the resource which, so to speak, utilises mediators, i.e., members of a functional group, for being processed*'. At first, feeding mode and/or foraging behaviour are integrated in the traits matrix associated with a functional group. Subsequently, other traits associated with ecosystem functioning may be added in order to characterize relationships between landscape features such as habitat fragmentation and ecosystem services such as pest control (Barbaro, Brockerhoff, Giffard, & van Halder, 2012; Tscharntke et al., 2007).

Other taxonomic groups, such as mammals (Flynn et al., 2009) or bats (Cisneros, Fagan, & Willig, 2015), have been studied and traits matrices have been collected in order to assess the functional diversity within these groups and the potential release of predation on insect pests when these groups are threatened by human activities and global change.

4. TROPHIC CASCADES FROM PLANTS TO INSECT PREDATORS

4.1 Trophic Cascade Concept

It is well recognized, even in terrestrial ecosystems, that predators (*s.l.*, i.e., predators, parasites, parasitoids and pathogens) can reduce herbivore abundance to the point that plants experience negligible levels of herbivory (Halaj & Wise, 2001; Schmitz, Hambäck, & Beckerman, 2000). The 'green world hypothesis' proposes that predators are food limited by the quantity of prey and then regulate herbivore populations. It indirectly helps plants to grow and reproduce (Hairston, Smith, & Slobodkin, 1960). This highlights the major 'top-down effect' that predators play in trophic interactions, which Oksanen, Fretwell, Arruda, and Niemela (1981) predict to be stronger in fertile environments where inorganic resources are unlimited. Trophic cascades may also occur in a 'bottom-up' fashion in ecosystems where inorganic resources are limited may not support high levels of herbivores and predators (Hairston et al., 1960; Polis & Strong, 1996). In both top-down and bottom-up controlled systems, trophic cascades have been shown for various taxa and ecosystems in terrestrial ecosystems (Mäntylä et al., 2011; Mooney et al., 2010), even if their strength has been found to be quite variable (Mooney & Linhart, 2006). In addition to the top-down effect of predators (*s.l.*), Polis (1999) listed five dominant factors that are supposed to limit the herbivore abundances and their associated top-down effect on vegetation. These are that (1) plants are not passive in reaction to their consumers and several adaptations of plants reduce the quality of resources for the consumer; (2) the quantity of plant resource for herbivores may occur and especially in nutrient limited environments; (3) animals are more affected than plants by some abiotic factors (water availability, temperature); (4) the spatial and temporal availability of resource (plants) varies for consumers (herbivores); and (5) negative interactions exist between herbivores (within and/or between species competition). All of these factors affect the biotic and abiotic context in which tritrophic interactions occur, and will therefore modify and influence the presence and strength of these interactions. Recently, several experiments and meta-analyses have been looking for describing this complexity and ask what factors determine cascade strength.

Within trophic cascades, those initiated by insectivorous birds have been particularly studied. Mäntylä et al. (2011) compiled about 30 experiments that tested the effect of bird exclosure and nestbox addition on herbivorous insects and consequences for plants. Globally, birds are beneficial to plants in multiple habitats and help to regulate damage from herbivores, even if consequences for plant fitness or growth are weaker. The ecosystem services provided by birds, and especially insectivorous birds which prey on arthropods, have recently been found to be particularly important in agricultural systems (or 'refund according to old studies in this subject' — see Whelan et al., 2008).

Within this area, and back to the initial concept of trophic cascades, several studies are based on the fact that the predator species or community such as avian community are specialized to one or several insect species associated to one plant species or a plant community (Polis, Sears, Huxel, Strong, & Maron, 2000). This excludes trophic interactions involving generalist predator or herbivorous species that can compete on a similar resource. Functional traits of species at each trophic level (e.g., behaviour for predator, host specialization for herbivores) or within communities (functional redundancy, species introduction or loss) are significantly responsible of an increase complexity of interactions within the ecosystem. Trophic cascades, first described as linear relationships between three successive levels, may be then defined at the whole community level and consider all the interactions within or between the trophic levels involved.

4.2 Complexity Within Tritrophic Systems

4.2.1 Predator Level

There is a strong increase of knowledge about the effect of plant traits and plant diversity on the upper trophic level of herbivores (Moreira, Abdala-Roberts, Rasmann, Castagneyrol, & Mooney, 2016). However, studies on the effects of predators on the structure and diversity of herbivore and plant communities have been focused on particular predator communities, rather than effects at each trophic level, such as by assessing the importance of top-down control for plant growth or fitness (Moreira et al., 2016).

Diversity within the trophic level of predators is supposed to have important effects on the levels below (Finke & Denno, 2005). Different predator species may synergistically prey on same herbivorous species, at different prey life stages for example, and then promote the indirect effect on primary producers. Conversely, they can compete for this same resource and their combined effects could be then limited or slightly additive. If one of the predators also predate an other one predator (intraguild predation), trophic cascades can be hindered and herbivorous species do not suffer or release from the loss of a predator species or particular community, process that has been shown in some ecosystems with spiders in salt marshes (Finke & Denno, 2005) or with insectivorous birds in pine and mistletoe (Mooney & Linhart, 2006). The meta-analysis conducted by Mooney et al. (2010) failed to correlate the predation by vertebrate insectivorous species and the weakening of trophic cascades. Moreover, trophic cascades are strengthened when predaceous vertebrate and arthropod species were more abundant, suggesting frequency-dependent response of top predators to the most abundant resource, predatory or herbivorous arthropods.

4.2.2 Herbivorous/Phytophagous Levels

A similar plant resource can be available for a wide range of phytophagous guilds, thus influencing the population dynamics of the host plant in different ways. In the frame of biological control towards the alien diffuse knapweed, *Centaurea diffusa* Lamk. (Asteraceae), an invasive species in North America, a total of 12 insect species were released from Eurasia between 1970 and 1990 (Smith, 2004). Based on the high seed production level of this knapweed species, most of the insect species released were seed feeders. Nonetheless, the alien weed continued to spread dramatically. Simulation models have shown that seed reduction is not sufficient if seedling survival is compensatory (Myers & Bazely, 2003). As a result, in the 1990s, insect species whose larvae feed on the root system, impacting directly knapweed rosettes before they bolt, and adults defoliating green parts, were selected for release. This phytophagous guild including the weevil species *Larinus* (Larinomesius) *minutus* (Gyllenhal, 1835) (Coleoptera: Curculioniade) was extremely efficient for regulating diffuse knapweed populations (Seastedt, Gregory, & Buckner, 2003), and the closely related spotted knapweed (*Centaurea stoebe* L.) (Story, Smith, Corn, & White, 2008). It is stated that knapweed density may not decrease significantly until the seed bank falls below a critical threshold.

4.2.3 Multiple Interactions at Primary Producer Level

Trophic cascades may change within a same community according to many plants characteristics. Differences in plant structure, as well as variation in chemical content of plants, are able to alter herbivores and also their predatory behaviours. These differences in plant biological traits can be due to environmental or genetic factors.

Focusing on target plants in tritrophic networks, Sipura (1999) compared relationships occurring on two species of willow (Salicaceae). Although *Salix myrsinifolia* Salisb. and *Salix phylicifolia* L. are morphologically very similar, *S. myrsinifolia* contain 50 times more phenolic compounds than *S. phylicifolia* does. When insectivorous birds were excluded, herbivores populations increased for both species of *Salix* but the magnitude of the effect and the associated damage were much higher on the species having the lowest rate of constitutive defences (*S. phylicifolia*). This process resulted in a strong trophic cascade on *S. phylicifolia*, which can be reinforced by specific attraction mechanisms for predators. Indeed, a plant resource with low secondary compounds is more palatable to herbivores and can support higher densities of both herbivores and predators alike. Moreover, a high level of damage on leaves creates a visual signal that will be detected by insectivorous birds during their foraging (e.g., Mäntylä et al., 2008).

In thermo-Atlantic pine forests, trophic cascades involving insectivorous birds, leaf-feeding insects and tree seedlings were identified, depending on the identity of the focal tree seedling. *Betula pendula* Roth showed lower leaf damages when bird predation occurred, whereas leaf damages for both oak species *Quercus ilex* L. and *Quercus robur* L. did not change according to the presence/absence of insectivorous birds (Giffard, Corcket, Barbaro, & Jactel, 2012).

The surrounding vegetation matters: trophic cascades involving tree seedlings, leaf eaters insects and insectivorous birds depend on the kind of forest stand. In the same temperate forests ecosystems, trophic cascades were significant in pine plantation only but not in broad leaves stands (Giffard et al., 2012). The structure of the understory vegetation is also important: in the same type of forest ecosystems, trophic cascades involving chewers being significant when tree seedlings were not surrounded by understory vegetation (Giffard, Barbaro, Jactel, & Corcket, 2013; see Fig. 2). In this figure, left diagram represents the positive indirect effect of insectivorous birds on seedlings through predation of generalist insects and the easier access of specialist insects on their host plants (focal plant species). Right diagram represents the disruption of the bird predation on generalist insects: plant neighbours reduce food accessibility for birds and/or shelter greater insect abundance which leads to increased damage on focal plant and divert the bird predation effect. At the opposite, specialist insects may be negatively influenced by the presence of surrounding vegetation: plant neighbours act as a barrier to the colonization of host plant.

4.2.4 Multiple Trophic Interactions Around Vine

As primary and secondary pests of grapevine, scale insects are among the main leaf consumers. According to the scale insect species and predators surrounding, one or several categories of parasitoids may coexist in vineyards (Fig. 3). In different temporal and spatial scales, oophagous or larval scale insect parasitoids will not be in competition, as they both occur at different time frames of the scale insect life cycle. Similarly, interactions between generalist predators such as spiders, lacewings and ladybugs are very limited as scale insects are always building up huge populations providing to

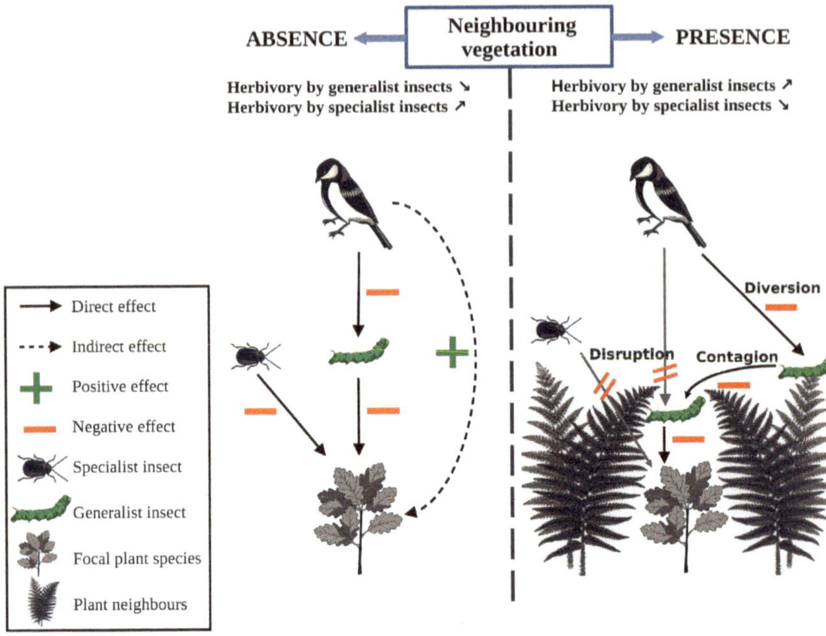

Figure 2 Conceptual diagram on the effects of the presence or absence of plant neighbourhood and the bird predation on generalist and specialist insects feeding on the focal plant species. When there is absence of neighbouring vegetation around focal plant, cascade trophic occurs (positive indirect effect of insectivorous birds on seedlings through predation of generalist insects). When there is presence of neighbouring vegetation, no trophic cascade by birds on focal plant occurs, due to the lack of accessibility of the resource. Plant neighbours also act as barrier to the colonization of host plant by specialist insects. See associated results in Giffard et al. (2013).

predators some kind of an unlimited source of feeding. Among other available ones there are others such as leafhoppers, whiteflies, thrips, etc. In this complex trophic food web, ants are playing a key role by tending scale insects, protecting them from the above-mentioned generalists predators, and receiving honeydew from scale insects in return (Gullan, 1997; Sforza, 2008) (Fig. 3). However, as being in direct contact with generalist predators and parasitoids of scale insects, ants are regulating their population densities. Presence of ants may negatively impact parasitoid densities such as parasitoid B (Fig. 3), but positively increase densities of predators by favouring the scale insect population density (Daane, Sime, Fallon, & Cooper, 2007). Thus, there is a nonnegligible effect of ants on the trophic web in general and this effect is mainly benefiting the plant itself (Styrsky & Eubanks, 2007).

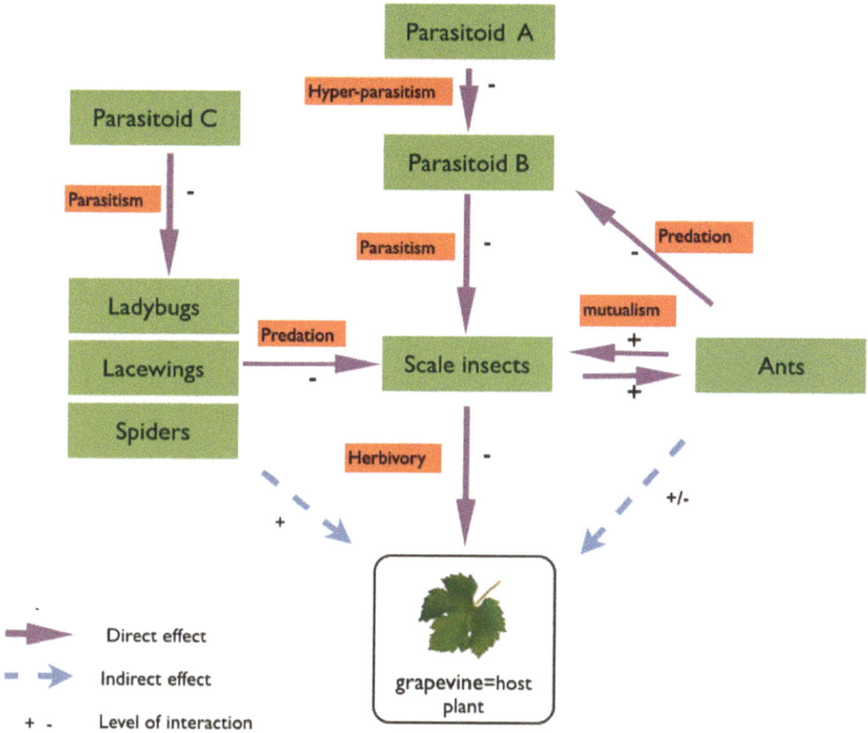

Figure 3 Scheme of a complex ecological web within a plant–insect model system involving grapevine crops, and scale insects (Hemiptera), their most important herbivores. Impacts of insects on different guilds are indicated by +(positive) and −(negative).

4.3 Trophic Cascade Control

Trophic cascades are controlled in their dynamic by several factors (Polis et al., 2000) that we can distinguish according to their biotic or abiotic nature, or to the endogenous or exogenous origin of these factors as regard to the ecosystem (Mooney, 2007). Biotic factors, usually endogenous, are biotic control within guilds or trophic levels, trophic control between trophic levels, quality and amount of available resources. Abiotic factors can be endogenous, such as habitat factors (microclimate), or exogenous such as landscape factors (Polis et al., 2000). A sharp understanding of the dynamic of trophic networks according to these different kind of factors is still lacking. Dynamic and magnitude of trophic cascades remain widely unpredictable.

4.3.1 Biotic Controls

Trophic interactions within the same trophic level are a first mechanism of control. Intraguild predation or cannibalism behaviours are reported (Polis et al., 2000), regulating carnivore's populations and decreasing the intensity of predation along the trophic chain. Such biotic controls lead to negative effects on plants since the trophic control of predators on herbivores populations is released.

The effects of biotic controls involving interactions between different trophic levels are less predictable. Herbivores may induce constitutive defences by plants, which alter the intensity of trophic cascade. Similarly, a more complex food web due to the multiplicity of organisms (i.e., a high diversity systems) is able to decrease the strength of trophic cascade: trophic links will be weakened but more numerous. Omnivorous behaviours contribute to the complexity of food webs, adding multiple trophic links due to the wide range of food interactions that can have such generalist consumers. The food consumption by omnivorous species will highly depend on the availability of their different potential resources, i.e., on the structure of the food web and not necessarily on intrinsic properties of omnivores. Influence of omnivorous species is then varying according to the studied cases.

Resources concentration (i.e., relative amount of consumed plants as compared to the total amount of vegetation; Root, 1973) is an important parameter for specialist herbivores. The more resource concentration is important, the more the foraging of herbivores will be easy and the more herbivory will be intense. Plant resource concentration will then reinforce trophic cascade effects. At the plant community scale, this kind of process suggest that low diversity systems are more susceptible for herbivory and experience strongest trophic cascades.

Biotic controls also include nontrophic interactions that are mentioned below in this chapter.

4.3.2 Controls by Physical and Chemical Factors of the Environment

Trophic cascades are under dependence of ecological factors influencing the amount or the quality of trophic resources and even the structure of landscape in which trophic interactions occur (Polis et al., 2000).

Nutrient availability for plants is certainly the most important factor, improving productivity of the ecosystem and the quality of plant tissues for herbivores. Nevertheless, the question of the interplay of bottom-up forces (fertility) and top-down pressure (predation) on herbivores is still a debating question, even for similar models involving tree saplings and insect herbivores. In trophic systems involving the cottonwood *Populus deltoides* Bartram ex Marshall, highest nutrient levels in soils increase areas of leaves but decrease the intensity of trophic cascade (i.e., the relative control of predators on insect herbivores; Hartvigsen, Wait, & Coleman, 1995). In trophic experiments involving *Quercus* spp., the predation intensity increases with fertilization but its importance remains low compared to the effect of nutrient availability per se: no trophic cascade was detected, whatever the nutrient level (Forkner & Hunter, 2000).

Water availability, another component of the fertility of ecosystems, may influence plant–insect interactions, mainly through water stress (Dhileepan, Setter, & McFadyen, 2000; Willis, Ash, & Groves, 1995).

Light is a complex abiotic factor, widely controlled by biotic component of ecosystems, especially plant canopies structure, and strongly related to other physical parameters such as temperature and air moisture. The impact of light availability and density of canopy on trophic cascades depends on their ability to alter plant biological traits (i.e., vegetation 'quality'), and the transmission of this bottom-up control of herbivore along the trophic chain. For *Vaccinium myrtillus* L. (Ericaceae) growing under dense canopies, leaves become more palatable because of their higher content in nitrogen and water, and because of their thinness. Although they experience relatively low density of herbivorous insect larvae (mainly Lepidopteraea), *V. myrtillus* shoots growing in close canopies are heavily damaged. Trophic cascade on *Vaccinium* in close forests is stronger, probably due to the higher consumption rate of palatable leaves per herbivore individual, and a more intensive predation by birds (Atlegrim, 1989).

At the wider scale, taxonomic and functional diversity is affected by landscape (Barbaro & van Halder, 2008), due to spatial structures, habitat availability or abiotic natural gradients (microclimate, fertility). These patterns of variation suggest potential changes in trophic cascades in relation to landscape. In fragmented broad leaves forests, populations of insectivorous birds are relatively numerous and their foraging activity is relatively high (Gonzalez-Gomez et al., 2006). This suggests that trophic cascades are strongest in fragmented forests. Similarly, bird insectivory is higher in fragmented native forests (Vásquez, Grez, Bustamante, & Simonetti, 2007).

5. APPLICATIONS OF TROPHIC CASCADES FOR MANAGEMENT

The introduction of exotic arthropod candidates for weed biocontrol, also called 'weed classical biocontrol', aims to introduce one or several specialized biocontrol agents (BCAs) from the native range to the introduced range to top down the spread of an alien plant species, considered as a target. The native range represents where the target weed is originating, and the introduced or invaded range where it has been accidentally/intentionally introduced. The distribution in invaded areas in the introduced range is always far larger than what is currently observed in the native range (thousands vs. tens of hectares). These invasions lead to a loss of native plant diversity. Once introduced in a new area and with the lack of herbivore pressure, alien plants develop new strategies to colonize the environment and become 'superweeds'. Based upon the Evolution of Increase Competitive Ability hypothesis, invasive plants are producing more biomass helping out to expand their colonizing surface (Blossey & Notzold, 1995). In addition, as invasive plants are usually introduced without their natural enemies (arthropods, micro-organisms), the main purpose of biocontrol is to restore the association plant—natural enemies in the introduced range with biocontrol candidates that are biologically selected during research programs conducted in the native range.

In order to understand the overall process, several steps must be followed before releasing selected BCAs. Phylogenetical, physiological and behavioural criteria will determine the set of plant species to be tested for evaluating the host specificity of a herbivore. In order to evaluate a seed feeder insect, a leaf miner, a root borer or any kind of phytophagous arthropods, a list of 30—50 + plant species closely related or not to the target plant is prepared. Since the mid-1970s, the methodology is improving, from the original centrifugal phylogenetic method (Wapshere, 1974), considering plants to be tested genetically closely related to the target, to a modernized approach taking into account ecological and biogeographic filters applied to ensure that plants to be tested included species with the highest risk profiles (Briese, 2005; Sheppard, Heard, & Briese, 2003). Tests, such as no-choice (one single plant species tested) and choice tests (several plant species tested together), are critical for evaluating specificity, development, including oviposition behaviour and impact of each potential biocontrol candidate. Multiple year tests are both conducted in controlled conditions and in the open, sometimes in the country of origin of the BCA tested. Once laboratory and field data are obtained and important criteria documented (specificity, direct impact on the target, easy to rear, multiple generations per year, etc.) researchers can move to the step of evaluating the possibility of release in the introduced range. This long process is conducted for each potential candidate.

Invasive plants are almost free of native enemies and then more susceptible to herbivory once one is introduced. This can be illustrated with scotch broom, *Cytisus scoparius* (L.) Link (Fabaceae), an example for which five or six herbivores on average are reported from Eurasia, and only one in the introduced range (Australia, New Zealand) (Memmott, Fowler, Paynter, Sheppard, & Syrett, 2000). Generally speaking, the number of natural enemies on invasive species at a larger scale is twice as large in the native range (Torchin, Lafferty, Dobson, McKenzie, & Kuris, 2003). The overall idea of biocontrol is to minimize the risk of adverse effects on short-, mid- and long-term timeframe. A recent review over 110 years of historical records on nontarget impacts of BCAs, found that >99% of the 512 agents released worldwide belonging to arthropods and plant pathogens had not shown adverse impact outside the target plant species (Suckling & Sforza, 2014).

6. NONTROPHIC INTERACTIONS IN FOOD WEBS

Organisms involved in trophic webs develop nontrophic interactions, each other or with their environment. These types of interactions may in turn affect the distribution and abundance of focal organisms as well as the food web itself. Nontrophic interactions include competition, mutualism or facilitation, as well as interference or behaviour-based processes. Indirect interactions (i.e., mediated by a third organism or ecological compartment such as atmosphere or soil) are widespread.

6.1 Positive Interactions of Plants Against Herbivory

A palatable plant can be protected from the predation of a herbivore through different type of associations with surrounding vegetation.

The palatable focal plants may share constitutive defences against herbivore by growing in the neighbourhood of vegetation, which have either physical adaptations to resist to herbivores (e.g., spiny stems or trichomes), or chemical compounds that are repellent or toxic for primary consumers (Callaway, 2007). Plants may also share active defences against herbivores: the surrounding vegetation may release in the air chemicals attracting predators, parasitoids or parasites of herbivorous insects (Callaway, 2007). This is a case of stimulation of top-down control of herbivores.

Associational defence allow a palatable plant to avoid herbivory by hiding in vegetation, whatever the palatable level of the surrounding plants (Callaway, 2007). Focal plants become then less detectable through olfactory or visual herbivore's foraging, They can avoid herbivory through resource dilution effects, which is a mechanism nonspecific to particular biological trait or defence adaptation of plants. Neighbouring vegetation may also change the microenvironment of the focal plant by shading or modifying microclimate, for example, and create conditions less favourable for herbivores. Whatever the process, these associational defences allow the palatable plant to avoid herbivores.

6.2 Susceptibility to Herbivory Due to Plant Neighbourhood

The presence of plant neighbourhood also may enhance herbivory level on a focal plant by attracting herbivores. Vicinity between plants may then protect or expose a focal plant to herbivores, depending on the identity of the neighbouring plants (Rousset & Lepart, 2002). Then, a focal plant may be overgrazed when surrounding by other plants more palatable, by species attracting herbivores through the emission of secondary chemical, by improving microenvironment for herbivores (Callaway, 2007).

Plant associations may show positive or negative effects according to the guild of herbivore involved. In pine forests, the shrub *Buxus sempervirens* L. (Buxaceae) attracts the predation on pine seeds by rodent such as *Apodemus sylvaticus* (L.) (Muridae), whereas *Buxus* improves recruitment of pines [*Pinus nigra* Arnold and *Pinus sylvestris* L. (Pinaceae)] by protection of seedlings against large herbivores such as sheeps (Boulant, Navas, Corcket, & Lepart, 2008).

6.3 Plant Competition and Herbivory

Patterns of insects herbivory may change according to the level of competition experienced by plants. In calcareous grasslands, the most competitive and late-successional grass species, *Brachypodium pinnatum* (L.) P. Beauv. (Poaceae), is the most negatively affected by grasshopper's herbivory (Corcket, Callaway, & Michalet, 2003). The intensity of competition occurring in plant communities depends on the relative density of vegetation (i.e., the relative amount of bare soil in ecosystems). When *Brachypodium* is surrounded by bare soil, plant competition is low but herbivory is important. This is probably due to the particular guild of grasshoppers, very sensitive to the increase in temperature and air aridity when vegetation cover decreases in thermophilic areas, and by their ability to identify more easily palatable plants when they are isolated (Corcket et al., 2003).

Trophic interactions also may alter below ground plant interactions. Larvae of the sulphur knapweed moth, *Agapeta zoegana* (Linneaus, 1767) (Lepidoptera: Cochylidae) and *Cyphocleonus achates* (Fahaeus, 1842)

(Coleoptera: Curculioniade), are used to control the population dynamic of an invasive species in north America, *Centaurea maculosa* Lam. (Asteraceae). The presence of these larvae induces the production by *Centaurea* of catechins. These biochemical compounds are released in soil and have allelopathic effects (i.e., release of chemicals/secondary metabolites from a plant or microbes that affect the growth of other plants) since they decrease the growth of native surrounding plants (Callaway, DeLuca, & Belliveau, 1999). By indirectly decreasing competitive ability of native species, the introduction of these below-ground herbivores favours the invasion of the exotic species.

The level of plant interactions occurring in vegetation may be related to the kind of defences against herbivores developed by plants, and by the susceptibility of plant species to herbivory (Mattson et al., 1988). In species-poor environments experiencing high plant competition intensities, the differentiation of antiherbivore defences is supposed to be low due to the scarcity of plant species. Plants tolerate herbivory since many conspecific plants coexist, having a similar probability to be eaten, and the effects of herbivores will be limited since they will not be able to alter plant competition levels in the neighbourhood of a focal plant. In species-rich ecosystems, plant consumption is more selective according to a wider range of palatability and defences developed by plants. In plant communities driven by competition relationships, herbivory will decrease the competitive ability of eaten plants and will indirectly favour neighbouring plants.

6.4 Competition Between Herbivores

The interaction between herbivores may be complex and seems to occur more for generalists than specialists as they are competing for primary resources. This intraguild competition as opposed to the principle of complementarity for using resources may decrease herbivore biomass in ecosystems (Thébault & Loreau, 2003). Although less common, intraguild competition does exist for specialist insects. Thus, two gall-making flies, *Urophora affinis* (Frauenfeld, 1857) and *Urophora quadrifasciata* (Meigen, 1826) (Diptera: Tephritidae), were released in North America for controlling spotted knapweed (*C. maculosa*). The two species were competing for the same resource: flower heads. After 14 years of postrelease monitoring, *U. quadrifasciata* was dominant (87%) over *U. affinis*, whereas the latter species was introduced 9 years before *U. quadrifasciata* (Mays & Kok, 2003). Strong interactions occurred in this species complex leading to the coexistence of the two *Urophora* species as 11.7% of flower heads hosted both *Urophora* flies. The choice of specialist phytophagous insects in weed biocontrol is done on purpose as it tends to minimize adverse effects on nontarget plants. At the interguild level, specialist phytophagous insects may induce nontrophic interactions, which may influence life trait parameters such as fecundity. As a result, interaction between *U. affinis* and curculionid beetles [*L. minutus* and *Bangasternus fausti* (Reitter, 1890)], also introduced into the United States for tackling spotted knapweed, is variable. Although weevils reproduction is not affected by the fly, the opposite occurs with about 80% reduction in fecundity when *L. minutus* is present, but no decrease is observed when *B. fausti* is present (Smith & Mayer, 2005).

6.5 Behaviour-Mediated Trophic Cascades

Predators can directly affect herbivorous prey and indirectly primary producer communities, through the reduction of herbivore density. However, the presence of predators does not only involve the direct consumption of prey: the herbivore prey foraging can be also strongly altered through a reduction of feeding time and/or shifts in diet and/or habitat selection. As a result, the behavioural mediation, also named as risk or nonconsumptive effects, could be as important as consumptive effects for the intensity and occurrence of trophic cascades (Schmitz, Beckerman, & O'Brien, 1997). The most popular example, even if not involving insect species, occurs in the Yellowstone National Park where wolves induce changes in elk behaviour that strongly impact aspen recruitment and all plant communities, since they have been reintroduced (Fortin et al., 2005). Behaviour of prey, herbivores in most studies, can be altered in various ways and function of the traits of the involved species (dispersal capacities, for example). In response to predator presence, herbivores may switch their habitat and then spatially consume the lower trophic level according to the density of their predators. The length of time herbivores devoted to feeding can be also altered (Beckerman, Uriarte, & Schmitz, 1997) or an increase in mortality can be observed due to the increase of starvation. Future studies should take into account behavioural responses of prey to predators, for example, experiments where predators are present but prevented to prey, and then assessing the survivorship and consequences for plant(s) not consumed by herbivores (Maguire, Nicole, Buddle, & Bennett, 2014).

ACKNOWLEDGEMENTS

The authors are very grateful to Dr. Dorothy Maguire for reading and commenting an earlier version of this article.

REFERENCES

Allesina, S., Alonso, D., & Pascual, M. (2008). A general model for food web structure. *Science, 320*, 658–661. http://dx.doi.org/10.1126/science.1156269.

Atlegrim, O. (1989). Exclusion of birds from bilberry stands: impact on insect larval density and damage to the bilberry. *Oecologia, 79*, 139. http://dx.doi.org/10.1007/BF00378251.

Baiser, B., Elhesha, R., & Kahveci, T. (2016). Motifs in the assembly of food web networks. *Oikos, 125*, 480–491. http://dx.doi.org/10.1111/oik.02532.

Barbaro, L., Brockerhoff, E. G., Giffard, B., & van Halder, I. (2012). Edge and area effects on avian assemblages and insectivory in fragmented native forests. *Landscape Ecology, 27*, 1451–1463. http://dx.doi.org/10.1007/s10980-012-9800-x.

Barbaro, L., Giffard, B., Charbonnier, Y., van Halder, I., & Brockerhoff, E. G. (2014). Bird functional diversity enhances insectivory at forest edges: a transcontinental experiment. *Diversity and Distributions, 20*, 149–159. http://dx.doi.org/10.1111/ddi.12132.

Barbaro, L., & van Halder, I. (2008). Linking bird, carabid beetle and butterfly life-history traits to habitat fragmentation in mosaic landscapes. *Ecography, 31*, 1–13. http://dx.doi.org/10.1111/j.1600-0587.2008.05546.x.

Beckerman, A. P., Uriarte, M., & Schmitz, O. J. (1997). Experimental evidence for a behavior-mediated trophic cascade in a terrestrial food chain. *Proceedings of the National Academy of Sciences of the United States of America, 94*, 10735–10738. http://dx.doi.org/10.1073/pnas.94.20.10735.

Blondel, J. (2003). Guilds or functional groups: does it matter? *Oikos, 100*, 223–231. http://dx.doi.org/10.1034/j.1600-0706.2003.12152.x.

Blossey, B., & Notzold, R. (1995). Evolution of increased competitive ability in invasive nonindigenous plants: a hypothesis. *Journal of Ecology, 83*(5), 887–889. http://dx.doi.org/10.2307/2261425.

Bouget, C., Brin, A., Tellez, D., & Archaux, F. (2015). Intraspecific variations in dispersal ability of saproxylic beetles in fragmented forest patches. *Oecologia, 177*, 911–920. http://dx.doi.org/10.1007/s00442-014-3162-9.

Boulant, N., Navas, M.-L., Corcket, E., & Lepart, J. (2008). Habitat amelioration and associational defence as main facilitative mechanisms in Mediterranean grasslands grazed by domestic livestock. *Ecoscience, 15*, 407–415. http://dx.doi.org/10.2980/15-3-3126.

Briese, D. T. (2005). Translating host-specificity test results into the real world: the need to harmonize the yin and the yang of current testing procedures. *Biological Control, 35*, 208–214. http://dx.doi.org/10.1016/j.biocontrol.2005.02.001.

Buddle, C. M., Spence, J. R., & Langor, D. W. (2000). Succession of boreal forest spider assemblages following wildfire and harvesting. *Ecography, 23*, 424–436. http://dx.doi.org/10.1111/j.1600-0587.2000.tb00299.x.

Callaway, R. M. (2007). *Positive interactions and interdependence in plant communities*. New York. USA: Springer.

Callaway, R. M., DeLuca, T. H., & Belliveau, W. M. (1999). Biological-control herbivores may increase competitive ability of the noxious weed *Centaurea maculosa*. *Ecology, 80*, 1196–1201. http://dx.doi.org/10.2307/177067.

Ceia, R. S., & Ramos, J. A. (2016). Birds as predators of cork and holm oak pests. *Agroforestry Systems, 90*, 159–176. http://dx.doi.org/10.1007/s10457-014-9749-7.

Cisneros, L. M., Fagan, M. E., & Willig, M. R. (2015). Effects of human-modified landscapes on taxonomic, functional and phylogenetic dimensions of bat biodiversity. *Diversity and Distributions, 21*, 523–533. http://dx.doi.org/10.1111/ddi.12277.

Coley, P. D. (1983). Herbivory and defense characteristics of tree species in a lowland tropical forest. *Ecological Monographs, 53*, 209–233. http://dx.doi.org/10.2307/1942495.

Corcket, E., Callaway, R. M., & Michalet, R. (2003). Insect herbivory and grass competition in a calcareous grassland: results from a plant removal experiment. *Acta Oecologica, 24*, 139–146. http://dx.doi.org/10.1016/s1146-609x(03)00064-x.

Daane, K. M., Sime, K. R., Fallon, J., & Cooper, M. L. (2007). Impacts of Argentine ants on mealybugs and their natural enemies in California's coastal vineyards. *Ecological Entomology, 32*, 583–596. http://dx.doi.org/10.1111/j.1365-2311.2007.00910.x.

Dhileepan, K., Setter, S. D., & McFadyen, R. E. (2000). Response of the weed *Parthenium hystephorus* (Asteraceae) to defoliation by the introduced biocontrol agent *Zygogramma bicolorata* (Coleoptera : Chrysomelidae). *Biological Control, 19*, 9–16. http://dx.doi.org/10.1006/bcon.2000.0847.

Díaz, S., Kattge, J., Cornelissen, J. H., Wright, I. J., Lavorel, S., Dray, S., ... Garnier, E. (2016). The global spectrum of plant form and function. *Nature, 529*, 167–171. http://dx.doi.org/10.1038/nature16489.

Díaz, S., Lavorel, S., McIntyre, S. U. E., Falczuk, V., Casanoves, F., Milchunas, D. G., ... Landsberg, J. (2007). Plant trait responses to grazing—a global synthesis. *Global Change Biology, 13*, 313–341. http://dx.doi.org/10.1111/j.1365-2486.2006.01288.x.

Finke, D. L., & Denno, R. F. (2005). Predator diversity and the functioning of ecosystems: the role of intraguild predation in dampening trophic cascades. *Ecology Letters, 8*, 1299–1306. http://dx.doi.org/10.1111/j.1461-0248.2005.00832.x.

Flynn, D. F. B., Gogol-Prokurat, M., Nogeire, T., Molinari, N., Richers, B. T., Lin, B. B., ... DeClerck, F. (2009). Loss of functional diversity under land use intensification across multiple taxa. *Ecology Letters, 12*, 22–33. http://dx.doi.org/10.1111/j.1461-0248.2008.01255.x.

Forkner, R. E., & Hunter, M. D. (2000). What goes up must come down? Nutrient addition and predation pressure on oak herbivores. *Ecology, 81*, 1588–1600. http://dx.doi.org/10.1890/0012-9658(2000)081[1588:wgumcd]2.0.co;2.

Fortin, D., Beyer, H. L., Boyce, M. S., Smith, D. W., Duchesne, T., & Mao, J. S. (2005). Wolves influence elk movements: behavior shapes a trophic cascade in Yellowstone National Park. *Ecology, 86*, 1320–1330. http://dx.doi.org/10.1890/04-0953.

Giffard, B., Barbaro, L., Jactel, H., & Corcket, E. (2013). Plant neighbours mediate bird predation effects on arthropod abundance and herbivory. *Ecological Entomology, 38*, 448–455. http://dx.doi.org/10.1111/een.12035.

Giffard, B., Corcket, E., Barbaro, L., & Jactel, H. (2012). Bird predation enhances tree seedling resistance to insect herbivores in contrasting forest habitats. *Oecologia, 168*, 415–424. http://dx.doi.org/10.1007/s00442-011-2089-7.

Gitay, H., Wilson, J. B., & Lee, W. G. (1996). Species redundancy: a redundant concept? *Journal of Ecology, 84*, 121–124. http://dx.doi.org/10.2307/2261706.

Gonzalez-Gomez, P. L., Estades, C. F., & Simonetti, J. A. (2006). Strengthened insectivory in a temperate fragmented forest. *Oecologia, 148*, 137–143. http://dx.doi.org/10.1007/s00442-005-0338-3.

Gray, S. M., & Banerjee, N. (1999). Mechanisms of arthropod transmission of plant and animal viruses. *Microbiology and Molecular Biology Reviews, 63*, 128–148. Retrieved from http://mmbr.asm.org/content/63/1/128.short.

Gullan, P. (1997). Adaptations in scale insects. *Annual Review of Entomology, 42*, 23–50.

Hairston, N. G., Smith, F. E., & Slobodkin, L. B. (1960). Community structure, population control, and competition. *The American Naturalist, 94*, 421–425. http://dx.doi.org/10.1146/annurev.ento.42.1.23.

Halaj, J., & Wise, D. H. (2001). Terrestrial trophic cascades: how much do they trickle? *The American Naturalist, 157*, 262–281. http://dx.doi.org/10.1086/319190.

Hartvigsen, G., Wait, D. A., & Coleman, J. S. (1995). Tri-trophic interactions influenced by resource availability: predator effects on plant performance depend on plant resources. *Oikos, 74*, 463–468. http://dx.doi.org/10.2307/3545991.

Lavorel, S., & Garnier, E. (2002). Predicting changes in community composition and ecosystem functioning from plant traits: revisiting the holy grail. *Functional Ecology, 16*, 545–556. http://dx.doi.org/10.1046/j.1365-2435.2002.00664.x.

Lavorel, S., McIntyre, S., Landsberg, J., & Forbes, T. D. A. (1997). Plant functional classifications: from general groups to specific groups based on response to disturbance. *Trends in Ecology and Evolution, 12*, 474–478. http://dx.doi.org/10.1016/S0169-5347(97)01219-6.

MacNally, R. (1994). Habitat-specific guild structure of forest birds in South-Eastern Australia: a regional scale perspective. *Journal of Animal Ecology, 63*, 988–1001.

Maguire, D. Y., Nicole, T., Buddle, C., & Bennett, E. M. (2014). Effect of fragmentation on predation pressure of insect herbivores in a north temperate deciduous forest ecosystem. *Ecological Entomology, 40*, 182–186. http://dx.doi.org/10.1111/een.12166.

Mäntylä, E., Alessio, G. A., Blande, J. D., Heijari, J., Holopainen, J. K., Laaksnonen, T., ... Klemola, T. (2008). From plants to birds: higher avian predation rates in trees responding to insect herbivory. *PLoS One, 3*, e2832.

Mäntylä, E., Klemola, T., & Laaksnonen, T. (2011). Birds help plants: a meta-analysis of top-down trophic cascades caused by avian predators. *Oecologia, 165*, 143–151. http://dx.doi.org/10.1371/journal.pone.0002832.

Mattson, W. J., Lawrance, R. K., Haack, R. A., Herms, D. A., & Charles, P. J. (1988). Defensive strategies of woody plants against different insect-feeding guilds in relation to plant ecological strategies and intimacy of association with insects. In W. J. Mattson, J. Levieux, & C. Bernard-Dagan (Eds.), *Mechanisms of woody plant defenses against insects. Search for patterns* (pp. 3–38). New York: Springer-Verlag.

Mays, W. T., & Kok, L. T. (2003). Population dynamics and dispersal of two exotic biological control agents of spotted knapweed, *Urophora affinis* and *U. quadrifasciata* (Diptera : Tephritidae), in southwestern Virginia from 1986 to 2000. *Biological Control, 27*, 43–52. http://dx.doi.org/10.1016/S1049-9644(02)00182-2.

McMeans, B. C., McCann, K. S., Tunney, T. D., Fisk, A. T., Muir, A. M., Lester, N., ... Rooney, N. (2016). The adaptive capacity of lake food webs: from individuals to ecosystems. *Ecological Monographs, 86*, 4–19. http://dx.doi.org/10.1890/15-0288.1.

Memmott, J., Fowler, S. V., Paynter, Q., Sheppard, A. W., & Syrett, P. (2000). The invertebrate fauna on broom, *Cytisus scoparius*, in two native and two exotic habitats. *Acta Oecologica, 21*, 213–222.

Mooney, K. A. (2007). Tritrophic effects of birds and ants on a canopy food web, tree growth, and phytochemistry. *Ecology, 88*, 2005–2014. http://dx.doi.org/10.1890/06-1095.1.

Mooney, K. A., Gruner, D. S., Barber, N. A., van Bael, S. A., Philpott, S. M., & Greenberg, R. (2010). Interactions among predators and the cascading effects of vertebrate insectivores on arthropod communities and plants. *Proceedings of the National Academy of Sciences of the United States of America, 107*, 7335–7340.

Mooney, K. A., & Linhart, Y. B. (2006). Contrasting cascades: insectivorous birds increase pine but not parasitic mistletoe growth. *Journal of Animal Ecology, 75*, 350–357. http://dx.doi.org/10.1111/j.1365-2656.2006.01054.x.

Moreira, X., Abdala-Roberts, L., Rasmann, S., Castagneyrol, B., & Mooney, K. A. (2016). Plant diversity effects on insect herbivores and their natural enemies: current thinking, recent findings, and future directions. *Current Opinion in Insect Science, 14*, 1–7.

Mougi, A., & Kondoh, M. (2016). Food-web complexity, meta-community complexity and community stability. *Scientific Reports, 6*, 24478. http://dx.doi.org/10.1038/srep24478.

Müller, H. (1989). Structural analysis of the phytophagous insect guilds associated with the roots of *Centaurea maculosa* Lam., *C. diffusa* Lam., and *C. vallesiaca* Jordan in Europe: 1. Field observations. *Oecologia, 78*, 41–52. http://dx.doi.org/10.1007/BF00377196.

Myers, J. H., & Bazely, D. R. (2003). *Ecology and control of introduced plants*. Cambridge, UK: Cambridge University Press. http://dx.doi.org/10.1007/s00606-005-0318-4.

Novotny, V., Miller, S. E., Baje, L., Balagawi, S., Basset, Y., Cizek, L., … Weiblen, G. D. (2010). Guild-specific patterns of species richness and host specialization in plant–herbivore food webs from a tropical forest. *Journal of Animal Ecology, 79*, 1193–1203. http://dx.doi.org/10.1111/j.1365-2656.2010.01728.x.

Oksanen, L., Fretwell, S. D., Arruda, J., & Niemela, P. (1981). Exploitation ecosystems in gradients of primary productivity. *The American Naturalist, 118*, 240–261. http://dx.doi.org/10.1086/283817.

Oksanen, L., & Ranta, E. (1992). Plant strategies along mountain vegetation gradients: a test of two theories. *Journal of Vegetation Science, 3*, 175–186. http://dx.doi.org/10.2307/3235678.

Paine, R. T. (1980). Food webs: linkage, interaction strength and community infrastructure. *Journal of Animal Ecology, 49*, 667–685. http://dx.doi.org/10.2307/4220.

Pimm, S. L., Lawton, J. H., & Cohen, J. E. (1991). Food web patterns and their consequences. *Nature, 350*, 669–674. http://dx.doi.org/10.1038/350669a0.

Poisot, T., Stouffer, D. B., & Gravel, D. (2015). Beyond species: why ecological interaction networks vary through space and time. *Oikos, 124*, 243–251. http://dx.doi.org/10.1111/oik.01719.

Polis, G. A. (1999). Why are parts of the world green? Multiple factors control productivity and the distribution of biomass. *Oikos, 86*, 3–15. http://dx.doi.org/10.2307/3546565.

Polis, G. A., Sears, A. L. W., Huxel, G. R., Strong, D. R., & Maron, J. (2000). When is a trophic cascade a trophic cascade? *Trends in Ecology and Evolution, 15*, 473–475.

Polis, G. A., & Strong, D. R. (1996). Food web complexity and community dynamics. *The American Naturalist, 147*, 813–842.

Prado, P. I., & Lewinsohn, T. M. (2004). Compartments in insect-plant associations and their consequences for community structure. *Journal of Animal Ecology, 73*, 1168–1178. http://dx.doi.org/10.1111/j.0021-8790.2004.00891.x.

Raymond, L., Sarthou, J.-P., Plantegenest, M., Gauffre, B., Ladet, S., & Vialatte, A. (2014). Immature hoverflies overwinter in cultivated fields and may significantly control aphid populations in autumn. *Agriculture Ecosystems and Environment, 185*, 99–105. http://dx.doi.org/10.1016/j.agee.2013.12.019.

Ribera, I., Dolédec, S., Downie, I. S., & Foster, G. N. (2001). Effect of land disturbance and stress on species traits of ground beetle assemblages. *Ecology, 82*, 1112–1129.

Root, R. B. (1967). The niche exploitation pattern of the blue-gray gnatcatcher. *Ecological Monographs, 37*, 317–350.

Root, R. B. (1973). Organization of a plant-arthropod association in simple and diverse habitat: the fauna of collards (*Brassica oleracea*). *Ecological Monographs, 43*, 95–124.

Rousset, O., & Lepart, J. (2002). Neighbourhood effects on the risk of an unpalatable plant being grazed. *Plant Ecology, 16*, 197–206.

Schmitz, O. J., Beckerman, A. P., & O'Brien, K. M. (1997). Behaviorally mediated trophic cascades: effects of predation risk on food web interactions. *Ecology, 78*, 1388–1399.

Schmitz, O. J., Hambäck, P. A., & Beckerman, A. P. (2000). Trophic cascades in terrestrial systems: a review of the effects of carnivore removals on plants. *The American Naturalist, 155*, 141–153.

Seastedt, T. R., Gregory, N., & Buckner, D. (2003). Effect of biocontrol insects on diffuse knapweed (*Centaurea diffusa*) in a Colorado grassland. *Weed Science, 51*, 237–245.

Sforza, R. (2008). Les cochenilles sur la vigne. In S. Kreiter (Vol. Ed.), *Les ravageurs de la vigne* (2nd ed., pp. 188–210). Bordeaux: Eds Feret.

Sheppard, A. W., Heard, T. A., & Briese, D. T. (2003). Workshop recommendations: the selection, testing and evaluation of weed biological control agents. In *Improving the selection, testing and evaluation of weed biological control agents* (pp. 89–100). CRC Technol. Ser. No. 7.

Simberloff, D., & Dayan, T. (1991). The guild concept and the structure of ecological communities. *Annual Reviews of Ecology and Systematics, 22*, 115–143.

Sipura, M. (1999). Tritrophic interactions: willows, herbivorous insects and insectivorous birds. *Oecologia, 121*, 537–545.

Smith, L. (2004). Impact of biological control agents on diffuse knapweed in central Montana. In J. M. Cullen, D. T. Briese, D. J. Kriticos, W. M. Lonsdale, L. Morin, & J. K. Scott (Eds.), *Proceedings of the XI symposium on biological control of weeds, April 27-May 2, 2003* (pp. 589–593). Canberra: Australia.

Smith, L., & Mayer, M. (2005). Field cage assessment of interference among insects attacking seed heads of spotted and diffuse knapweed. *Biocontrol Science and Technology, 15*, 427–442.

Southwood, T. R., Wint, G. R. W., Kennedy, C. E. J., & Greenwood, S. R. (2004). Seasonality, abundance, species richness and specificity of the phytophagous guild of insects on oak (*Quercus*) canopies. *European Journal of Entomology, 101*, 43–50.

Story, J. M., Smith, L., Corn, J. G., & White, L. J. (2008). Influence of seed head-attacking biological control agents on spotted knapweed reproductive potential in Western Montana over a 30-year period. *Environmental Entomology, 37*, 510–519.

Styrsky, J. D., & Eubanks, M. D. (2007). Ecological consequences of interactions between ants and honeydew-producing insects. *Proceedings of the Royal Entomological Society of London Series A, General Entomology, 274*, 151–164.

Suckling, D. M., & Sforza, R. F. H. (2014). What magnitude are observed non-target impacts from weed biocontrol? *PLoS One, 9*, e84847. http://dx.doi.org/10.1371/journal.pone.0084847.

Thébault, E., & Loreau, M. (2003). Food-web constraints on biodiversity–ecosystem functioning relationships. *Proceedings of the National Academy of Sciences of the United States of America, 100*, 14949–14954.

Torchin, M. E., Lafferty, K. D., Dobson, A. P., McKenzie, V. J., & Kuris, A. M. (2003). Introduced species and their missing parasites. *Nature, 421*, 628–630.

Tscharntke, T., Bommarco, R., Clough, Y., Crist, T. O., Kleijn, D., Rand, T. A., … Vidal, S. (2007). Conservation biological control and enemy diversity on a landscape scale. *Biological Control, 43*, 294–309.

Vásquez, P. A., Grez, A. A., Bustamante, R. O., & Simonetti, J. A. (2007). Herbivory, foliar survival and shoot growth in fragmented populations of *Aritotelia chilensis*. *Acta Oecologica, 31*, 48–53.

Wapshere, A. J. (1974). A strategy for evaluating the safety of organisms for biological control. *Annals of Applied Biology, 77*, 201–211.

Whelan, C. J., Wenny, D. G., & Marquis, R. J. (2008). Ecosystem services provided by birds. *Annals of the New York Academy of Sciences, 1134*, 25–60.

Williams, S. E., & Hero, J. M. (1998). Rainforest frogs of the Australian Wet Tropics: guild classification and the ecological similarity of declining species. *Proceedings of the Royal Society B: Biological Sciences, 265*, 597–602. http://dx.doi.org/10.1098/rspb.1998.0336.

Willis, A. J., Ash, J. E., & Groves, R. H. (1995). The effects of herbivory by a mite, *Aculus hyperici*, nutrient deficiency on growth in *Hypericum* species. *Australian Journal of Botany, 43*, 305–316.

Wootton, K. L., & Stouffer, D. B. (2016). Many weak interactions and few strong; food-web feasibility depends on the combination of the strength of species' interactions and their correct arrangement. *Theoretical Ecology, 9*, 185–195. http://dx.doi.org/10.1007/s12080-015-0279-3.

CHAPTER FIVE

Chemical Signatures in Plant–Insect Interactions

B. Frérot*[,1], E. Leppik*, A.T. Groot[§], M. Unbehend[¶], J.K. Holopainen[||]
*INRA, UMR 1392, Institut d'Ecologie et des Sciences de l'Environnement de Paris, Versailles, France
[§]University of Amsterdam, Amsterdam, Netherlands
[¶]Max Planck Institute for Chemical Ecology, Jena, Germany
[||]University of Eastern Finland, Kuopio, Finland
[1]Corresponding author: E-mail: brigitte.frerot@versailles.inra.fr

Contents

Advances in Botanical Research, Volume 81
ISSN 0065-2296
http://dx.doi.org/10.1016/bs.abr.2016.10.003

Abstract

Chemical signals are important cues throughout the life of an insect especially for mate location and for prey and host finding. The chemical signal, whether pheromone or plant volatile organic compound (VOC), remains specific because of the mixture, of the ratio of the components in mixture and of the release quantity. The plasticity of pheromone emissions is now studied in several insect species in relation to geographic variation, host plant specialization and chemical and light environment. The actual vision is that the pheromone composition is likely to be more plastic than previously assumed. The perception of the environmental odorscape produced by living plants and animals together addressed the question on the specific detection of the pheromone signal in the atmospheric blend of molecules. In agrobiocoenosis, the cultivated plants produce a specific odorscape. The insects rely on plant VOCs to locate the crop or the host plant, after which specific mixtures act as oviposition stimulants. The insect responses to host plants and their odours vary with the physiological status of both actors: the plant and the insect. Chemical signals released by plants vary with plant physiology, diel periodicity, climatic factors and pollution, and these signals can be species or even variety specific. Many of plants signalling compounds detected by insects have important roles as warning signals, which can also function in plant–plant communication.

1. INTRODUCTION

The chemical signals or semiochemicals manage the relations between organisms belonging to the animal kingdom and plant kingdom, via air, water or soil. They are classified in different categories according to their nature, the effects and which is the beneficiary of the relation (the releaser or the receiver). The chemicals involved in an intraspecific relation are pheromones by opposition to the allelochemicals that control the interspecific relation (Dicke & Sabelis, 1988; Nordlund & Lewis, 1976).

The pheromones are classified according to the action on the conspecific. When it acts immediately on the behaviour, they are releasers and when it induces a long term physiological change they are called modifiers (Wilson, 1963).

The allelochemicals are classified on the base of the beneficiary of the released chemical: when the benefit is for the releaser they are named allomones and when it is for the receiver, they are kairomones. Nowadays, the studies of the allelochemicals are expanding in the field of plant–insect interactions.

The chemical signature is a concept that defines the specificity and the originality of the semiochemicals: pheromone and allelochemical. The chemical signature can be considered as a specific blend more or less complex, involving often ubiquist compounds or/and original compounds. The pheromone chemical signature is rather simple made of one to at least five compounds. The concept becomes more complex for the chemical signature produced by plants and at the worst when the chemical signature of a plant acts on the pheromone production or on the perception.

The chemical signature characterized a specific mixture of chemical that acts on another organism. The specificity relies on the composition, the ratio of the components and the dose released. The originality and the diversity of the components are poor compared to the diversity of the fauna and flora. The combinations of these compounds make the chemical signature of an individual.

In insects, the chemical signals play an important role for each steps of the life. Intraspecific recognition is well-documented for reproduction widely mediated in insects by pheromone. The receiver perceived the signal by the olfactory receptors located on the antennae. The signal, a very precise mixture of molecules produced in a particular ratio and released in very low amount (10^{-8}; 10^{-9} gr), is analyzed at the level of the brain in antennal lobes and converted in a behavioural or a physiological change.

The pheromones are mostly synthetized de novo (in Lepidoptera) but can also be sequestered from the food (in Coleoptera; in Lepidoptera male pheromone). In Lepidoptera, the fatty acids synthetized de novo in the epithelial cells of the pheromone gland undergo actions of enzymes that shorten the carbohydrate chain, change the position of double bonds then reduce and esterify (Bjostad, Wolf, & Roelofs, 1987). The pheromone molecules are rather simple: carbohydrate chain of 10–18 carbons; with one, two, rarely more; double bond. The chain has a functional group, acetate, aldehyde, alcohol on the first carbon. Rare examples exist with a triple bond and with epoxyde function. This molecular construction allowed many different possibilities according to the position of the double bonds along the chain. The different functional groups contribute to increase the diversity of molecules. This particularity and the complexity of the blend of two, three or more compounds insure for most of the species the specificity of the pheromone.

The chemical structure, especially the chain length, has followed evolution and can be used as an added criterion in systematic (Descoins & Frérot, 1979; Roelofs & Brown, 1982). In Coleoptera, the gut bacteria used the terpenoids ingested to biosynthesize the pheromone. As a consequence, the same molecules are shared by several species of bark beetles. In this group, the attraction of conspecifics is achieved by the combination of the pheromone and tree odours. The signal became an original chemical signature because of this association (Blum, 1987).

The universality of the pheromone components in the animal kingdom is documented by the examples of bumblebees and the elephant. Males of some bumblebee species mark a territory by depositing on plant products secreted by the labial glands. The molecules involved in this behaviour are the same as those identified as sex pheromone in some noctuid moths (Berström et al., 1985; Descoins et al., 1984). In mammals certain molecules found in urine inform conspecific on the physiological and social status. One of the compounds, (Z)-7-dodecen-1-yl acetate (Z7-12:Ac), identified in urine of elephant females is a component of the mating pheromones for numerous species of Lepidoptera (Rasmussen, Lee, Roelofs, Zhang, & Daves, 1996).

The interspecific communication and especially the interaction between plants and insects followed a schema closely related to the intraspecific communication. The insects are tuned to a specific signal released by the plants. The signal could be more complex than the pheromone blend in number of components. The chemicals originate from primary and secondary metabolites and are bio-synthetized by the plants. The diversity of the plant signals depends of the botanical family but is also achieved by combining molecules more or less common to the plant kingdom.

2. PLASTICITY AND SPECIFICITY OF THE CHEMICAL INFORMATION

2.1 Plasticity and Specificity in Pheromone Communication

In most nocturnal moths, females are the pheromone signallers and males are the responders. The female signal is exclusively chemical and produced in a pheromone gland de novo every night. Females of each moth species produce a species-specific sex pheromone blend, which is determined by the combination of, in general, two–five pheromone compounds that are emitted in species-specific ratios.

The predominant role of moth sexual communication is species recognition (i.e., to not attract or be attracted by other closely related species with a similar pheromone blend) (Butlin, Hewitt, & Webb, 1985; Droney, Musto, Mancuso, Roelofs, & Linn, 2012; Löfstedt, 1993; Paterson, 1985).

As these communications are generally thought to be mostly important to distinguish between species and not in intraspecific sexual selection, they are generally hypothesized to be under stabilizing selection and to exhibit low intraspecific variation. As males are behaviourally fine-tuned to their species-specific pheromone blend (Cossé et al., 1995; Linn, Young, Gendle, Glover, & Roelofs, 1997), a mutation that alters the female pheromone blend is likely to lower her reproductive fitness (Butlin & Trickett, 1997; Zhu, Chastain, Spohn, & Haynes, 1997). Therefore, the means by which novel signals in sexual communication can evolve, in the face of selection against such change, is still an evolutionary mystery, especially as moths are one of the most diverse group of animals on earth, with ~120.000 of 160.000 Lepidoptera species (Bazinet, Cummings, Mitter, & Mitter, 2013).

To resolve the dilemma of how the female signal and male preference can concomitantly change, the genes that are responsible for intraspecific variation in and interspecific divergence of the pheromone systems should be more extensively identified (Groot, Dekker, & Heckel, 2016). So far, only five such genes have been identified (Albre, Steinwender, & Newcomb, 2013; Fujii et al., 2011; Gould et al., 2010; Lassance et al., 2010; Leary et al., 2012). In the moth species studied until now, the genomic regions involved in female pheromone production and male response do not appear to overlap (i.e., they are located on different chromosomes). Without a genetic association between signal and response, genetic changes seem to have evolved independently, which makes Fisherian runaway selection (a hypothesized genetic sexual selection mechanism for the evolution of exaggerated male ornamentation) unlikely. Therefore, other evolutionary scenarios have been proposed for moth sexual communication, including stabilizing and directional selection, the asymmetric tracking hypothesis (Phelan, 1992) and the rare male hypothesis (Roelofs & Rooney, 2003) (see Groot et al., 2016 for a more detailed description). Once the genes underlying variation in moth sex pheromone signals and responses are eventually known, more powerful tests based on comparisons of allele frequencies among populations and over time will be possible.

Although moth sex pheromones are generally assumed to be under stabilizing selection, there are numerous descriptive studies of geographic variation in female sex pheromone blends, male responses and attraction to artificial lures formulated to mimic females (Cossé et al., 1995; Gries, Schaefer, Gries, Liska, & Gotoh, 2001; Groot et al., 2009; Linn et al., 1997; McElfresh & Millar, 1999, 2001; Unbehend et al., 2014). Geographic variation in moth sexual communication systems could result in reproductive isolation and subsequently may lead to speciation (Coyne & Orr, 2004; Löfstedt, 1993; Smadja & Butlin, 2009; Symonds & Elgar, 2008). The causes of geographic variation may be different selection forces in different regions, stochastic events or habitat variation, which includes abiotic factors, such as temperature, relative humidity and day length. In moths, pheromone production has been shown to be affected by host plant volatiles (McNeil & Delisle, 1989a; Raina, Jackson, & Severson, 1997) (see Section 2.1), sucrose nutrition (Foster, 2009) and temperature (Raina, 2003). However, these factors affect the quantity rather than the quality of the pheromone blend. Biotic factors in the habitat that likely affect chemical communication are the presence and abundance of species with similar chemical cues because they may either affect the signal-to-noise ratio (e.g., Eizaguirre, Albajes, Lopez, Sans, & Gemeno, 2007; Gemeno, Sans, Lopez, Albajes, & Eizaguirre, 2006; Haynes, Gemeno, Yeargan, Millar, & Johnson, 2002; Sole et al., 2008) or generate communication interference (e.g., Butlin, 1995; Cardé, Cardé, Hill, & Roelofs, 1977; Groot et al., 2006; McElfresh & Millar, 1999, 2001). Both signal-to-noise ratio and communication interference would result in selection for females with the most distinct, optimized pheromone blend (i.e., negative frequency-dependent selection). Such local natural selection forces may alternate directions or be unidirectional, similar to what has been found for beak sizes in the Galapagos finches (Grant & Grant, 2002). Only when specific local environmental conditions persist, selection forces from the environment may result in directional or divergent selection. Variation related to selection across a persistent environmental gradient has been referred to as the "selection hypothesis" (Groot et al., 2009).

Even though the finding of variation often leads to the conclusion that differences are due to selection that may lead to speciation, as in the Galapagos finches, not all variation may be subject to selection. An alternative, more speculative hypothesis to explain variation in chemical communication signals, is that females and males exhibit phenotypic plasticity in sexual signalling, and experience — either by immature stages or by early adults — shapes the expressed phenotype. Early adult (postimaginal) experience has been shown to be an important factor in female oviposition preference (Barron, 2001; Jaenike, 1983, 1990; Van Emden et al., 1996), and preexposure of male moths to sex pheromone blends has been shown to affect their subsequent responses to sex pheromone (Anderson et al., 2007; Anderson, Sadek, & Hansson, 2003; Anderson, Sadek, Larsson, Hansson, Thoming, 2013). Whether preexposure to pheromone blends and other semiochemicals could also cause females to alter the blend that they produce has only been investigated in one species so far (Groot, Classen, Staudacher, Schal, & Heckel, 2010).

Plasticity in female sex pheromone composition in moths can be expected because of a number of characteristics. First of all, females produce their pheromone de novo every night (e.g., Jurenka, 2004; Rafaeli, 2005), which may allow adjustment of time and temporal patterning of pheromone release (calling) in relation to environmental conditions (Lim & Greenfield, 2007; Schal & Carde, 1986). Also, in many moth species females can perceive their own species' female pheromone compounds (Den Otter, Schuil, & Sandervanoosten, 1978; Groot, Gemeno, Brownie, Gould, & Schal, 2005; Hillier, Kleineidam, & Vickers, 2006; Schneider, Schulz, Priesner, Ziesmann, & Francke, 1998), so that they are likely to perceive conspecific as well as heterospecific sex pheromone, at least if their own pheromone blend overlaps with that of other species. In addition, pheromone receptors have been found in sensilla on the ovipositor of a female moth (Widmayer, Heifetz, & Breer, 2009), suggesting that these receptors might allow a feedback mechanism onto the gland that might affect pheromone emission. Hence, it may be possible that females can vary their biosynthesized as well as emitted pheromone blend to some extent, depending on the prevailing olfactory cues in their habitat. This hypothesis has been referred to as the "experience hypothesis" (Groot, Classen, et al., 2010; Groot et al., 2009).

In many species, the attractiveness of a potential mate is determined by the quality of a sexual signal (Domb & Pagel, 2001; Groot et al., 2014; Scheuber, Jacot, & Brinkhof, 2003), which may be affected by immune defence responses, as generally hypothesized by the Zahavi handicap principle (i.e., hypothesis proposed to explain how evolution may lead to "honest" or reliable signalling between animals which have an obvious motivation to bluff or deceive each other). The level of sexual attraction may signal the level and extent of health. However, if sexual attractiveness and immunity compete for the same resource pool, they may negatively affect each other. Very little data exist on whether and how immune responses affect sexual communication in moths. In the tobacco budworm, *Heliothis virescens* (Fabricius, 1777) (Lepidoptera: Noctuidae), the sex pheromone blend shifted towards an unattractive blend when females were infected with the entomopathogenic bacterium *Serratia entomophila* (Barthel, Staudacher, Schmaltz, Heckel, & Groot, 2015), suggesting that at least in this species sex pheromone production is condition dependent.

2.2 Relevant Examples of Intraspecific Variation in Moth Sexual Communication

2.1.1 *Pheromone Races of* Ostrinia nubilalis *(Fig. 1A)*

The existence of pheromonal races within the same species is known since 1975 (Klun, 1975), who identified two pheromone races in the European corn borer *Ostrinia nubilalis* (Hübner, 1796) (Lepidoptera: Crambidae), the E– and the Z-strain. The sex pheromone of the two *O. nubilalis* strains consists of two pheromone components, i.e., (Z)-11-tetradecenyl acetate (Z11-14:Ac) and (E)-11-tetradecenyl acetate (E11-14:Ac), which are produced in a 97:3 Z:E ratio in the Z-strain, while E-strain females emit a 1:99 Z:E mixture (Klun, 1975; Kochansky, Cardé, Liebherr, & Roelofs, 1975; Roelofs et al., 1987). Males of each strain are highly attracted to females of their own strain (Linn, 1997; Roelofs et al., 1987), which leads to assortative mating in the field (i.e., a form of nonrandom mating in which pair bonds are established on the basis of phenotype) (Dopman, Robbins, & Seaman, 2010; Klun & Huettel, 1988; Malausa et al., 2005). Following the first evidence for pherotypes in US population of *O. nubilalis*, a more recent

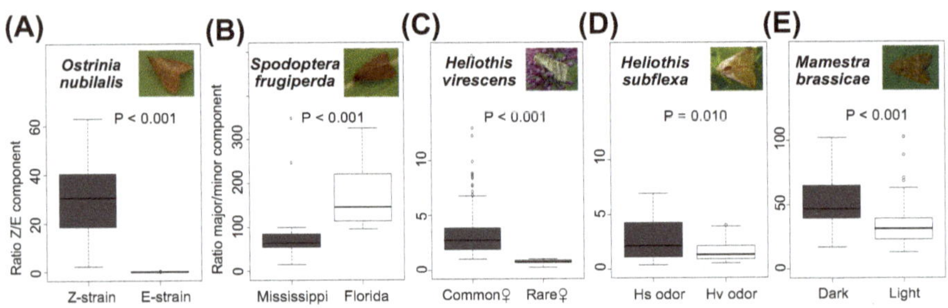

Figure 1 Intraspecific variation in the pheromone composition of female moths. (A) Pheromone polymorphism of *Ostrinia nubilalis* Z- and E-strain females. Values on the y-axes show the relative percentages of (Z)-11-tetradecenyl acetate (Z11-14: Ac)/(E)-11-tetradecenyl acetate (E11-14:Ac). (B) Geographic variation in the relative amount of (Z)-9-tetradecenyl acetate (Z9-14: Ac)/(Z)-7-dodecenyl acetate (Z7-12: Ac) in *Spodoptera frugiperda* corn-strain females from Mississippi and Florida. (C) Within-population variation in the relative amount of (Z)-l1-hexadecenal (Z11-16: Ald)/hexadecenal (16:Ald) in common and rare *Heliothis virescens* females. (D) Phenotypic plasticity in the pheromone composition (relative amount of Z11-16: Ald/(Z)-11-hexadecenyl acetate (Z11-16:Ac)) of *Heliothis subflexa* (Hs) females that were reared in the presence of Hs odour or *H. virescens* (Hv) odour. (E) Light-dependent phenotypic plasticity of the pheromone composition (relative amount of Z11-16:Ac/(Z)-9-hexadecenyl acetate (Z9-16:Ac) + (Z)-11-hexadecenol (Z11-16: OH)) of *Mamestra brassica* females that were either extracted under normal night conditions (Dark) or exposed to artificial green, red, and white light (Light).

study demonstrates that in France the two pheromone populations were also present but segregated by the host plant. The Z population was specialized on maize whereas the E population was on hop and mugwort, whatever was the geographic location (Pelozuelo, Malosse, Genestier, Guenego, & Frérot, 2004). Genetic approach evidenced a very low gene flow between the two populations (Bethenod et al., 2005). Thus, although hybridization is still possible (Liebherr & Roelofs, 1975), the Z- and E-strains seem to be in process of speciation and are isolated enough to be considered sibling species (Cardé et al., 1978; Malausa et al., 2007).

2.1.2 Geographic Variation in Spodoptera frugiperda *(Fig. 1B)*

The fall armyworm, *Spodoptera frugiperda* (J.E. Smith, 1797) (Lepidoptera: Noctuidae), consists of two genetically and behaviourally distinct strains, the corn- and rice-strain, that appear to be undergoing ecological speciation in sympatry (Dumas et al., 2015; Kergoat et al., 2012). One of the prezygotic isolation barriers is differentiation in sexual communication (Groot, Marr, Heckel, & Schoefl, 2010). Strain-specific as well as geographic variation in the sex pheromone composition has been found in this species (Batista-Pereira et al., 2006; Groot et al., 2009; Lima & McNeil, 2009). Genetic analysis showed that only 9% of the strain-specific pheromone variation was explained by genetic differences (Unbehend, 2013), which suggest that environmental conditions influence the female pheromone phenotype to a large degree. In addition, males of the two strains were similarly attracted to the blends of both strains, and there was more geographic than strain-specific variation in male response (Unbehend et al., 2014). Thus, this prezygotic isolation barrier in itself does not seem strong enough to keep the two strains separated. However, if environmental conditions affect the female signal and male response in the same direction, geographic differentiation in sexual communication may ensue.

2.1.3 Within-Population Variation in Heliothis virescens *(Fig. 1C)*

In the moth *H. virescens*, consistent high phenotypic variability in the female sex pheromone blend was found within each of four geographically distant populations (Groot et al., 2014). The within-population variation found in *H. virescens* is in ratio of unsaturated to saturated pheromone compounds; common females produce mostly unsaturated compounds while rare females produce higher relative amounts of saturated pheromone compounds. Selection experiments in the laboratory showed a genetic basis of this within-population variation, while field experiments showed that females producing a higher relative proportion of the saturated compounds are less attractive to males (Groot et al., 2014). Interestingly, *H. virescens* females expressing the unusual phenotype (high relative amounts of saturated compounds) were found in all populations, across regions and years, albeit at low frequency (Groot et al., 2014). Such a polymorphism may be maintained through balancing selection, in this case specifically through heterozygote advantage, as heterozygote females produce significantly more of the critical sex pheromone component through which males are attracted. Thus, selection need not act solely in a purifying role, eliminating genetic variation in signal—response systems as is generally assumed but can also act to maintain genetic variation in them.

2.1.4 Phenotypic Plasticity in Heliothis subflexa Due To Varying Chemical Environments (Fig. 1D)

In the moth, *Heliothis subflexa* (Guenée, 1852) (Lepidoptera: Noctuidae), one compound in the sex pheromone blend was found to have a dual function: the acetate (Z)-11-hexadecenyl acetate (Z11-16: Ac) not only enhances the attraction of conspecific males but also inhibits the attraction of the closely related heterospecific *H. virescens* males (Groot et al., 2006; Vickers & Baker, 1997). In these two closely related species, not only geographical but also temporal variation in the pheromonal signals was found (Groot et al., 2007, 2009). The geographic and temporal variation in the sex pheromone of *H. subflexa* was correlated with *H. virescens*; when *H. virescens* was highly abundant, *H. subflexa* females contained significantly more of this acetate than when *H.virescens* was much less abundant (Groot et al., 2009). In the laboratory, when *H. subflexa* females emerged and remained in the odour of *H. virescens* for three days, these females contained significantly more acetate compared to *H. subflexa* females that emerged in control odour or in the sex pheromone odour of conspecific females (Groot, Classen, et al., 2010). Thus, early-adult experience of different chemical environments affects the sex pheromone composition in *H. subflexa* females. Because a higher acetate level increases the attraction of conspecific males (Groot et al., 2006), this behavioural adjustment may lead to assortative mating.

2.1.5 *Phenotypic Plasticity in* Mamestra brassicae *Due To Varying Light Regimes (Fig. 1E)*

The cabbage moth, *Mamestra brassicae* (Linnaeus, 1758) (Lepidoptera: Noctuidae), is a common species in (sub)urban areas of western Europe, and adult moths are frequently exposed to artificial light at night, as they are nocturnal and attracted to sources of artificial light. Van Geffen et al. (2015) tested the effect of low levels of artificial night lighting with different spectral compositions on the amount and composition of the sex pheromone produced by *M. brassicae* females. This experiment showed that artificial light at night not only strongly reduced the total amount of sex pheromone, but also changed the chemical composition of the pheromone blend. The range of variation that was found in the pheromone composition of *M. brassicae* under different light conditions is similar to or even higher than that found in the geographic variation of, for example, *H. virescens*, *H. subflexa* (Groot et al., 2009), *Agrotis ipsilon* (Hufnagel, 1766)(Lepidoptera: Noctuidae) (Gemeno, Lutfallah, & Haynes, 2000), *Agrotis segetum* (Denis & Schiffermüller, 1775) (Lepidoptera: Noctuidae) (Wu, Cottrell, Hansson, & Löfstedt, 1999), and *Cydia pomonella* (Linnaeus, 1758) (Lepidoptera: Tortricidae)(Dumenil et al., 2014). Hence, varying only one environmental factor, in this case light, can already change the sex pheromone blend significantly.

2.3 Plasticity and Specificity in Plant Volatile Organic Compounds

2.3.1 Major Groups of Plant Volatiles

Plants are capable of emitting wide variety volatile organic compounds (VOCs). Because of their biological origin these compounds are often called biogenic VOCs (BVOCs) to separate them from volatile compounds of anthropogenic origin. Volatiles can be emitted from different plant organs including leaves, flowers, stem and root system and they may have different ecological functions such as activating defences in neighbouring plants, attraction of pollinators by flower volatiles or attraction of parasitoids and predators of defoliating herbivores by leaf volatiles (Holopainen & Gershenzon, 2010).

Majority of VOC emitted by plant originate from three major biosynthesis pathways: (1) terpenes form most diverse group of VOCs, (2) oxylipins are fatty acid–derived six-carbon C6 compounds and (3) shikimate and benzoic acid pathways produce benzenoids that are aromatic compounds such as methyl salicylate and indole (Maffei, 2010). In addition, there are several low molecular weights, C1 and C2, compounds, such as methanol, ethanol, formaldehyde, and acetaldehyde methane and ethylene synthesized via other biosynthetic routes (Loreto & Schnitzler, 2010). Most of common plant VOCs can be classified into alkanes, alkenes, alcohols, esters, aldehydes, and ketones (Maffei, 2010). Furthermore, the breakdown products of glucosinolates are very important group of VOC in the family

Brassicaceae (e.g., Blande, Holopainen, & Niinemets, 2014; Pinto, Blande, et al., 2007).

Terpenes represent highly diverse group of volatile chemicals found from plants and they are also most intensively studied. Isoprene, synthesized in plant chloroplast in light and temperature-dependent way (Loreto & Schnitzler, 2010), is the single compound emitted from the vegetation in the atmosphere in the highest rate, but not all plant species emit this compound (Laothawornkitkul, Taylor, Paul, & Hewitt, 2009). Isoprene unit (C5) is the basic structure of terpenes. Other major volatile terpene (isoprenoid) groups are monoterpenes (C10) and sesquiterpenes (C15) (Holopainen & Gershenzon, 2010). Recently even some volatile diterpenes (C20) have been detected from plant headspace of tobacco (Jud et al., 2016), although diterpenes are mostly nonvolatile compounds. Herbivore-inducible homoterpenes (E)-4,8, dimethyl-1,3,7-nonatriene (C11, DMNT) and 4,8,12-trimethyltrideca-1,3,7,11-tetraene (C16, TMTT) (Dicke, 2009) are important signalling compounds in plant—carnivore communication as their emissions are induced by herbivore feeding.

Fatty acid—derived C6 oxylipins are often called green leaf volatiles (GLVs) as typically they are smelled by humans after cutting of grass. They are rapidly emitted from stressed plants, particularly from those suffering mechanical damage (Brilli et al., 2011) or insect feeding, which are causing membrane damage in plant cells. The emission rate of some GLV compounds, such as leaf aldehydes (*E*)-3-hexenal and (*E*)-2-hexenal, are following strictly the feeding activity of Lepidopteran larvae having the emission peak within 2—3 min from the start of feeding activity. Leaf alcohol (Z)-3-hexenol and (Z)-3-hexenyl acetate have the emission peaks nearly 10 min later (Maja et al., 2014; Simpraga, Takabayashi, & Holopainen, 2016). This succession of GLV compounds in plant VOC emissions has chemical information for herbivorous insects to detect the sites in plant that are already occupied by other herbivores. Value for predatory and parasitoid insects is also high as their can use typical GLV profiles of current feeding and older damage in their orientation behaviour to separate the older damage from fresh damage. This maximizes their probability to find host larvae. Furthermore, the proportion of GLV compounds in volatile bouquet is an important tool in detection of pest attack and its extension in crops when using VOC-based profiling (Jansen et al., 2011).

Quite a few benzenoid compounds are emitted by plants (Misztal et al., 2015), although some compounds such as methyl salicylate are induced in higher rate (e.g., by mites - Blande, Holopainen, & Li, 2010, by aphids - Blande, Korjus, & Holopainen, 2010; by plant pathogens - Shulaev, Silverman, & Raskin, 1997) and have very important function in plant to plant communication (Blande, Holopainen, et al., 2010; Shulaev et al., 1997).

2.3.2 Plant Physiological Functions of VOCs and Diel Periodicity

Secondary metabolites are generally considered to lack primary functions for plant physiology and act more for plant defence and communication. However, many of the terpenes (isoprene, monoterpenes) are synthesised in chloroplasts and they have crucial role to maintain plant photosynthesis under variable environmental conditions [e.g., by stabilizing chloroplast membranes under heat and pollution stress (Loreto & Schnitzler, 2010)]. This suggests that chloroplast-originating terpenes have very primary functions for plant physiology. Turnover of freshly fixed carbon to isoprenoid emissions from the chloroplasts could be unexpectedly fast in sufficient light conditions. Isoprene is the volatile compound studied most intensively in the association with plant photosynthesis. Delwiche and Sharkey (1993) found that freshly photosynthetically fixed labelled ^{13}C carbon can be detected from isoprene emitted from oak leaves just 9 min later. Under high temperature and solar radiation conditions photosynthesis is activated and will better provide fresh carbon for isoprene production. However, drought stress causes some limitation for photosynthesis as closure of stomata does not allow carbon uptake. Recent studies have shown that isoprene synthesis and emission still continues under drought stress, but carbon source can earlier fix carbon possibly from respiration or starch breakdown (Loreto & Schnitzler, 2010). While stomata are closed, plant volatiles can be emitted by passive diffusion from cells to inter cellular gas space and from there by diffusion through leaf epidermis. Recently it has been suggested that specific transporter molecules might transfer VOC molecules in plant tissues (Widhalm, Jaini, Morgan, & Dudareva, 2015). In general, emission rates of isoprene can be substantially higher in photosynthetically active mature leaves than in young and old leaves (Niinemets, Sun, & Talts, 2015).

Diel periodicity of leaf volatiles, showing highest emission rates of VOC during day time and low emission rates at night can be directly linked to their very strong coupling of photosynthetic activity of foliage during day time. However, under 24h sunshine subarctic conditions some plant species can keep their VOC emission rates at midnight on the same level as during daytime, while in colder high arctic condition plants seem to show diel periodicity under sunny nights probably due to colder night temperature (Lindwall, Faubert, & Rinnan, 2015). Floral scents are known for the strong diurnal activity. Scent volatiles are often produced in petal lobes of flowers (Pichersky & Dudareva, 2007), but tissue-specific scents for the stigma and stamens has been found (Burdon, Raguso, Kessler, & Parachnowitsch, 2015). Most of the floral scent volatiles are terpenoids or phenylpropanoid/benzenoid compounds, but their proportions are changing during the progress of flowering. Total floral scent emissions might be highest during the day, but also composition has differences between night and day. Burdon et al. (2015) found that day emissions dominated by monoterpenes while sesquiterpenes and other aliphatic compounds dominated night emissions. Typically, floral scent emissions peak often during the highest activity hours of the efficient pollinator species showing importance of coevolution in plant pollinator interaction (Burdon et al., 2015).

2.3.3 Genotypic Diversity of Plant VOCs

Plant volatile emission from flowers, fruits and vegetative parts vary strongly between plant species, but also between different genotypes within a species. Even the dominating VOC emitted by a plant species may differ between different locations. In small pine seedlings the provenance of seed material collected from a 1000 km strongly affects, if the volatile monoterpene pool is dominated by α-pinene of Δ3-carene (Nerg et al., 1994). In agricultural plants similar variation can be found. In three-week-old carrot seedlings, myrcene is the dominating monoterpene, but in the comparison of four cultivars the second most common monoterpene was sabinene in one cultivar and limonene in three others (Kainulainen, Nissinen, Piirainen, Tiilikkala, & Holopainen, 2002). In older seedlings the order changed and the sesquiterpene β-caryophyllene became the second most common terpene. Cultivar splendid with highest limonene and β-caryophyllene content was preferred by a generalist plant bug, *Lygus rugulipennis* Poppius, 1911 (Heteroptera: Miridae), in oviposition tests (Kainulainen et al., 2002).

2.3.4 Relationship of Plant VOCs With Climate Changes and Pollution

As long as there has been human activity leading to burning of the fuels, the levels of pollutants have increased in the atmosphere. Most important atmospheric pollutants affecting biogenic VOCs and other hydrocarbons by reacting with them in the atmosphere are ozone (O_3) and OH and NO_3 radicals (McFrederick, Kathilankal, & Fuentes, 2008). These compounds have increased substantially in the atmosphere since late 1800s and model calculation shows that this might have affected behaviour of pollinators (McFrederick et al., 2008). There are also natural sources of these emissions, such as volcanic eruptions, wild fires and lightning, but these sources are mostly local and temporal and not so widely distributed as human activities. Recent years' elevated concentrations of atmospheric carbon dioxide (CO_2) and global climate warming–related CO_2 have become even more important factors to affect ecosystems and crop production (Forkel et al., 2016).

Important for the understanding of plant communication with conspecifics and other organisms using VOCs is the fact that most of the herbivore-inducible VOCs and other volatile signalling compounds have rather short lifetime in the atmosphere with reactive air pollutants. Oxidation and other photochemical reactions in the atmosphere will lead degradation of compound originally emitted by plants (Pinto, Blande, Souza, Nerg, & Holopainen, 2010). Reduced atmospheric lifetime means also reduced signalling distance by the VOCs produced by plants (Simpraga et al., 2016). On the other hand, we do not know enough how important the specific ratio of emitted compounds is for the volatile signatures that plants and animals use to detect the specific VOC composition emitted by plants (Bruce, Wadhams, & Woodcock, 2005). Recently, it has been suggested that the ratio of the original plant-emitted volatile compound and the semivolatile degradation products might inform, e.g., for approaching parasitic wasp, the distance to the plant where the host larvae are feeding (Simpraga et al., 2016).

When signalling compounds are destroyed in the polluted atmosphere, it has been found to reduce this signalling value of VOCs in plant–plant communication (Blande, Holopainen, et al., 2010), in floral pollinator cues (Farre-Armengol et al., 2016), and in attraction of natural enemies on herbivore-damaged plants (Himanen et al., 2009; Pinto, Nerg, & Holopainen, 2007). Blande, Holopainen, et al. (2010) observed that in ozone-rich air (80 ppb O_3) spider mite–induced VOCs from lima bean–induced defences (production of extra floral nectar) in intact plants only in 20 cm distance while in ambient air response was induced also in 70 cm distance. Homoterpenes *E*-DMNT, TMTT and monoterpene β-ocimene were the induced compound, which were significantly reduced at 80 ppb ozone. Elevated ozone concentrations also reduced several herbivore-induced VOCs in cabbage and lima bean plants damaged by *Plutella xylostella* (Linnaeus, 1758) (Lepidoptera: Plutellidae) and *Tetranychus urticae* Koch, 1836 (Trombidiformes: Tetranychidae), respectively (Pinto, Blande, et al., 2007). However, this did not reduce the efficiency of parasitoid wasps and predatory mites to orientate towards herbivore-damage plant in olfactometric tests. It was concluded that parasitoid wasp probably detected lower concentration of some key volatiles of damaged plant than the used CG—MS system (Pinto, Blande, et al., 2007). This result also suggests that some of the individual herbivore-induced volatile compounds even in trace concentration might be more important cues for natural enemies than the changes in ratios of emitted compounds.

Elevated CO_2 and other greenhouse gas emissions to the atmosphere are the main reason for global climate warming. As CO_2 is a major atmospheric gas taken up by plants and it has significant effects on plant growth and carbon—nitrogen balance. Elevated atmospheric CO_2 concentrations have also effects on constitutive and induced emissions of plant VOCs. In their literature analysis, Penuelas and Staudt (2010) found that the increases in atmospheric CO_2 led lower emission rates of all type of VOC, but warming had an opposite effect with increases of VOC emissions, most clearly in the emissions of different terpenes. They also found that the results for biogenic VOC emissions in response to increasing atmospheric CO_2 concentrations varied depending on the species, phenology and other environmental conditions. Explanation to reduced VOC emissions could be that increasing CO_2 concentration might uncouple isoprene emission from photosynthesis (the carbon source for VOC synthesis) and inhibit isoprenoid (terpenes) emissions at leaf level (Penuelas & Staudt, 2010). As elevating temperature activate biosynthesis of VOCs and warming increases diffusion of VOCs from plant tissues to the atmosphere (Widhalm et al., 2015) the numerous observations of elevated emission rates of isoprenoids under warming is a logical consequence.

2.3.5 Mechanism of Plant—Plant Communication

Plant volatiles in plant—plant communication are expected to evolve from the use of volatile compounds as molecular signals in unitary plant development (Holopainen & Blande, 2013). This view is also supported by results, which have shown plant—plant signalling to be more efficient between clonal cuttings of the same plant than between nonclonal conspecifics (Karban, Shiojiri, Ishizaki, Wetzel, & Evans, 2013). This indicates a degree of self or kin recognition to occur in receiver plants. Although currently we do not know any specific plant organs aimed to sense the volatile signals received by plants (Simpraga et al., 2016), plants have dense network of stomata in their leaves with a capacity of gas transfer and known of uptake of atmospheric plant volatiles (Niinemets, Fares, Harley, & Jardine, 2014).

This allows for volatile compound an access to cellular interspaces of leaf mesophyll cells. It is known that plants have several salicylic acid-binding proteins (Manohar et al., 2015) which could be related (e.g., to perception volatile methyl salicylate signals). On the other hand, glycosylation of key volatile signalling compounds, such as glycosylation of (Z)-3-hexenol in receiver plant, is proposed to be one of the mechanisms involved in the reception volatile signalling in plants (Sugimoto et al., 2014).

Evidence of communication by VOCs within plant is gathered especially from woody plant species (Baldwin, Halitschke, Paschold, von Dahl, & Preston, 2006) where volatile signalling between damaged plant parts and intact branches results in intact branches being better protected against subsequent attacks by herbivores (Frost et al., 2007; Heil & Karban, 2010; Shiojiri, Karban, & Ishizaki, 2009). Signalling between individual conspecific plants has shown that volatiles emitted by a herbivore-attacked plant will cause activation of various defence in the receiver plants. These include, for example, activation of defence genes (priming) resulting in more vigorous response in primed plants under herbivore-attacked when compared to nonprimed plants (Arimura et al., 2000; Frost et al., 2008). Also increased emission rates of typical herbivore-induced volatile compounds that are known to attract natural enemies of herbivores are found in the neighbouring intact lima plants when focal plant is attacked by spider mites (Blande, Holopainen, et al., 2010). Induced production of extra floral nectar, which is known to keep the predatory mites on plant foliage, was also found from these neighbouring plants (Blande, Holopainen, et al., 2010). Current evidence in literature suggests that plants have a capacity to activate efficient defences in neighbouring plants, but applications for pest control are still needed. An important result in most of the studies is that the plant–plant communication by volatile signals seems to be efficient usually at less than 1 m distance.

3. PLANT–INSECT CHEMICAL INTERACTION IN REPRODUCTION

The host plant location is more often the burden of the mated females excepted for the specialized species for which the host plant is the "rendezvous" place for mating. In polyphagous insects an extensive host range is associated with a less delicacy link between the possible host plants and the insect. The host plant chemical signals become less necessary for reproduction and mate finding relies exclusively on pheromone signal.

3.1 Host Plant Chemical Signal and Reproduction

For phytophagous insect, plant is principally the food resource for the larval stage and occasionally for adults in some species. Host plants are also a place where adults mate. For some species the larval host plant is required for mating whereas not for others. In this case, the adults move to other places and vegetation sometimes completely different from the native host plant.

Some literature highlighted the role of host plant chemical on reproduction. The host plants release compounds affecting several important steps of reproductive behaviour, such as pheromone production (Raina, Kingan, & Mattoo, 1992), and pheromone releasing behaviour (Landolt & Phillips, 1997; McNeil & Delisle, 1989a, 1989b; Tamhankar, 1994).

A research question arising from the past decades is: how the insects find a partner in a constant changing olfactory environment or in the middle of the host plant complex odours? Few studies have demonstrated the action of host plants on male behaviour attraction. In most cases, modifications in the male attraction are due to increase of female pheromone production (Reddy & Guerrero, 2004). Emelianov, Drès, Baltenweiler, and Mallet (2001) and Emelianov, Simpson, Narang, and Mallet (2003) described female calling behaviour improvement of *Zeiraphera diniana* Guénée, 1845 (Lepidoptera: Tortricidae) in relation with host plant. The two host races of *Z. diniana* develop on larch and pine respectively. The male attraction is improved by female calling on the host of conspecific. Males find more females on host plant or a synergy occur between calling behaviour, pheromone production and host plant volatiles. That being, reproductive isolation and host plant adaptation can be associated with mate finding. For a polyphagous noctuid feeding on maize no change in male and female reproductive behaviour was found in presence and absence of the host plant (Félix, Smail, & Frérot, 2013), whereas a synergistic effect on male attraction behaviour was shown in the tortricid moth *Eupoecilia ambiguella* (Hübner, 1796) (Lepidoptera: Tortricidae) (Schmidt-Büsser, von Arx, & Guerin, 2009) although the species is not specialized on grape.

Electrophysiology experiments on *Heliothis zea* (Boddie, 1850) (Lepidoptera: Noctuidae) males showed that olfactory receptor neuron response is synergized when plant compounds are presented in a mixture with major female pheromone component. The increase in male attraction to *H. zea* female placed on host plant is probably also due to synergy between pheromone components and plants volatiles (Ochieng et al., 2002). Namiki, Iwabuchi, and Kanzaki (2008) have shown the synergy at the level of antennal lobe neurons. The main information available about the processing of mixtures of plant odours and pheromone at the level of the antennal lobe is that the neurones located at that level of the brain are not so specific than postulated. It is now accepted that several peripheral neurones respond as well to a pheromone component and a plant odour compound (Anton & Hansson, 1995; Trone, Anfora, Bengtsson, Witzgall, & Ignell, 2010).

Recently on *H. virescens*, an extensive study including behaviour and neuroperception studies demonstrate that under natural conditions the olfactory system of the male moth appears to be well adapted to follow the female pheromone without interference from plant-emitted odours (Badeke, Haverkamp, Hansson, & Sachse, 2016).

What is less clear for most of the insect species is whereabout do they mate? It has been demonstrated that the *O. nubilalis* however specialized on maize or on other plants; whatever the pheromone strain it was, mate in grassy area out of the host plant spot (Showers, Reed, Robinson, & Derozari, 1976). Scouting in maize field during mating period reveals the absence of both males and females whereas pairs were observed in the "rendezvous" grassy area (Ponsard et al., 2004). *M. brassicae*, a noctuid moth specialized on cabbages for oviposition and larval feeding did not mate in the field planted with cabbages. Pheromone trapping of males is less efficient in the cabbage field than in wooden bush. Scouting after artificial release of males and females in the field showed that the adult insects did not stay in the cabbage field (Frérot, unpublished data), only the mated females come back for oviposition. Few examples on mating location are available whereas a foisonnante literature appeared during the last decades on the role of host plant VOCs on pheromone perception by male and on female pheromone production. Such basic knowledge is missing for most of the pest insects and will be a challenge for developing efficient treatment with sex pheromone. The basic knowledge of the mating procedure is indeed fundamental for using mating disruption or mass trapping of males. For instance there is no reports on mating place for *Cydia pomonella*, nor for *Lobesia botrana* (Denis & Schiffermüller, 1775) (Lepidoptera: Tortricidae) main pest in orchards and vineyard respectively. Could we imagine that they do not mate in their respective cultivated areas where mating disruption is applied?

3.2 Species for Which the Host Plant Is the "Rendezvous" Place

Phytophagous insects are generally adapted to their native host plant via behavioural and/or physiological adjustments. Some of them become through evolution highly specialized to a host plant family and by extension to a single host within a family.

To illustrate the case, the leek moth *Acrolepiopsis assectella* (Walker, 1864) (Lepidoptera: Yponomeutidae) is an example. Thibout (1974) and Auger and Thibout (1983) demonstrated the importance of the host plant in the mate recognition process. The host plant induced pheromone production and act on mating performances. The host plant is a "rendezvous" place. Such a thigh link between a plant and an insect is not rare and plenty of examples are available within all the insect orders. Another case where the plant attracts either one or both sexes and the pheromone become a coattractant of the plant signal as in *Rhynchophorus palmarum* (Linnaeus, 1758) (Coleoptera: Curculionidae) where the host plant volatiles synergize the sex pheromone attraction (Rochat, Gonzalez, Mariau, Villanueva, & Zagatti, 1991).

3.3 Species for Which Host Plants Are Sex Pheromone Precursor

In Lepidoptera, the female sex pheromone is biosynthetised de novo and bears no relation with the host plant. In some species males carry hair pencils associated with glands that produced a pheromone used in courtship behaviour. The male pheromone originated from sequestration of alkaloids or other compounds from larvae food as in some Arctiid moths (Conner, Eisner, Vander Meer, Guerrero, & Meinwald, 1981) or from male foraging intakes on specific plants like in nymphalids (Meinwald, 1986; Pliske, 1975). The reproduction is linked to the availability of host plants. In Coleoptera, the bark beetles need to feed on the host plant to produce the sex and aggregation pheromone (Blomquist et al., 2010). For such insects the reproduction relies on the availability of the host plants.

4. PLANT–INSECT CHEMICAL INTERACTION IN HOST FINDING FOR OVIPOSITION

In most of the species that did not rely on trial-error strategy, mated females have to locate the suitable host plant for their brood. Few examples reported active host seeking by larvae. The choosy behaviour of the ovipositing female is well described in several species and there is no doubt that the mated females in some species are attracted by the host plant. After landing, they exhibited sophisticated probing of the substratum leading to evaluation of the chemical signal and the physical properties (Calatayud et al., 2008; Frérot & Robert, 1998).

Locating a host plant is crucial for herbivore to find the suitable site for the brood to develop on. The sensory cues that elicit or inhibit host location have an important role in survival of the offspring (Renwick & Chew, 1994). When an insect is searching for host plant, it may use different senses, including olfaction, vision, tactile and gustation (Bernays & Chapman, 1994). At the first stages of selection, olfaction and vision are the most important senses because they operate at long distances whereas at short distance the gustatory and tactile cues become more important.

The oviposition choice is based on a complex set of external and internal stimuli and responses in which plant volatiles play an essential role (Miller & Strickler, 1984; Visser, 1986). They are emitted by plants and diffused in the air and mixed with volatiles of different sources to form an aerial soup (Cardé & Willis, 2008) in which the insects are capable to locate their potential host at distance even when the host plant is hidden among an array of other plants and buried in the background noise (Schröder & Hilker, 2008).

Plant cues that guide gravid females include long distance attractants that act during search of oviposition site and close range attraction. Compounds with host plant specific distribution as well as ubiquitous plant volatiles and primary metabolites have been found to act on the oviposition choice (Simmonds, 2001). By now, there are two major hypotheses for host plant recognition based on olfactory cues: (a) plant-specific odour recognition and (b) ratio-specific odour recognition.

In the plant specific odour recognition, the host plant recognition relies on highly specific volatiles cues that are not found in unrelated plant species. For numerous specialist insects the host plant recognition is guided by chemoreception of token stimuli (Dethier, 1982; Fraenkel, 1959). *Pieris* butterflies are specialist on cruciferous plants and use glucosinolates, secondary plant metabolites chemotaxonomically characteristic for this plant family, as token stimuli (Huang & Renwick, 1994). This nonvolatile glucosinolates were shown to stimulate the oviposition of other crucifer-adapted insect species (Giamoustaris & Mithen, 1995; Griffiths et al., 2001; Hopkins et al., 1997; Mewis, Ulrich, & Schnitzler, 2002). On the contrary, rejection of nonhost plant for oviposition is linked to contact chemoreception of secondary plant metabolites that have repulsive action on females. Compounds present in some plant family that are token stimuli to specialist insects can have inhibitory effect to nonassociated generalist (Huang & Renwick, 1993).

In the ratio-specific odour recognition the host plant recognition is provided by a specific ratio of mostly ubiquitous volatiles and not by a single compounds or a class of plant-specific volatiles. Wide range of herbivore insects are tuned to detect ubiquitous plant volatiles and the specificity of the signal relies on recognition of particular blends of compounds. The Colorado potato beetle, *Leptinotarsa decemlineata* (Say, 1824) (Coleoptera: Chrysomelidae), uses a ratio-specific blend of green leaf volatiles to locate its host plant and a subtle alternation of the blend ratios switches off attraction (Visser & Avé, 1978). Moreover the blend composition of host plant compounds is critical since ratio-specific blends are more attractive than individual compounds (Piñero, Galizia, & Dorn, 2008).

Herbivore insects are attracted or repelled by volatiles cues emitted from plants (Foster & Harris, 1997) but also by compounds released by others herbivores. In some case, the primary cue that leads to oviposition site is not the plant volatiles, but the chemical signal released by a conspecific insect. In many bark beetles species (i.e., "true weevil" family Curculionidae), males choose the host plant and attract females at distance with a pheromone, thus the females are lured by the interplay of pheromones and host plant cues (Wood, 1982). Some female flies, e.g., the cabbage root fly, *Delia radicum* (Linnaeus, 1758) (Diptera: Anthomyiidae), mark the host plant with an oviposition deterring substance to discourage other females from laying eggs on the same host (De Jong & Städler, 2001). As well, it was shown also that the gravid female in search of oviposition site, avoid actively host plants that release volatiles induced by conspecific larvae (De Moraes, Mescher, & Tumlinson, 2001). On the contrary, some species like the leaf miner species, *Liriomyza trifolii* (Burgess, 1880) (Diptera: Agromyzidae), and the beetle *Plagiometriona clavata* (Fabricius, 1798) (Coleoptera: Chrysomelidae) showed a strong oviposition preference for conspecific-infested plants (Facknath, 2012; Viswanathan, Narwani, & Thaler, 2005),

4.1 Case of Insects That Do Not Mate on the Host Plant

In many herbivorous lepidopteran species, the imago does not feed and uses the energy accumulated from the food eaten by the larval stage. Depending on the species, the adult may live from few days to few weeks.

The adult stage is very short compared to the overall lifecycle of the insect and the main concern of the insects at this stage is reproduction and oviposition. In some species, adults do not stay on the host plant after emerging from the pupae stage. They leave the larval host plant and mate in a different biotope. For example, the European corn borer (*O. nubilalis*) adults leave the maize field and mate in dense grassy areas where they rest and hide during the day (Showers et al., 1976). After mating, only the nocturnal-behaving gravid females fly back to maize field to lay eggs. Field observations clearly show that the gravid *O. nubilalis* females reach maize fields by oriented flights from the resting area, flying up the prevailing wind carrying maize field odours (Leppik & Frérot, 2014). As the females mate far from their host plant, they have a formidable task to find a suitable oviposition site in a complex and changing odorscape. The gravid females discriminate the right host plant at a precise time of the night just after the sunset.

The chemical signature of the maize field evolves through diel change linked with photosynthetic activity between the light and dark time. In the maize field odorscape, the ratios of monoterpenes, sesquiterpenes and some green leaf volatiles change between the day and night and consequently the chemical signal encounter by nocturnal flying insect is specific to their oviposition period (Leppik, Tammaru, & Frérot, 2014). Navigating in a complex and ever-changing odorscape, gravid females need to be constantly tuned into their host plant volatiles and avoid the background odours emitted by other plants. A quick view to what is released by a woody area compared with maize field evidences specific blend released by each biotope (Leppik & Frérot, 2014).

O. nubilalis is present in Europe with two pherotypes specialized on different host plants (Pelozuelo et al., 2004). Z-pherotype feed and lay eggs on maize (*Zea mays* L.), whereas the E-pherotype does on wild plants such as mugwort (*Artemisia vulgaris* L.) or hop (*Humulus lupulus* L.). These three plants have a distinctive volatile signature based on ubiquitous volatiles present in specific ratios (Leppik & Frérot, 2012). Gravid females, based on plant volatile cues specificity, make the host plant selection. The Z-pherotypes females are attracted to the maize volatile blend, whereas the E-pherotype is attracted to the mugwort and hop volatile blend (Bengtsson et al., 2006; Leppik & Frérot, 2012; Molnár, Tóth, Fejes-Tóth, Dekker, & Kárpáti, 2015). Wide range of herbivore insects are tuned into detecting ubiquitous plant volatiles and the specificity of the signal relies on recognition of particular blends of compounds (Bruce & Pickett, 2011). Moreover the blend composition of the host plant is critical because ratio-specific blends are more attractive than individual compounds (Piñero et al., 2008). The use of VOC in oviposition site recognition is the most prevalent mechanism in most herbivore insects that rely on gravid females to locate and choose the host plant.

4.2 Case of Specialized Insects that Mate and Oviposit on the Host Plant

The chemical signature of the plant changes with the development stages. Therefore, the airborne host plant signals encountered by the insects that colonize the crop at young leaf stage or at flowering stage are different. The different development stages of the plant may act on the different phases of specialized herbivore life such as host location, mating and oviposition. The broad bean weevil, *Bruchus rufimanus* (Boheman, 1833) (Coleoptera: Chrysomelidae), is a specialist pest of *Faba* bean (*Vicia faba* L.). The insects hibernate in woody areas away from Faba bean fields (Balachowsky, 1962). Early in the spring, the weevils come out of hibernation places and have to locate the host plant newly planted. The first weevils observed in the field coincide with the development of flower buds. Behavioural tests in the wind tunnel confirm that the weevils are only attracted to their host plant at the flower bud stage, not before at leaf stage (Leppik, Pinier, & Frérot, 2014). For oviposition, only the pod stage of the *Faba* bean is attractive to gravid females. Chemical analyses on the *Faba* bean VOCs show a clear separation of leaf, flower and pod stage. The chemical signature of all the three development stages of Faba bean is composed of ubiquitous plant volatiles; the specificity is ensured by change of ratios in the blend (Leppik, Pinier, et al., 2014). The nonattractive leaf stage is dominated mainly by monoterpenes. The flowering stage, when weevils arrive massively in the field, is characterized by a blend of monoterpenes and sesquiterpenes (Bruce, Martin, Smart, & Pickett, 2011). At the pod stage when the females search for oviposition sites, the chemical signature of *Faba* bean is an original blend of green leaf volatiles and monoterpenes (Leppik & Frérot, unpublished data).

5. CONCLUSION

All the advances in understanding the plasticity of the chemical signal released by insects and plants were made possible because of the technological contribution and the improvement of the analytical equipment.

For example, solid phase microextraction (SPME) and headspace gas chromatography mass spectrometry (HS-GC-MS) have allowed the detection of Lepidoptera pheromone blend from a single female (Frérot, Malosse, & Cain, 1997) leading to the vision of individuals belonging to defined populations. The computer and dedicated software allow a quick identification by comparison of compounds spectra with the assistance of improved database.

In conclusion, significant variation in moth sex pheromone blends can be found, not only geographically, but also after being in different experimental conditions, showing that the pheromone composition is likely to be more plastic than previously assumed (Butlin & Trickett, 1997; Groot, Classen, et al., 2010; Löfstedt, 1993). Such variations can directly affect reproduction (e.g., when attraction of conspecific males is reduced), or when other closely related and/or sympatrically occurring moth species with similar pheromone blends are attracted. Such mating reductions or interactions could result in either divergence of moth populations or convergence through hybrid speciation (Harrison, 2012). The plasticity of the pheromone signal makes the mate recognition system able to evolve and thus to preserve the assortative mating even in case of accidental introduction of alien insects in a specific habitat. Plasticity also contributes to adaptation to new host plant and to diversification of species via the specific mate recognition system.

The chemical signals produced by the plants and acting on the insect behaviour are based on the same principle as pheromones. The insects respond to a specific blend composed of a mixture of compounds being released in different quantities. The chemical profile is therefore made of major and minor components and changes in ratio of these compounds have significant information value. The concentration is also a component of the efficient chemical signal as at longer distances from the emission source concentration of many plant VOCs in the atmosphere decreases fast. The insects are tuned to these specific blends and as for pheromone they respond very quickly by a behaviour. The plant chemical signals evolve with the plant development, with the diel periodicity, with the environmental conditions and atmospheric pollution and are often species and variety specific.

The understanding of what are the key compounds for insects may be used by breeders to select new varieties missing the chemical information attractive for herbivorous species but also enhance information on what attracts natural enemies of pest insects on the crop plants.

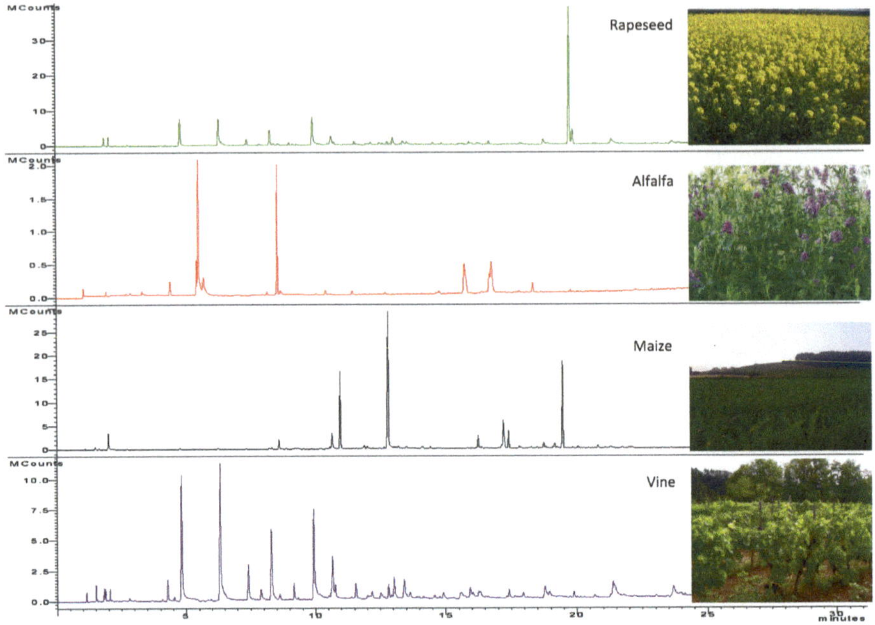

Figure 2 Odorscapes of different agrobiocoenosis. The volatile organic compounds were collected in different open fields for 4 h using solid phase microextraction (SPME) (Divinylbenzene/Carboxen/Polydimethylsiloxane, Supelco). SPME was analyzed by gas chromatography–mass spectrometry (Bruker Scion 436-GC linked to a Bruker Scion SQ detector equipped with an Rxi1-5SilMS, 30 m × 0.32 mm i.d.).

Odorscape Definition

Odorscape is a new concept that considers the general atmospheric bouquet of VOCs emitted from all the organisms: the flora, the fauna, the soil, bacteria and fungus, developing in a biotope.

Odorscape is the odorant description of a landscape. An agrobiocoenose in which biodiversity is reduced can be defined by an odorscape made of a specific mixture of VOCs (Fig. 2). Specificity of the odorscape relies on the chemical structures of the VOCs and on the ratio of the compounds forming the mixture. The dose is also specific, with landscape releasing different amount of VOCs.

Odorscape is characterised by a blend made of one or more main compounds associated with several minor compounds. The action on insects depends of all the components. The odorscape complexity is related with the plant species, the physiological stages of the plants, the biodiversity.

Technique of the Odorscape Collection

Odorscape can be collected by dynamic collection on adsorbent or by passive adsorption on SPME. SPME-HS/GC–MS is relatively simple nondestructive sampling method for collecting and characterizing the composition of the plant volatiles. Since the sampling is done on undamaged plants in situ, the SPME-HS analysis gives a realistic picture of plant volatile profile released by plants and allowed comparison of biocoenosis chemical signature. The volatile profile obtained by this method is pertinent for many ecological applications. For an absolute view of the odorscape or of the plant volatile profile, the fibres need to be calibrated with a series of chemicals belonging to different classes, i.e., alcohol, aldehyde, terpenes and sesquiterpenes. As well the mass spectrometry detection should be known for each relevant chemical.

Methodological Considerations of Plant VOC Sampling and Analysis

Sampling of VOCs from living plant material should be conducted so that the sampling method does not change the authentic emission profiles or concentration of emitted compounds significantly. This does not allow, for example, picking the flowers, leaves or twigs from plants for VOC sampling. The studied plant or part should be enclosed in a sampling cuvettes made of glass or inert plastic or even inside transparent plastic bag (Stewart-Jones & Poppy, 2006) so that the enclosed organ is keeping its normal physiological status during VOC sampling. The sampling environment can be the natural environment in the field or if potted, plants can be transferred into the laboratory where the environmental condition can be controlled better. Vuorinen et al. (2005) compared VOC emission of the same branches of potted silver birch seedlings before and after detachment from the seedling and found that detached branches (base in a water container after detachment) had significant increase in emission rates GLV compounds. For instance, emission of (*Z*)-3-hexenol was 13 and 23 fold higher and emission of (*Z*)-3-hexenyl acetate was 5 and 28 fold higher after detachment in two different birch genotypes, respectively. However, detachment did not have significant effect on mono-, sesqui- or homoterpene emission rates from the same plants (Vuorinen et al., 2005). If focussing just on analysis of special group or compounds, such as terpenes of essential oil, the separation of specific organs from the original plant does not necessarily change the scent composition significantly, but it does not represent the full volatile signature, which has ecological relevance.

Tholl et al. (2006) listed plant VOC sampling methods in three categories: (1) Static headspace sampling, (2) Dynamic headspace sampling, these both are for gas chromatography–mass spectrometry (GC–MS) sampling and (3) fast, GC-independent analysis [e.g., for proton transfer reaction –mass spectrometry (PTR-MS)]. Main difference between static and dynamic headspace sampling is

(Continued)

that dynamic sampling has controlled air flow in the sampling cuvette, which allows to quantify the emission rates, i.e., calculation of VOC emission per leaf area or dry weight. This allows, for example, calculation of proportion of VOC fluxes of total net carbon exchange, which can account for up to 5–10% particularly in stressed condition (Penuelas & Staudt, 2010) and comparison of the emission. Dynamic head space sampling could disadvantage the low molecular weight compounds such as isoprene, eluated by the air flow passing through the adsorbent trap. Also the replacement air in dynamic sampling needs additional filtering to exclude volatile contaminants and an ozone scrubber to remove atmospheric ozone, which can otherwise degrade most reactive plant VOCs in the sample tube. Static headspace sampling is suitable for qualitative analyses of, for example, flower emissions. Static, noncirculated air allows enrichment of many compounds, which might stay below detection level in dynamic sampling. In the third category, PTR-MS sampling, the cuvettes have often higher air flow rate than dynamic sampling for GC–MS. PTR-MS instrument is connected directly to the sampling cuvette or bag for online sampling with real time analysis having few seconds time resolution (Brilli et al., 2011; Maja et al., 2014). This is important for proper detection of some volatile compounds that are highly reactive in the atmosphere after release from the plant.

Also GC–MS sampling can be done with near real time in fast GC 1–10 min time resolution (Materic et al., 2015). More common in static and dynamic sampling is off-line/storing sampling, which allows storing of VOC samples adsorbent for several weeks and transportation of samples for long distances to the GC–MS analysis laboratory. In static headspace sampling with the SPME method is used. It is a fast and simple method allowing collection of volatiles at detection limits in the ppbv (parts per billion by volume) range (Tholl et al., 2006). SPME is based on ad/absorption and desorption of volatiles inside the static on an inert fibre coated with different types of ad/absorbents.

VOCs are most conveniently collected in situ from undamaged plants. Nonetheless, sometimes it is necessary to collect volatiles from plant parts or organs, for example, to distinguish the VOCs profiles of reproductive organs and vegetative parts. In that case, volatiles are collected either from cut plant parts or preferably in situ from enclosed plant organs to avoid supplementary emission of volatiles due to injuring effects.

The plant is tightly enclosed into a *Teflon* bag or a vessel to form a headspace. The air surrounding the plant is static, i.e., with no airflow. As the plant continues to release volatile compounds, the static air around the plant becomes enriched with volatile compounds. A small hole is pierced to the *Teflon* bag and an SPME fibre is inserted into the headspace. The volatile compounds emitted by plant are passively adsorbed on the fibre coating according with the affinity of the fibre for the compounds. Following equilibration between the SPME fibre and the enriched headspace, the SPME fibre is retracted and packed in aluminium foil or analyzed directly in GC–MS. When properly stored, SPME samples can be analyzed days later without significant loss of volatile compounds. The adsorbed volatiles are thermally desorbed from the SPME fibre without further preparation in a common GC injector. Thermal desorption of VOCs from the SPME fibre eliminates the need for solvents and allowed detection of very volatile compounds, usually undetected due to the solvent peak. Furthermore, the technique is extremely sensitive and enables to detect trace compounds with detection limits down to parts per trillion (ppt) (pg/mL) levels for certain compounds. The same SPME fibres can be reused nearly 100 times.

SPME has some inconveniences: the fibres provide semiquantitative information for a fraction of the plant volatiles, since the adsorbed amount depends on the fibre-coating affinity for the compound and the coating-free sites where compounds are adsorbed (Pawliszyn, 1997), in addition to their concentration in the plant headspace. As a result, the plant volatile composition may misrepresent some volatiles and over represent others. However, by desorbing the entire volatile sample into the injector, no repeated injections of the sample are possible.

In dynamic sampling for GC–MS sample, air is sucked through adsorbent powder inside a glass or steel tube. The sample air-flow through sample tube allows calculation of air volume and concentration of compound in that air volume. Both SPME and adsorbent tube sampling methods use thermodesorption systems to transfer VOC sample for final GC–MS analysis (Tholl et al., 2006).

REFERENCES

Albre, J., Steinwender, B., & Newcomb, R. D. (2013). The evolution of desaturase gene regulation involved in sex pheromone production in leafroller moths of the genus Planotortrix. *Journal of Heredity, 104*, 627–638.

Anderson, P., Hansson, B. S., Nilsson, U., Han, Q., Sjoholm, M., Skals, N., & Anton, S. (2007). Increased behavioral and neuronal sensitivity to sex pheromone after brief odor experience in a moth. *Chemical Senses, 32*, 483–491.

Anderson, P., Sadek, M. M., & Hansson, B. S. (2003). Pre-exposure modulates attraction to sex pheromone in a moth. *Chemical Senses, 28*, 285–291.

Anderson, P., Sadek, M. M., Larsson, M., Hansson, B. S., & Thoming, G. (2013). Larval host plant experience modulates both mate finding and oviposition choice in a moth. *Animal Behaviour, 85*, 1169–1175.

Anton, S., & Hansson, B. S. (1995). Sex pheromone and plant associated processing in antennal lobe interneurons of male *Spodoptera littoralis* (Lepidoptera, Noctuidae). *Journal of Comparative Physiology, 176*, 773–789.

Arimura, G., Ozawa, R., Shimoda, T., Nishioka, T., Boland, W., & Takabyashi, J. (2000). Herbivory-induced volatiles elicit defence genes in lima bean leaves. *Nature, 406*, 512–515.

Auger, J., & Thibout, E. (1983). Spécificité des substances non volatiles des Allium responsables de la ponte de la teigne du poireau, Acrolepiopsis assectella (Lepidoptera). *Entomologia Experimentalis et Applicata, 34*, 71–77.

Badeke, E., Haverkamp, A., Hansson, B. S., & Sachse, S. (2016). A challenge for a male noctuid moth? Discerning the female sex pheromone against the background of plant volatiles. *Frontiers in Physiology, 7*, 143.

Balachowsky, A. S. (1962). *Entomologie appliquée à l'agriculture. Tome I. Coléoptères*. Paris: Masson et Cie Editeurs.

Baldwin, I. T., Halitschke, R., Paschold, A., von Dahl, C. C., & Preston, C. A. (2006). Volatile signaling in plant-plant interactions: "talking trees" in the genomics era. *Science, 311*, 812–815.

Barron, A. B. (2001). The life and death of Hopkins' host-selection principle. *Journal of Insect Behavior, 14*, 725–737.

Barthel, A., Staudacher, H., Schmaltz, A., Heckel, D. G., & Groot, A. T. (2015). Sex-specific consequences of an induced immune response on reproduction in a moth. *BMC Evolutionary Biology, 15*, 282. http://dx.doi.org/10.1186/s12862-015-0562-3.

Batista-Pereira, L. G., Stein, K., de Paula, A. F., Moreira, J. A., Cruz, I., Figueiredo Mde, L., ... Corrêa, A. G. (2006). Isolation, identification, synthesis, and field evaluation of the sex pheromone of the brazilian population of *Spodoptera frugiperda*. *Journal of Chemical Ecology, 32*, 1085–1099.

Bazinet, A. L., Cummings, M. P., Mitter, K. T., & Mitter, C. W. (2013). Can RNA-Seq resolve the rapid radiation of advanced moths and butterflies (Hexapoda: Lepidoptera: Apoditrysia)? An exploratory study. *PLoS One, 8*, e82615.

Bengtsson, M., Karpati, Z., Reuveny, H., Yang, Z., Witzgall, P., & Szöcs, G. (2006). Flight tunnel responses of Z strain European Corn Borer females to corn and hemp plants. *Environmental Entomology, 35*, 1238–1243.

Bergström, G., Appelgren, M., Svensson, B. G., Ågren, L., Descoins, C., Frerot, B., ... Lettere, M. (1985). Marking pheromones of *Megabombus sylvarum* (L.) and *M. ruderarius* (Müller) males (Hymenoptera: Apidae). *Apidologie, 16*, 57–58.

Bernays, E. A., & Chapman, R. F. (1994). *Host-plant selection by phytophagous insects*. US: Springer.

Bethenod, M. T., Thomas, Y., Rousset, F., Frerot, B., Pelozuelo, L., Genestier, G., & Bourguet, D. (2005). Genetic isolation between two sympatric host plant races of the European corn borer, *Ostrinia nubilalis* Hubner. II: assortative mating and host-plant preferences for oviposition. *Heredity, 94*, 264–270.

Bjostad, L., Wolf, W., & Roelofs, W. L. (1987). Pheromone biosynthesis in lepidopterans: desaturation and chain shortening. In G. D. Prestwich, & G. L. Blomquist (Eds.), *Pheromone biochemistry* (pp. 77–120). New-York: Acad. Press.

Blande, J. D., Holopainen, J. K., & Li, T. (2010). Air pollution impedes plant-to-plant communication by volatiles. *Ecology Letters, 13*, 1172–1181.

Blande, J. D., Holopainen, J. K., & Niinemets, Ü. (2014). Plant volatiles in a polluted atmosphere: stress response and signal degradation. *Plant, Cell & Environment, 37*, 1892–1904.

Blande, J. D., Korjus, M., & Holopainen, J. K. (2010). Foliar methyl salicylate emissions indicate prolonged aphid infestation on silver birch and black alder. *Tree Physiology, 30*, 404–416.

Blomquist, G. J., Figueroa-Teran, R., Aw, M., Song, M. M., Gorzalski, A., Abbott, N. L., ... Tittiger, C. (2010). Pheromone production in bark beetles. *Insect Biochemistry and Molecular Biology, 40*, 699–712.

Blum, M. S. (1987). Biosynthesis of arthropod exocrine compounds. *Annual Review of Entomology, 32*, 381–413.

Brilli, F., Ruuskanen, T. M., Schnitzhofer, R., Mueller, M., Breitenlechner, M., Bittner, V., ... Hansel, A. (2011). Detection of plant volatiles after leaf wounding and darkening by proton transfer reaction "time-of-flight" mass spectrometry (PTR-TOF). *PLoS One, 6*, e20419.

Bruce, T. J., Martin, J. L., Smart, L. E., & Pickett, J. A. (2011). Development of semiochemical attractants for monitoring bean seed beetle, *Bruchus rufimanus*. *Pest Management Science, 67*, 1303–1308.
Bruce, T. J., & Pickett, J. A. (2011). Perception of plant volatile blends by herbivorous insects-finding the right mix. *Phytochemistry, 72*, 1605–1611.
Bruce, T. J., Wadhams, L. J., & Woodcock, C. M. (2005). Insect host location: a volatile situation. *Trends in Plant Science, 10*, 269–274.
Burdon, R. C. F., Raguso, R. A., Kessler, A., & Parachnowitsch, A. L. (2015). Spatiotemporal floral scent variation of Penstemon digitalis. *Journal of Chemical Ecology, 41*, 641–650.
Butlin, R. K. (1995). Reinforcement – an idea evolving. *Trends in Ecology & Evolution, 10*, 432–434.
Butlin, R. K., Hewitt, G. M., & Webb, S. F. (1985). Sexual selection for intermediate optimum in *Chorthippus brunneus* (Orthoptera, Acrididae). *Animal Behaviour, 33*, 1281–1292.
Butlin, R., & Trickett, A. J. (1997). Can population genetic simulations help to interpret pheromone evolution? In R. T. Cardé, & A. K. Minks (Eds.), *Insect pheromone research: New directions* (pp. 548–562). New York: Chapman and Hall.
Calatayud, P. A., Ahuya, P. O., Wanjoya, A., Le Rüe, B., Silvain, J. F., & Frérot, B. (2008). Importance of plant physical cues in host acceptance for oviposition by *Busseola fusca*. *Entomologia Experimentalis et Applicata, 126*, 233–243.
Cardé, R. T., Cardé, A. M., Hill, A. S., & Roelofs, W. L. (1977). Sex pheromone specificity as a reproductive isolating mechanism among sibling species *Archips argyrospilus* and *A. mortuanus* and other sympatric tortricine moths (Lepidoptera, Tortricidae). *Journal of Chemical Ecology, 3*, 71–84.
Cardé, R. T., Roelofs, W. L., Harrison, R. G., Vawter, A. T., Brussard, P. F., Mutuura, A., & Munroe, E. (1978). European corn borer-pheromone polymorphism or sibling species. *Science, 199*, 555–556.
Cardé, R. T., & Willis, M. A. (2008). Navigational strategies used by insects to find distant, wind-borne sources of odor. *Journal of Chemical Ecology, 34*, 854–866.
Conner, W. E., Eisner, E., Vander Meer, R. K., Guerrero, A., & Meinwald, J. (1981). Precopulatory sexual interaction in an Arctiid moth (*Utetheisa ornatrix*): role of a pheromone derived from dietary alkaloids. *Behavioral Ecology and Sociobiology, 9*, 227–235.
Cossé, A. A., Campbell, M. G., Glover, T. J., Linn, C. E., Todd, J. L., Baker, T. C., & Roelofs, W. L. (1995). Pheromone behavioral responses in unusual male European corn borer hybrid progeny not correlated to electrophysiological phenotypes of their pheromone-specific antennal neurons. *Experientia, 51*, 809–816.
Coyne, J. A., & Orr, H. A. (2004). *Speciation*. MA, Sunderland: Sinauer Associates, Inc.
De Jong, R., & Städler, E. (2001). Sensilla on cabbage root fly tarsae sensitive to egg-associated compounds. *Chemoecology, 11*, 145–147.
De Moraes, C. M., Mescher, M. C., & Tumlinson, J. H. (2001). Caterpillar-induced nocturnal plant volatiles repel conspecific females. *Nature, 410*, 577–580.
Delwiche, C. F., & Sharkey, T. D. (1993). Rapid appearance of ^{13}C in biogenic isoprene when $^{13}CO_2$-C is fed to intact leaves. *Plant, Cell & Environment, 16*, 587–591.
Den Otter, C. J., Schuil, H. A., & Sandervanoosten, A. (1978). Reception of host-plant odors and female sex-pheromone in *Adoxophyes orana* (Lepidoptera, Tortricidae) – electrophysiology and morphology. *Entomologia Experimentalis et Applicata, 24*, 570–578.
Descoins, C., & Frérot, B. (1979). Sex pheromone specificity in the tortricid fauna of apple-orchards. In F. J. Ritter (Ed.), *Chemical ecology: Odour communication in animals* (pp. 181–185). Amsterdam: Elsevier North-Holland Publish. Co.
Descoins, C., Frérot, B., Gallois, M., Lettere, M., Bergström, G., Appelgren, M., … Ågren, L. (1984). Identification des composés de la phéromone de marquage produite par les glandes labiales des mâles de *Megabombus pascuorum* (Scopoli) (Hymenoptera, Apidae). *Nova Acta Reg. Soc. Sci. Upsal., 3*, 149–152. Serie VC.

Dethier, V. G. (1982). Mechanism of host-plant recognition. *Entomologia Experimentalis et Applicata, 31*, 49–56.

Dicke, M. (2009). Behavioural and community ecology of plants that cry for help. *Plant Cell and Environment, 32*, 654–665.

Dicke, M., & Sabelis, M. W. (1988). Infochemical terminology: Based on cost-benefit analysis rather than origin of compounds? *Functional Ecology, 22*, 131–139.

Domb, L. G., & Pagel, M. (2001). Sexual swellings advertise female quality in wild baboons. *Nature, 410*, 204–206.

Dopman, E. B., Robbins, P. S., & Seaman, A. (2010). Components of reproductive isolation between North American pheromone strains of the European corn borer. *Evolution, 64*, 881–902.

Droney, D. C., Musto, C. J., Mancuso, K., Roelofs, W. L., & Linn, C. E. (2012). The response to selection for broad male response to female sex pheromone and its implications for divergence in close-range mating behavior in the European corn borer moth, *Ostrinia nubilalis. Journal of Chemical Ecology, 38*, 1504–1512.

Dumas, P., Legeai, F., Lemaitre, C., Scaon, E., Orsucci, M., Labadie, K., … d'Alencon, E. (2015). *Spodoptera frugiperda* (Lepidoptera: Noctuidae) host-plant variants: two host strains or two distinct species? *Genetica, 143*, 305–316.

Dumenil, C., Judd, G. J. R., Bosch, D., Baldessari, M., Gemeno, C., & Groot, A. T. (2014). Intraspecific variation in female sex pheromone of the codling moth *Cydia pomonella. Insects, 5*, 705–721.

Eizaguirre, M., Albajes, R., Lopez, C., Sans, A., & Gemeno, C. (2007). Inhibition of pheromone response in *Sesamia nonagrioides* by the pheromone of the sympatric corn borer, *Ostrinia nubilalis. Pest Management Science, 63*, 608–614.

Emelianov, I., Drès, M., Baltenweiler, W., & Mallet, J. (2001). Host-induced assortative mating in host races of larch budmoth. *Evolution, 55*, 2002–2010.

Emelianov, I., Simpson, F., Narang, P., & Mallet, J. (2003). Host choice promotes reproductive isolation between host races of the larch budworm *Zeiraphera diniana. Journal of Evolutionary Biology, 16*, 208–218.

Facknath, S. (2012). Study of herbivore response to the presence of conspecifics and heterospecifics in an insect-host plant relationship. *University of Mauritius Research Journal, 18*, 250–274.

Farre-Armengol, G., Penuelas, J., Li, T., Yli-Pirila, P., Filella, I., Llusia, J., & Blande, J. D. (2016). Ozone degrades floral scent and reduces pollinator attraction to flowers. *New Phytologist, 209*, 152–160.

Félix, A. E., Smail, T., & Frérot, B. (2013). Host plants and reproductive behaviour in the African maize borer *Busseola fusca* (Fuller 1901). *Annales de la Société Entomologique de France, 49*, 68–72.

Forkel, M., Carvalhais, N., Roedenbeck, C., Keeling, R., Heimann, M., Thonicke, K., … Reichstein, M. (2016). Enhanced seasonal CO_2 exchange caused by amplified plant productivity in northern ecosystems. *Science, 351*, 696–699.

Foster, S. (2009). Sugar feeding via trehalose haemolymph concentration affects sex pheromone production in mated *Heliothis virescens* moths. *Journal of Experimental Biology, 212*, 2789–2794.

Foster, S. P., & Harris, M. O. (1997). Behavioral manipulation methods for insect pest-management. *Annual Review of Entomology, 42*, 123–146.

Fraenkel, G. (1959). The raison d'être of secondary plant substances. *Science, 129*, 1466–1470.

Frérot, B., Malosse, C., & Cain, A. H. (1997). Solid phase microextraction (SPME) a new tool in pheromone identification in Lepidoptera. *Journal of High Resolution Chromatography, 20*, 340–342.

Frérot, B., & Robert, P. (1998). Oviposition behaviour of *Sesamia nonagrioides* Lefebvre (Lepidoptera: Noctuidae). Effect of conspecific eggs. *Annales de la Socitété Entomologique de France, 34*, 189–194.

Frost, C. J., Appel, M., Carlson, J. E., De Moraes, C. M., Mescher, M. C., & Schultz, J. C. (2007). Within-plant signalling via volatiles overcomes vascular constraints on systemic signalling and primes responses against herbivores. *Ecology Letters, 10*, 490–498.

Frost, C. J., Mescher, M. C., Dervinis, C., Davis, J. M., Carlson, J. E., & De Moraes, C. M. (2008). Priming defense genes and metabolites in hybrid poplar by the green leaf volatile cis-3-hexenyl acetate. *New Phytologist, 180*, 722–733.

Fujii, T., Namiki, S., Abe, H., Sakurai, T., Ohnuma, A., Kanzaki, R., Katsuma, S., Ishikawa, Y., & Shimada, T. (2011). Sex-linked transcription factor involved in a shift of sex-pheromone preference in the silkmoth *Bombyx mori*. *Proceedings of the National Academy of Sciences of the United States of America, 108*, 18038–18043.

Gemeno, C., Lutfallah, A. F., & Haynes, K. F. (2000). Pheromone blend variation and cross-attraction among populations of the black cutworm moth (Lepidoptera: Noctuidae). *Annals of the Entomological Society of America, 93*, 1322–1328.

Gemeno, C., Sans, A., Lopez, C., Albajes, R., & Eizaguirre, M. (2006). Pheromone antagonism in the European corn borer moth *Ostrinia nubilalis*. *Journal of Chemical Ecology, 32*, 1071–1084.

Giamoustaris, A., & Mithen, R. (1995). The effect of modifying the glucosinolate content of leaves of oilseed rape (*Brassica napus* ssp. *oleifera*) on its interaction with specialist and generalist pests. *Annals of Applied Biology, 126*, 347–363.

Gould, F., Estock, M., Hillier, N. K., Powell, B., Groot, A. T., Ward, C. M., … Vickers, N. J. (2010). Sexual isolation of male moths explained by a single pheromone response QTL containing four receptor genes. *Proceedings of the National Academy of Sciences of the United States of America, 107*, 8660–8665.

Grant, P. R., & Grant, B. R. (2002). Unpredictable evolution in a 30-year study of Darwin's finches. *Science, 296*, 707–711.

Gries, G., Schaefer, P. W., Gries, R., Liska, J., & Gotoh, T. (2001). Reproductive character displacement in *Lymantria monacha* from northern Japan? *Journal of Chemical Ecology, 27*, 1163–1176.

Griffiths, D. W., Deighton, N., Birch, A. N. E., Patrian, B., Baur, R., & Städler, E. (2001). Identification of glucosinolates on the leaf surface of plants from the Cruciferae and other closely related species. *Phytochemistry, 57*, 693–700.

Groot, A. T., Classen, A., Staudacher, H., Schal, C., & Heckel, D. G. (2010). Phenotypic plasticity in sexual communication signal of a noctuid moth. *Journal of Evolutionary Biology, 23*, 2731–2738.

Groot, A. T., Dekker, T., & Heckel, D. G. (2016). The genetic basis of pheromone evolution in moths. *Annual Review of Entomology, 61*, 99–117.

Groot, A. T., Gemeno, C., Brownie, C., Gould, F., & Schal, C. (2005). Male and female antennal responses in *Heliothis virescens* and *H. subflexa* to conspecific and heterospecific sex pheromone compounds. *Environmental Entomology, 34*, 256–263.

Groot, A. T., Horovitz, J. L., Hamilton, J., Santangelo, R. G., Schal, C., & Gould, F. (2006). Experimental evidence for interspecific directional selection on moth pheromone communication. *Proceedings of the National Academy of Sciences of the United States of America, 103*, 5858–5863.

Groot, A. T., Inglis, O., Bowdrigde, S., Santangelo, R. G., Blanco, C., Lopez, J. D., … Schal, C. (2009). Geographic and temporal variation in moth chemical communication. *Evolution, 63*, 1987–2003.

Groot, A. T., Marr, M., Heckel, D. G., & Schoefl, G. (2010). The roles and interactions of reproductive isolation mechanisms in fall armyworm (Lepidoptera: Noctuidae) host strains. *Ecological Entomology, 35*, 105–118.

Groot, A. T., Santangelo, R. G., Ricci, E., Brownie, C., Gould, F., & Schal, C. (2007). Differential attraction of *Heliothis subflexa* males to synthetic pheromone lures in eastern US and western Mexico. *Journal of Chemical Ecology, 33*, 353–368.

Groot, A. T., Schoefl, G., Inglis, O., Donnerhacke, S., Classen, A., Schmalz, A., … Heckel, D. G. (2014). Within-population variability in a moth sex pheromone blend: genetic basis and behavioral consequences. *Proceedings of the Royal Society B-Biological Sciences, 281*, 20133054.

Harrison, R. G. (2012). The language of speciation. *Evolution, 66*, 3643–3657.

Haynes, K. F., Gemeno, C., Yeargan, K. V., Millar, J. G., & Johnson, K. M. (2002). Aggressive chemical mimicry of moth pheromones by a bolas spider: how does this specialist predator attract more than one species of prey? *Chemoecology, 12*, 99–105.

Heil, M., & Karban, R. (2010). Explaining evolution of plant communication by airborne signals. *Trends in Ecology & Evolution, 25*, 137–144.

Hillier, N. K., Kleineidam, C., & Vickers, N. J. (2006). Physiology and glomerular projections of olfactory receptor neurons on the antenna of female *Heliothis virescens* (Lepidoptera : Noctuidae) responsive to behaviorally relevant odors. *Journal of Comparative Physiology A-Neuroethology Sensory Neural and Behavioral Physiology, 192*, 199–219.

Himanen, S. J., Nerg, A., Nissinen, A., Pinto, D. M., Stewart, C. N., Jr., Poppy, G. M., & Holopainen, J. K. (2009). Effects of elevated carbon dioxide and ozone on volatile terpenoid emissions and multitrophic communication of transgenic insecticidal oilseed rape (*Brassica napus*). *New Phytologist, 181*, 174–186.

Holopainen, J. K., & Blande, J. D. (2013). Where do herbivore-induced plant volatiles go? *Frontiers in Plant Science, 4*, 185.

Holopainen, J. K., & Gershenzon, J. (2010). Multiple stress factors and the emission of plant VOCs. *Trends in Plant Science, 15*, 176–184.

Hopkins, R. J., Birch, A. N. E., Griffiths, D. W., Baur, R., Städler, E., & McKinlay, R. G. (1997). Leaf surface compounds and oviposition preference of Turnip root fly *Delia floralis*: the role of glucosinolate and nonglucosinolate compounds. *Journal of Chemical Ecology, 23*, 629–643.

Huang, X., & Renwick, J. A. A. (1993). Differential selection of host plants by two Pieris species: the role of oviposition stimulants and deterrents. *Entomologia Experimentalis et Applicata, 68*, 59–69.

Huang, X., & Renwick, J. A. A. (1994). Relative activities of glucosinolates as oviposition stimulants for *Pieris rapae* and *P. napi oleracea*. *Journal of Chemical Ecology, 20*, 1025–1037.

Jaenike, J. (1983). Induction of host preference in Drosophila melanogaster. *Oecologia, 58*, 320–325.

Jaenike, J. (1990). Host specialization in phytophagous insects. *Annual Review of Ecology and Systematics, 21*, 243–274.

Jansen, R. M. C., Wildt, J., Kappers, I. F., Bouwmeester, H. J., Hofstee, J. W., & van Henten, E. J. (2011). Detection of diseased plants by analysis of volatile organic compound emission. *Annual Review of Phytopathology, 49*, 157–174.

Jud, W., Fischer, L., Canaval, E., Wohlfahrt, G., Tissier, A., & Hansel, A. (2016). Plant surface reactions: an opportunistic ozone defence mechanism impacting atmospheric chemistry. *Atmospheric Chemistry and Physics, 16*, 277–292.

Jurenka, R. (2004). Insect pheromone biosynthesis. In S. Schulz (Ed.), *Chemistry of pheromones and other semiochemicals I* (pp. 97–131). Berlin Heidelberg: Springer-Verlag.

Kainulainen, P., Nissinen, A., Piirainen, A., Tiilikkala, K., & Holopainen, J. K. (2002). Essential oil composition in leaves of carrot varieties and preference of specialist and generalist sucking insect herbivores. *Agricultural and Forest Entomology, 4*, 211–216.

Karban, R., Shiojiri, K., Ishizaki, S., Wetzel, W. C., & Evans, R. Y. (2013). Kin recognition affects plant communication and defence. *Proceedings of the Royal Society B-Biological Sciences, 280*, 20123062.

Kergoat, G. J., Prowell, D. P., Le Rü, B. P., Mitchell, A., Dumas, P., Clamens, A. L., … Silvain, J.-F. (2012). Disentangling dispersal, vicariance and adaptive radiation patterns: a case study using armyworms in the pest genus *Spodoptera* (Lepidoptera: Noctuidae). *Molecular Phylogenetics and Evolution, 65*, 855–870.

Klun, J. A. (1975). Insect sex pheromones – intraspecific pheromonal variability of *Ostrinia nubilalis* (Lepidoptera, Pyralidae) in North America and Europe. *Environmental Entomology, 4*, 891–894.

Klun, J. A., & Huettel, M. D. (1988). Genetic regulation of sex pheromone production and response – interaction of sympatric pheromonal types of European corn borer, *Ostrinia nubilalis* (Lepidoptera, Pyralidae). *Journal of Chemical Ecology, 14*, 2047–2061.

Kochansky, J., Cardé, R. T., Liebherr, J., & Roelofs, W. L. (1975). Sex pheromone of the European corn borer *Ostrinia nubilalis* (Lepidoptera Pyralidae) in New-York USA. *Journal of Chemical Ecology, 1*, 225–231.

Landolt, P. J., & Phillips, T. W. (1997). Host plant influences on sex pheromone behavior of phytophagous insects. *Annual review of entomology, 42*, 371–391.

Laothawornkitkul, J., Taylor, J. E., Paul, N. D., & Hewitt, C. N. (2009). Biogenic volatile organic compounds in the earth system. *New Phytologist, 183*, 27–51.

Lassance, J. M., Groot, A. T., Liénard, M. A., Binu, A., Borgwardt, C., Andersson, F., … Löfstedt, C. (2010). Allelic variation in a fatty-acyl reductase gene causes pheromone divergence in European corn borer races. *Nature, 466*, 486–489.

Leary, G. P., Allen, J. E., Bunger, P. L., Luginbill, J. B., Linn, C. E., Macallister, I. E., … Wanner, K. W. (2012). Single mutation to a sex pheromone receptor provides adaptive specificity between closely related moth species. *Proceedings of the National Academy of Sciences of the United States of America, 109*, 14081–14086.

Leppik, E., & Frérot, B. (2012). Volatile organic compounds and host-plant specialization in European corn borer E and Z pheromone races. *Chemoecology, 22*, 119–129.

Leppik, E., & Frérot, B. (2014). Maize field odorscape during the oviposition flight of the European corn borer. *Chemoecology, 24*, 221–228.

Leppik, E., Pinier, C., & Frérot, B. (2014). Paysage chimique d'une agrobiocenose: un exemple de la féverole et son ravageur spécialist *Bruchus rufimanus*. *AFPP – Proceeding de la Dixième conférence internationale sur les ravageurs en agriculture, 10*, 585–595. http://www.fredon-corse.com/standalone/5/004F60b5P6PTvHsJ7g5P3I67.pdf.

Leppik, E., Tammaru, T., & Frérot, B. (2014). A view of diel variation of maize odorscape. *American Journal of Plant Sciences, 05*, 811–820.

Liebherr, J., & Roelofs, W. (1975). Laboratory hybridization and mating period studies using 2 pheromone strains of *Ostrinia nubilalis*. *Annals of the Entomological Society of America, 68*, 305–309.

Lim, H., & Greenfield, M. D. (2007). Female pheromonal chorusing in an arctiid moth. *Utetheisa ornatrix*. *Behavioral Ecology, 18*, 165–173.

Lima, E. R., & McNeil, J. N. (2009). Female sex pheromones in the host races and hybrids of the fall armyworm, *Spodoptera frugiperda* (Lepidoptera: Noctuidae). *Chemoecology, 19*, 29–36.

Lindwall, F., Faubert, P., & Rinnan, R. (2015). Diel variation of biogenic volatile organic compound emissions- A field study in the sub, low and high arctic on the effect of temperature and light. *PLoS One, 10*, e0123610.

Linn, C. E. (1997). Neuroendocrine factors in the photoperiodic control of male moth responsiveness to sex pheromone. In R. T. Cardé, & A. K. Minks (Eds.), *Insect pheromone research: New directions* (pp. 194–209). US: Springer.

Linn, C. E., Young, M. S., Gendle, M., Glover, T. J., & Roelofs, W. L. (1997). Sex pheromone blend discrimination in two races and hybrids of the European corn borer moth, *Ostrinia nubilalis*. *Physiological Entomology, 22*, 212–223.

Löfstedt, C. (1993). Moth pheromone genetics and evolution. *Philosophical Transactions of the Royal Society B: Biological Sciences, 340*, 167–177.

Loreto, F., & Schnitzler, J. P. (2010). Abiotic stresses and induced BVOCs. *Trends in Plant Science, 15*, 154–166.
Maffei, M. E. (2010). Sites of synthesis, biochemistry and functional role of plant volatiles. *South African Journal of Botany, 76*, 612–631.
Maja, M. M., Kasurinen, A., Yli-Pirilä, P., Joutsensaari, J., Klemola, T., Holopainen, T., & Holopainen, J. K. (2014). Contrasting responses of silver birch VOC emissions to short- and long-term herbivory. *Tree Physiology, 34*, 241–252.
Malausa, T., Bethenod, M. T., Bontemps, A., Bourguet, D., Cornuet, J.-M., & Ponsard, S. (2005). Assortative mating in sympatric host races of the European corn borer. *Science, 308*, 258–260.
Malausa, T., Dalecky, A., Ponsard, S., Audiot, P., Streiff, R., Chaval, Y., & Bourguet, D. (2007). Genetic structure and gene flow in French populations of two *Ostrinia* taxa: host races or sibling species? *Molecular Ecology, 16*, 4210–4222.
Manohar, M., Tian, M., Moreau, M., Park, S. W., Choi, H. W., Fei, Z., … Shi, K. (2015). Identification of multiple salicylic acid-binding proteins using two high throughput screens. *Frontiers in Plant Science, 5*, 777.
Materic, D., Bruhn, D., Turner, C., Morgan, G., Mason, N., & Gauci, V. (2015). Methods in plant foliar volatile organic compounds research. *Applications in Plant Sciences, 3*, 1500044.
McElfresh, J. S., & Millar, J. G. (1999). Geographic variation in sex pheromone blend of *Hemileuca electra* from Southern California. *Journal of Chemical Ecology, 25*, 2505–2525.
McElfresh, J. S., & Millar, J. G. (2001). Geographic variation in the pheromone system of the saturniid moth *Hemileuca eglanterina*. *Ecology, 82*, 3505–3518.
McFrederick, Q. S., Kathilankal, J. C., & Fuentes, J. D. (2008). Air pollution modifies floral scent trails. *Atmospheric Environment, 42*, 2336–2348.
McNeil, J. N., & Delisle, J. (1989a). Are host plants important in pheromone-mediated mating systems of lepidoptera? *Experientia, 45*, 236–240.
McNeil, J. N., & Delisle, J. (1989b). Host plant pollen influences calling behavior and ovarian development of the sunflower moth, *Homoeosoma electellum*. *Oecologia, 80*, 201–205.
Meinwald, J. (1986). The insect as synthetic chemist: chemical aspects of defense, courtship and mate selection. *Engineering and Science, 49*, 14–19.
Mewis, I., Ulrich, C., & Schnitzler, W. H. (2002). The role of glucosinolates and their hydrolysis products in oviposition and host-plant finding by cabbage webworm, *Hellula undalis*. *Entomologia Experimentalis et Applicata, 105*, 129–139.
Miller, J. R., & Strickler, K. L. (1984). Finding and accepting host plants. In W. J. Bell, & R. T. Cardé (Eds.), *Chemical ecology of insects* (pp. 127–157). US: Springer.
Misztal, P. K., Hewitt, C. N., Wildt, J., Blande, J. D., Eller, A. S. D., Fares, S., … Guenther, A. B. (2015). Atmospheric benzenoid emissions from plants rival those from fossil fuels. *Scientific Reports, 5*.
Molnár, B. P., Tóth, Z., Fejes-Tóth, A., Dekker, T., & Kárpáti, Z. (2015). Electrophysiologically-active maize volatiles attract gravid female European corn borer, *Ostrinia nubilalis*. *Journal of Chemical Ecology, 41*, 997–1005.
Namiki, S., Iwabuchi, S., & Kanzaki, R. (2008). Representation of a mixture of pheromone and host plant odor by the antennal lobe projection neurons of the silkmoth *Bombyx mori*. *Journal of Comparative Physiology A, 194*, 501–515.
Nerg, A., Kainulainen, P., Vuorinen, M., Hanso, M., Holopainen, J., & Kurkela, T. (1994). Seasonal and geographical variation of terpenes, resin acids and total phenolics in nursery grown seedlings of scots pine (*Pinus sylvestris* L). *New Phytologist, 128*, 703–713.
Niinemets, U., Fares, S., Harley, P., & Jardine, K. J. (2014). Bidirectional exchange of biogenic volatiles with vegetation: emission sources, reactions, breakdown and deposition. *Plant Cell and Environment, 37*, 1790–1809.

Niinemets, U., Sun, Z., & Talts, E. (2015). Controls of the quantum yield and saturation light of isoprene emission in different-aged aspen leaves. *Plant Cell and Environment, 38*, 2707–2720.

Nordlund, D. A., & Lewis, W. J. (1976). Terminology of chemical releasing stimuli in intraspecific and interspecific interaction. *Journal of Chemical Ecology, 2*, 211–216.

Paterson, H. E. H. (1985). The recognition concept of species. In E. S. Vrba (Ed.), *Species and speciation* (pp. 21–30). Pretoria: Transvaal Museum Monograph No.4.

Pawliszyn, J. (1997). *Solid phase microextraction: Theory and practice*. John Wiley & Sons.

Pelozuelo, L., Malosse, C., Genestier, G., Guenego, H., & Frérot, B. (2004). Host-plant specialization in pheromone strains of the European corn borer *Ostrinia nubilalis* in France. *Journal of Chemical Ecology, 30*, 335–352.

Penuelas, J., & Staudt, M. (2010). BVOCs and global change. *Trends in Plant Science, 15*, 133–144.

Phelan, P. L. (1992). Evolution of sex pheromones and the role of asymmetric tracking. In B. D. Roitber, & M. B. Isman (Eds.), *Insect chemical ecology: An evolutionary approach* (pp. 265–314). New York: Chapman and Hall.

Pichersky, E., & Dudareva, N. (2007). Scent engineering: toward the goal of controlling how flowers smell. *Trends in Biotechnology, 25*, 105–110.

Piñero, J. C., Galizia, C. G., & Dorn, S. (2008). Synergistic behavioral responses of female oriental fruit moths (Lepidoptera: Tortricidae) to synthetic host plant-derived mixtures are mirrored by odor-evoked calcium activity in their antennal lobes. *Journal of Insect Physiology, 54*, 333–343.

Pinto, D. M., Blande, J. D., Nykanen, R., Dong, W. X., Nerg, A. M., & Holopainen, J. K. (2007). Ozone degrades common herbivore-induced plant volatiles: does this affect herbivore prey location by predators and parasitoids? *Journal of Chemical Ecology, 33*, 683–694.

Pinto, D. M., Blande, J. D., Souza, S. R., Nerg, A., & Holopainen, J. K. (2010). Plant volatile organic compounds (VOCs) in ozone (O_3) polluted atmospheres: the ecological effects. *Journal of Chemical Ecology, 36*, 22–34.

Pinto, D. M., Nerg, A., & Holopainen, J. K. (2007). The role of ozone-reactive compounds, terpenes, and green leaf volatiles (GLVs), in the orientation of *Cotesia plutellae*. *Journal of Chemical Ecology, 33*, 2218–2228.

Pliske, T. (1975). Courtship behaviour and use of chemical communication by males of certain species of Ithomine butterflies (Nymphalidae: Lepidoptera). *Annals of the Entomological Society of America, 68*, 935–942.

Ponsard, S., Bethenod, M.-T., Bontemps, A., Pélozuelo, L., Souqual, M.-C., & Bourguet, D. (2004). Carbon stable isotopes: a tool for studying the mating, oviposition, and spatial distribution of races of European corn borer, *Ostrinia nubilalis*, among host plants in the field. *Canadian Journal of Zoology, 82*, 1177–1185.

Rafaeli, A. (2005). Mechanisms involved in the control of pheromone production in female moths: recent developments. *Entomologia Experimentalis et Applicata, 115*, 7–15.

Raina, A. K. (2003). Pheromone production in corn earworm: effect of temperature and humidity. *Southwestern Entomologist, 28*, 115–120.

Raina, A. K., Jackson, D. M., & Severson, R. F. (1997). Increased pheromone production in wild tobacco budworm (Lepidoptera: Noctuidae) exposed to host plants and host chemicals. *Environmental Entomology, 26*, 101–105.

Raina, A. K., Kingan, T. G., & Mattoo, A. K. (1992). Chemical signals from host plant and sexual behavior in a moth. *Science, 255*, 592–594.

Rasmussen, L. E. L., Lee, T. D., Roelofs, W. L., Zhang, A., & Daves, G. D., Jr. (1996). Asian elephants and Lepidoptera have a common sex pheromone. *Nature, 379*, 684.

Reddy, G. V. P., & Guerrero, A. (2004). Interactions of insect pheromones and plant semiochemicals. *Trends in Plant Sciences, 9*, 253–261.

Renwick, J. A. A., & Chew, F. S. (1994). Oviposition behavior in Lepidoptera. *Annual Review of Entomology, 39*, 377–400.

Rochat, D., Gonzalez, A. V., Mariau, D., Villanueva, A. G., & Zagatti, P. (1991). Evidence for male-produced aggregation pheromone in American palm weevil, *Rhynchophorus palmarum* (L.) (Coleoptera: Curculionidae). *Journal of Chemical Ecology, 17*, 1221–1230.

Roelofs, W. L., & Brown, R. (1982). Pheromones and evolutionary relationships of Tortricidae. *Annual Review of Ecology and Systematics, 13*, 395–422.

Roelofs, W., Glover, T., Tang, X. H., Sreng, I., Robbins, P., Eckenrode, C., ... Bengtsson, B. O. (1987). Sex pheromone production and perception in European corn borer moths is determined by both autosomal and sex-linked genes. *Proceedings of the National Academy of Sciences of the United States of America, 84*, 7585–7589.

Roelofs, W. L., & Rooney, A. P. (2003). Molecular genetics and evolution of pheromone biosynthesis in Lepidoptera. *Proceedings of the National Academy of Sciences of the United States of America, 100*, 9179–9184.

Schal, C., & Carde, R. T. (1986). Effects of temperature and light on calling in the tiger moth *Holomelina lamae* (Freeman) (Lepidoptera, Arctiidae). *Physiological Entomology, 11*, 75–87.

Scheuber, H., Jacot, A., & Brinkhof, M. W. G. (2003). Condition dependence of a multicomponent sexual signal in the field cricket *Gryllus campestris*. *Animal Behaviour, 65*, 721–727.

Schmidt-Büsser, D., von Arx, M., & Guerin, P. M. (2009). Host plant volatiles serve to increase the responses of the male European grape berry moth, *Eupoecilia ambiguella*, to their sex pheromone. *Journal of Comparative Physiology, 195*, 853–864.

Schneider, D., Schulz, S., Priesner, E., Ziesmann, J., & Francke, W. (1998). Autodetection and chemistry of female and male pheromone in both sexes of the tiger moth *Panaxia quadripunctaria*. *Journal of Comparative Physiology A-Neuroethology Sensory Neural and Behavioral Physiology, 182*, 153–161.

Schröder, R., & Hilker, M. (2008). The relevance of background odor in resource location by insects: a behavioral approach. *Bioscience, 58*, 308–316.

Shiojiri, K., Karban, R., & Ishizaki, S. (2009). Volatile communication among sagebrush branches affects herbivory: timing of active cues. *Arthropod-Plant Interactions, 3*, 99–104.

Showers, W. B., Reed, G. L., Robinson, J. F., & Derozari, M. B. (1976). Flight and sexual activity of the European corn borer. *Environmental Entomology, 5*, 1099–1104.

Shulaev, V., Silverman, P., & Raskin, I. (1997). Airborne signalling by methyl salicylate in plant pathogen resistance. *Nature, 385*, 718–721.

Simmonds, M. S. (2001). Importance of flavonoids in insect–plant interactions: feeding and oviposition. *Phytochemistry, 56*, 245–252.

Simpraga, M., Takabayashi, J., & Holopainen, J. K. (2016). Language of plants: where is the word? *Journal of Integrative Plant Biology, 58*, 343–349.

Smadja, C., & Butlin, R. K. (2009). On the scent of speciation: the chemosensory system and its role in premating isolation. *Heredity, 102*, 77–97.

Sole, J., Sans, A., Riba, M., Rosa, E., Bosch, M. P., Barrot, M., ... Guerrero, A. (2008). Reduction of damage by the Mediterranean corn borer, *Sesamia nonagrioides*, and the European corn borer, *Ostrinia nubilalis*, in maize fields by a trifluoromethyl ketone pheromone analog. *Entomologia Experimentalis et Applicata, 126*, 28–39.

Stewart-Jones, A., & Poppy, G. (2006). Comparison of glass vessels and plastic bags for enclosing living plant parts for headspace analysis. *Journal of Chemical Ecology, 32*, 845–864.

Sugimoto, K., Matsui, K., Iijima, Y., Akakabe, Y., Muramoto, S., Ozawa, R., ... Nobuke, T. (2014). Intake and transformation to a glycoside of (Z)-3-hexenol from infested neighbors reveals a mode of plant odor reception and defense. *Proceedings of the National Academy of Sciences of the United States of America, 111*, 7144–7149.

Symonds, M. R. E., & Elgar, M. A. (2008). The evolution of pheromone diversity. *Trends in Ecology & Evolution, 23*, 220–228.
Tamhankar, A. J. (1994). Influence of host presence, age and mating on the calling behaviour of female Earias vitella (Fabricius). *International Journal of Tropical Insect Science, 15*, 275–280.
Thibout, E. (1974). Influences respectives de la plante-hôte et de la copulation sur la longévité, la ponte, la production ovarienne et la fertilité des femelles d'*Acrolepia assectella. Annales de Zoologie: Ecologie Animale, 6*, 81–96.
Tholl, D., Boland, W., Hansel, A., Loreto, F., Rose, U., & Schnitzler, J. (2006). Practical approaches to plant volatile analysis. *Plant Journal, 45*, 540–560.
Trone, F., Anfora, G., Bengtsson, M., Witzgall, P., & Ignell, R. (2010). Coding interaction of sex pheromone and plant volatile signals in the antennal lobe of codling moth *Cydia pomonella. Journal of Experimental Biology, 213*, 4291–4303.
Unbehend, M. (2013). *Variation in sexual communication and its role in divergence of two host strains of the noctuid moth* Spodoptera frugiperda (Ph.D. thesis). Amsterdam: Institute for Biodiversity and Ecosystem Dynamics (IBED), University of Amsterdam http://hdl.handle.net/11245/1.398457.
Unbehend, M., Haenniger, S., Vásquez, G., Juarez, M. L., Reisig, D., McNeil, J. N., … Groot, A. T. (2014). Geographic variation in sexual attraction of *Spodoptera frugiperda* corn- and rice-strain males. *PLoS One, 9*, e89255.
Van Emden, H. F., Sponagl, B., Baker, T., Ganguly, S., & Douloumpaka, S. (1996). Hopkins 'host selection principle', another nail in its coffin. *Physiological Entomology, 21*, 325–328.
Van Geffen, K. G., Groot, A. T., Van Grunsven, R. H. A., Donners, M., Berendse, F., & Veenendaal, E. M. (2015). Artificial night lighting disrupts sex pheromone in a noctuid moth. *Ecological Entomology, 40*, 401–408.
Vickers, N. J., & Baker, T. C. (1997). Chemical communication in heliothine moths. 7. Correlation between diminished responses to point source plumes and single filaments similarly tainted with a behavioral antagonist. *Journal of Comparative Physiology A-Neuroethology Sensory Neural and Behavioral Physiology, 180*, 523–536.
Visser, J. H. (1986). Host odor perception in phytophagous insects. *Annual Review of Entomology, 31*, 121–144.
Visser, J. H., & Avé, D. A. (1978). General green leaf volatiles in the olfactory orientation of the Colorado beetle. *Leptinotarsa decemlineata. Entomologia Experimentalis et Applicata, 24*, 738–749.
Viswanathan, D. V., Narwani, A. J., & Thaler, J. S. (2005). Specificity in induced plant responses shapes patterns of herbivore occurrence on *Solanum dulcamara. Ecology, 86*, 886–896.
Vuorinen, T., Nerg, A., Vapaavuori, E., & Holopainen, J. K. (2005). Emission of volatile organic compounds from two silver birch (*Betula pendula* Roth) clones grown under ambient and elevated CO_2 and different O_3 concentrations. *Atmospheric Environment, 39*, 1185–1197.
Widhalm, J. R., Jaini, R., Morgan, J. A., & Dudareva, N. (2015). Rethinking how volatiles are released from plant cells. *Trends in Plant Science, 20*, 545–550.
Widmayer, P., Heifetz, Y., & Breer, H. (2009). Expression of a pheromone receptor in ovipositor sensilla of the female moth (*Heliothis virescens*). *Insect Molecular Biology, 18*, 541–547.
Wilson, E. O. (1963). Pheromones. *Scientific American, 208*, 100–111.
Wood, D. L. (1982). The role of pheromones, kairomones, and allomones in the host selection and colonization behavior of bark beetles. *Annual Review of Entomology, 27*, 411–446.
Wu, W., Cottrell, C. B., Hansson, B. S., & Löfstedt, C. (1999). Comparative study of pheromone production and response in Swedish and Zimbabwean populations of turnip moth, *Agrotis segetum. Journal of Chemical Ecology, 25*, 177–196.
Zhu, J. W., Chastain, B. B., Spohn, B. G., & Haynes, K. F. (1997). Assortative mating in two pheromone strains of the cabbage looper moth, *Trichoplusia ni. Journal of Insect Behavior, 10*, 805–817.

CHAPTER SIX

The Plant as a Habitat for Entomophagous Insects

L. Kaiser*[,1], P. Ode[§], S. van Nouhuys[¶], P.-A. Calatayud*[,||], S. Colazza[#], A.-M. Cortesero, A. Thiel[§§], J. van Baaren[¶¶]**

*IRD UMR EGCE (Evolution, Génome, Comportement, Ecologie), CNRS-IRD-Université Paris-Saclay, Gif-sur-Yvette, France
[§]Colorado State University, Fort Collins, CO, United States
[¶]University of Helsinki, Helsinki, Finland
[||]IRD c/o icipe, Nairobi, Kenya
[#]University of Palermo, Palermo, Italy
**UMR 1349 IGEPP (Institute for Genetics, Environment and Plant Protection), Université de Rennes 1, Rennes, France
[§§]University of Bremen, FB 02, Institute of Ecology, Bremen, Germany
[¶¶]UMR-CNRS 6553 Ecobio, Université de Rennes 1, Université Bretagne-Loire, Rennes, France
[1]Corresponding author: E-mail: laure.kaiser-arnauld@egce.cnrs-gif.fr

Contents

Advances in Botanical Research, Volume 81
ISSN 0065-2296
http://dx.doi.org/10.1016/bs.abr.2016.09.006

Abstract

Populations of herbivorous insects are naturally consumed by other predacious or predatory insect species. These entomophagous insects are thus plant-dwelling organisms that use the plant for several vital functions and are affected by plant traits at the evolutionary, organism and population levels. Many entomophagous species are used for the biological control of insect pests worldwide. The aim of this chapter is to provide an exhaustive review of mechanisms underlying the interactions between plants and entomophagous insects, including those governing life history traits at the individual level, as well as those acting on population and community structure and dynamics. We detail how properties of host-infested plants determine parasitism behaviour, development (in the case of parasitoids) and nectar consumption by adult entomophagous insects. We detail how plants respond to and benefit from natural enemies attacking insect herbivores. We also illustrate how plant architecture, the vegetation communities and their climatic correlates can influence predator and parasitoid behaviour and populations. This chapter considers the biology and ecology of the interactions and mentions some implications for the biological control of plant pests.

1. INTRODUCTION

Herbivorous insects are consumed by predatory and parasitic ('entomophagous' or 'carnivorous') arthropods. These entomophagous arthropods, mainly insects, spiders and mites, visit plants not only to find hosts or prey, but they may also mate, feed and develop on plants. Various plant traits can affect entomophagous arthropods (Bottrell, Barbosa, & Gould, 1998; Kaiser, Couty, & Cortesero, 2013). The importance of the plant in the evolution and diversification of entomophagous species is evident in their searching and oviposition behaviours, which are adapted to the vegetational structure inhabited, and often created or transformed by their hosts. There are thus spectacular adaptations of the third trophic level to the first. For instance, ichneumonid wasps of the genus *Megarhyssa* that attack wood-boring larvae possess an ovipositor equipped with a hydraulic pressure system and a saw-like terminal part to drill through the wood and sting the host within (Gauld & Bolton, 1988; Quicke, 2015). Different strategies have evolved in response to similar constraints exerted by the host habitat. For instance, parasitoids of grain borers that are adapted to locate their host within seeds have different ways to reach the host. *Dinarmus basalis* (Rondani, 1877) (Chalcidoidea: Pteromalidae) pierces the seed coat with its ovipositor, just above the cavity inhabited by the host, whereas *Eupelmus vuilleti* (Crawford, 1913) (Chalcidoidea: Eupelmidae) inserts its ovipositor inside the tiny tunnel bored by the host larva (Jaloux, 2004). Similarly, stem borer Lepidoptera larvae can be attacked through the stem by *Iphiaulax* spp., which are relatively large wasps equipped with long ovipositors (Quicke, 1988, 2015) (Fig. 1), or by much smaller species like *Cotesia sesamiae* (Cameron, 1906) which enters the tunnel and walks up to the host larva (Kimani-Njogu & Overholt, 1997). Beyond these obvious adaptations, the reader of this chapter will discover that entomophagous insects can respond behaviourally and physiologically to a wide array of plant traits often modified by the herbivorous host to insure their reproductive success.

The plant is also a direct resource, providing water and nectar, especially to adult entomophagous arthropods. Besides these trophic interactions, the plant also offers a physical refuge (e.g., trichomes, domatia and shelter) from unfavourable weather conditions and higher level predators.

Most studies of multitrophic interactions involve chemically mediated effects of plants on higher trophic levels; e.g., the emission and perception of plant volatiles by entomophagous insects, and the physiological response of entomophagous insects to plant compounds that mediate resistance to phytophagous insects. These interactions have been particularly well studied in the case of insect parasitoids that lay eggs in or on an insect host which later dies, following the larval development of the parasite. This intimate connection between life histories of host and parasitoid, presumably driven by reciprocal selection between host immune defences and virulence of the parasitoids, often leads to extremely specialized host—parasite interactions. It is within Hymenoptera that the largest number of parasitoid species (parasitic wasps) is found. To a lesser extent, multitrophic interactions have also been documented in the case of parasitoid flies (especially the tachinids), predatory insects (e.g., hoverflies, lacewings, ladybugs and hemipteran bugs) and predatory mites. Much literature has been published on responses of the third trophic level to plant resistance mechanisms since 1990 (Agrawal, 2000a, 2000b; Ode, 2006; Turlings & Wäckers, 2004). More recently, plant-mediated multitrophic interactions, i.e., involving several herbivores, pathogens, primary and hyper parasitoids, have been the focus of several studies (Gish, De Moraes, & Mescher, 2015; Hare, 2011; Mooney & Singer, 2012; Poelman & Dicke, 2014). Several books have been dedicated to the biology and evolution of insect parasitoids (Godfray, 1994; Quicke, 1997, 2015; Wajnberg, Bernstein, & Van Alphen, 2007).

The aim of this chapter is to provide an exhaustive review of the mechanisms underlying the interactions between plants and entomophagous insects, those governing adult and larval life history traits at the individual

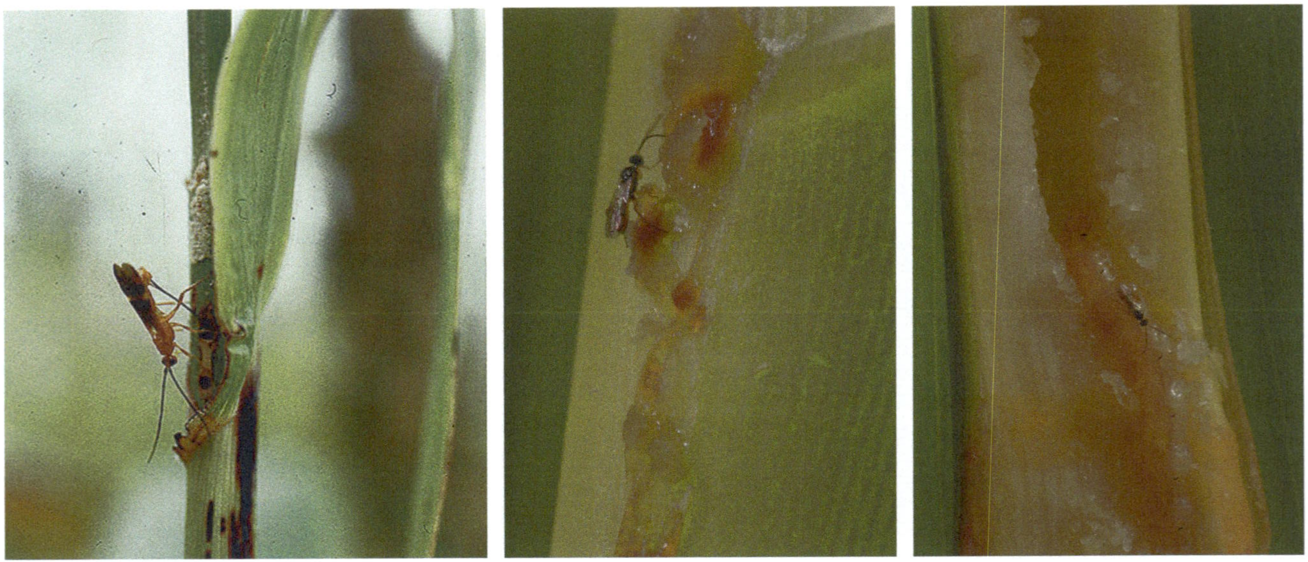

Figure 1 Left: *Iphiaulax* species are relatively large Hymenoptera that can either insert their long ovipositor in the host larval tunnel, as seen here on a sorghum stem bored by *Chilo partellus* (Lepidoptera: Crambidae), or drill through the stem to oviposit on the host larva. Middle: on a maize stalk, tiny *Cotesia sesamiae* examines faeces of its stem borer host left at the entrance of the larval tunnel. Right: once in the tunnel, *C. sesamiae* oviposits in the larval body, here *Sesamia nonagrioides* Lefebvre (Lepidoptera Noctuidae). *Courtesy of Iphiaulax: J. van der Berg;* C. sesamiae: © *2016 « Les nouveaux guerriers des champs » un film de Claude-Julie Parisot pour Galaxie Presse & Arte France.*

level and those acting on population and community structure and dynamics. We detail how properties of host-infested plants determine parasitism behaviour, larval development (in the case of parasitoids) and nectar consumption by adult entomophagous insects. We detail how plants respond to and benefit from natural enemies attacking insect herbivores. We also illustrate how plant architecture, the vegetation communities and their climatic correlates can influence predator and parasitoid behaviour and populations. This chapter considers the biology and ecology of the interactions; it mentions some implications for the biological control of plant pests, which are topics further developed in other books and reviews (e.g., Van Driesche, Hoddle, & Center, 2008).

2. THE PLANT: PLACE OF PREDATION AND PARASITISM

The plant is the most frequent place of predation and parasitism of herbivorous insects, with the exception of predation and parasitism of hosts that are no longer on or in the plant, such as parasitism of host pupae that undergo metamorphosis in the soil. How carnivorous insects select herbivorous host insects has been mainly studied in insect parasitoids, because host selection behaviour directly determines the developmental success of these species. Host selection typically involves the following sequence of behaviours (Godfray, 1994; Nordlund, Jones, & Lewis, 1981): searching for the appropriate host habitat (often, the plant), usually by remote-oriented flight; searching on the plant for the host by an exploration accompanied by antennal examinations of leaf area and faeces of the host; examination of the host; and finally, egg-laying.

Although the use of plant visual cues such as shape and colour has been demonstrated in some parasitoids (Wäckers & Lewis, 1999), most species rely on chemical cues during host searching and acceptance.

2.1 In-flight Search for Host and Prey

Insect parasitoids and predators can use various chemical cues as a reliable source of information about the presence of host or prey, which are often small and inconspicuous (Wajnberg & Colazza, 2013). Among all possible chemical cues, the volatile organic compounds (VOCs) emitted by plants in response to herbivory have long been recognized as playing an important role for the in-flight searching by natural enemies (Dicke, 2016; Erb, Robert, Hibbard, & Turlings, 2011; Heil, 2008). This indirect defence mechanism was documented for the first time in the 1990s by two pioneering and independent studies: one conducted on the parasitoid *Cotesia marginiventris* (Cresson, 1865) (Hymenoptera: Braconidae), which parasitizes a maize pest caterpillar (Turlings, Tumlinson, & Lewis, 1990); the other on the mite *Phytoseiulus persimilis* Athias-Henrio 1957 (Acarina: Phytoseiidae), which is a predator of apple pest mites (Dicke, Sabelis, & De Jong, 1988). Since then, several comprehensive reviews have been published on the attraction of parasitoids and predators towards VOCs, showing that it is a widespread ecological phenomenon recorded for at least 50 plant species belonging to about 30 different families, including both monocots and dicots (Mumm & Dicke, 2010). In several cases, insect parasitoids and predators can recognize specific plant volatile blends induced by their herbivore hosts (Dicke & Baldwin, 2010; Heil & Karban, 2010). This indicates that plants have specific responses to herbivory, depending on the identity of the herbivore (Dicke, 2016). A growing body of literature has documented the dynamic and specific nature of odours produced as well as their biosynthetic pathways (Ode, 2013; Schuman & Baldwin, 2016), and it is now recognized that plant VOCs can be induced either by feeding (named herbivore-induced plant volatiles (HIPVs)) or by egg-laying (named oviposition-induced plant volatiles (OIPVs)). The role of HIPVs has been extensively documented in larval parasitoids (Hare, 2011). For example, in a tritrophic system comprising maize, the Lepidoptera *Spodoptera* spp. and several larval endoparasitoids, damage induced by *Spodoptera* larvae stimulates the release of green leaf volatiles and the accumulation of two plant hormones, jasmonic acid (JA) and ethylene (Carroll, Schmelz, Meagher, & Teal, 2006). These plant hormones are responsible for the emission of indole, terpenoids and other compounds, which attract larval parasitoids, and this attraction can increase the fitness of the attacked plants. Volatile emission is not limited to the part of the plant attacked. The emission spreads throughout the plant, including roots, through the release of internal plant hormones, such as JA, salicylic acid (SA) and their volatile derivatives, methyl jasmonate and methyl salicylate. This spreading extends even to neighbouring plants, through air- or soil-transmitted chemicals. The reaction of plant defence can be induced by salivary enzymes of the herbivores. In particular, chewing insects (e.g., caterpillars) tend to induce the JA-signalling pathway, while phloem feeders (e.g., aphids) can induce both JA and −SA pathways, depending on insect species (Hare, 2011).

Many endoparasitoids of herbivorous insects do not immediately kill their hosts (termed ‘koinobiotic parasitoids’), so plants continue to be injured even after successful parasitism (De Rijk, Dicke, & Poelman, 2013). The emission of OIPVs upon herbivore egg-laying can be an effective indirect defence strategy that attracts egg parasitoids, which prevent the pest from hatching (Pashalidou et al., 2015). Furthermore, OIPVs are highly reliable and detectable for egg parasitoids (Colazza et al., 2004; Hilker & Fatouros, 2015). Evidence of this 'early herbivore alert' was provided in a pioneering study by Meiners and Hilker (2000). They showed that oviposition by elm leaf beetles *Xanthogaleruca luteola* (Müller, 1766) (Coleoptera: Chrysomellidae) on elm *Ulmus minor*, caused wounding of the plant tissue, which induced OIPVs that attracted the specialist egg parasitoid *Oomyzus gallerucae* (Fonscolombe, 1832) (Hymenoptera: Eulophidae).

Recent studies indicate that plant volatiles can mediate community interactions between parasitized and unparasitized herbivores (Poelman, Zheng, Zhang, Heemskerk, & Cortesero, 2011), and even at the fourth trophic level (parasitoids of parasitoids). For example, Poelman et al. (2012) showed that *Lysibia nana* (Gravenhorst, 1829) (Hymenoptera: Ichneumonidae), a solitary hyperparasitoid that attacks pupae of braconid wasps, is able to locate its host using HIPVs from cabbage plants attacked by *Pieris rapae* (L.) (Lepidoptera: Pieridae) that were parasitized by *Cotesia glomerata* (Linnaeus, 1758) (Hymenoptera: Braconidae).

In a complex natural environment, in-flight search by parasitoids for their hosts may be impeded by the presence of different herbivores on different food plants, or by herbivores in different feeding guilds that induce different VOC blends in the same plant (Cusumano, Weldegergis, Colazza, Dicke, & Fatouros, 2015). For example, the attraction of the egg parasitoid, *Trissolcus basalis* (Wollaston, 1858) (Hymenoptera: Scelionidae) towards *Nezara viridula* (Linnaeus, 1758)-egg induced volatiles produced by the fava bean *Vicia faba* is disrupted by the nonhost beetle, *Sitona lineatus* (Linnaeus, 1758) (Coleoptera: Curculionidae) feeding on the roots or leaves (Moujahed et al., 2014). VOC emission can also be induced by plant pathogen development, and its effects on insect parasitoids have received recent and growing interest, especially because the identification of biochemical defence pathways induced by pathogens and herbivores is making rapid progress (Appel et al., 2014; Pieterse, Van der Does, Zamioudis, Leon-Reyes, & Van Wees, 2012). Depending on the plant as well as the pathogen and herbivore species, the blend and timing of volatile emission vary and may interfere with responses of predators and parasitoids, which may decrease (Desurmont, Xu, & Turlings, 2016), remain unchanged or increase (Ponzio, Gols, Weldegergis, & Dicke, 2014). Similar to the interaction between herbivores in different feeding guilds, such interference can be interpreted as the outcome of crosstalk between the pathways induced by each plant enemy (Appel et al., 2014).

Because HIPVs and OIPVs are central to the ability of natural enemies to find hosts, their use in the manipulation of parasitoid host searching behaviour in biological controls is promising (Meiners & Peri, 2013). However, whether their use can improve the effective biological control of plant pests, it is still an open question, which requires more study taking into account crop-specific aspects and landscape context (Gish et al., 2015; Kaplan, 2012a; Traperò, Wilson, Stiller, & Wilson, 2016).

2.2 Searching for Hosts While on Plants and Host Acceptance

Host acceptance behaviour by parasitoids depends on the perception of several host criteria, including species identity, growth stage, health status and size, among many others. The abundant studies on host searching and acceptance mechanisms allow us to establish a general trend; the importance of direct plant signals (e.g., VOCs emitted by infested plants) for parasitoids decreases when approaching the host, whereas the host-specific contact sensory cues become increasingly important (e.g., chemical, visual, vibrational) (Godfray, 1994). Although parasitoids may use nonchemical information sources (e.g., physical cues) to locate their target host, they are differentially attracted mostly on the basis of specificity of host-derived kairomones (semiochemicals that benefit the receiver to the detriment of the emitter; Nordlund & Lewis, 1976; Dicke & Sabelis, 1988). These chemicals constitute reliable cues for the foraging parasitoid, especially in the final stages of host location, recognition and acceptance (Godfray, 1994; Afsheen, Wang, Li, Zhu, & Lou, 2008). Parasitoids exploit both volatile and nonvolatile contact kairomones to differentiate host and nonhost species as well as host developmental stage. These kairomones are emitted by a specific herbivore stage (egg, larva, pupa or adult) or by-product (e.g., frass, exuvia, mandibular gland secretion or defensive secretion) of its specific host (see Afsheen et al. (2008) for review) (Table 1).

The kairomones identified so far include various chemical groups but are predominately hydrocarbons such as aldehydes, esters and terpenoids (Table 1). Rani, Kumari, Sriramakrishna, and Sudhakar (2007) proposed that compounds that have a high number of carbon atoms might act as contact stimulants, whereas the chemicals with less than 10 carbon atoms are more likely to be volatile and might attract parasitoids to the vicinity of the host.

Table 1 Diversity of Kairomones Involved in Host Recognition and Acceptance by Parasitoids and Their Sources

References	Compound	Source	Category	Parasitoid Species	Host Species
Ananthakrishnan, et al. (1991). *Journal of Biosciences*, *16*, 111–119.	Hexatriacontane, pentacosane, docosane 2,6,10-dodecatrienal-3,7,11-trimethyl	Moth scale	Egg	*Trichogramma chilonis*	*Heliothis armigera*
Arakaki, et al. (2011). *Applied Entomology and Zoology*, *46*, 195–200.	(6Z, 9Z, 11S, 12S)-11, 12-epoxyhenicosa-6,9-diene	Sex pheromone	Egg	*Telenomus euproctidis*	*Orgyia postica*
Bénédet, et al. (1999). *Journal of Insect Physiology*, *45*, 375–384.	Four glycopolypeptides	Silk cocoon	Pupal	*Diadromus pulchellus*	*Acrolepiopsis assectella*
Burks, & Nettles (1978). *Environmental Entomology*, *7*, 897–900.	Cuticular extracts	Cuticle of the larvae	Larval	*Eucelatoria*	*Heliothis virescens*
Boo, & Yang (2000). *Journal of Chemical Ecology*, *26*, 359–375.	*Z*11-16:Ac E12-14:Ac	Sex pheromone Sex pheromone	Egg Egg	*T. chilonis* *T. chilonis*	*Helicoverpa assulta* *Ostrinia funacalis*
Calatayud, et al. (2001). *Journal of Chemical Ecology*, *27*, 2203–2217.	*O*-caffeoylserine	Host cover	Larval	*Acerophagus coccois*, *Aenasius vexans*	*Phenacoccus herreni*
Colazza, et al. (2007). *Journal of Chemical Ecology*, *33*, 1405–1420.	*n*-nonadecane	Adult tarsi and scutella	Egg	*Trissolcus basalis*	*Nezara viridula*

(Continued)

Table 1 Diversity of Kairomones Involved in Host Recognition and Acceptance by Parasitoids and Their Sources—cont'd

References	Compound	Source	Category	Parasitoid Species	Host Species
DeLury, et al. (1999). *Journal of Chemical Ecology*, *25*, 2419–2431.	Heptanal, octanal, nonanal, decanal undecan-2-one, dodecanal, pentadecan-2-one, (*Z*)-6-pentadecen-2-one, (*Z*)-9-hexadecenal, (*Z*)-6-heptadecen-2-one, 3,7,11-trimethyl-2E,6E, 10-dodecatrien-l-ol acetate	Scales	Egg	*Ascogaster quadridentata*	*Cydia pomonella*
Fatouros, et al. (2005). *Journal of Insect Behavior*, *20*, 53–65.	Benzyl cyanide	Antiaphrodisiac	Egg	*Trichogramma brassicae*	*Pieris brassicae*
Gauthier, et al. (2004). *Journal of Insect Physiology*, *50*, 1065–1074.	Polypeptides	Silk cocoon	Pupal	*D. pulchellus*	*A. assectella*
Millar, & Hare (1993). *Journal of Chemical Ecology*, *19*, 1721–1736.	*O*-caffeoyltyrosine	Host cover	Larval	*Aphytis melinus*	*Aonidiella aurantii*
Hilker, et al. (2000). *Journal of Chemical Ecology*, *26*, 2591–2601.	Acetate and propionate of (2S,3R,7R)-3,7-dimethyl- 2-tridecanol (2S,3S,7S)-3,7-dimethyt-2-pentadecyl acetate	Sex pheromone	Egg	*Chrysonotomyia ruforum*	*Diprion pini*
Jones, et al. (1971). *Science*, *17*, 842–843.	13-Methy-lhentriacontane	Frass, larvae	Larval	*Microplitis croceipes*	*Heliothis* zea

Jones, et al. (1973). *Environmental Entomology*, *2*, 593–596.	Docosane, tricosane	Wing scales	Egg	*Trichogramma evanescens*	*H. zea*
Kuwahara, et al. (1983). *Agricultural and Biological Chemistry*, *47*, 1929–1931.	2-Palmitoyl- and 2-oleoyl-cyclohexane- 1, 3- dione	Frass		*Venturia canescens*	*Plodia interpunctella*
Lewis, et al. (1982). *Journal of Chemical Ecology*, *8*, 1323–1331.	(*Z*)-9-Hexadecenal	Sex pheromone	Egg	*Trichogramma pretiosum*	*H. zea*
Lou, et al. (1999). In J. Du (Ed) *Proceedings of first Asia–Pacific conference on chemical ecology*, Shanghai, China. November 1–4, 1999.	Palm oil	Adult, nymph	Egg	*Anagrus nilaparvatae*	*Nilaparvata lugens*
Lou, & Cheng (2001). *Entomologia Experimentalis et Applicata*, *101*, 59–67.	Palm oil	Adult, nymph	Egg	*A. nilaparvatae*	*Sogatella furcifera*
Mattiacci, et al. (1993). *Journal of Chemical Ecology*, *19*, 1167–1181.	α^{β} – unsaturated aldehyde, (*E*)-2-decenal	Defensive metathoracic gland	Egg	*T. basalis*	*Nezara viridula*
Mizutani (2006). *Japanese Journal of Applied Entomology and Zoology*, *50*, 87–99.	(*E*)-2-hexenyl (*Z*)-3-hexenoate (*E*2H*Z*3H)	Aggregation pheromone	Egg	*Ooencyrtus nezarae*	*Riptortus clavatus*

(Continued)

Table 1 Diversity of Kairomones Involved in Host Recognition and Acceptance by Parasitoids and Their Sources—cont'd

References	Compound	Source	Category	Parasitoid Species	Host Species
Ma, et al. (1992). *Annals of the Entomological Society of America*, *85*, 72–79.	11 free amino acids including serine and glutamic acid	Frass, oral secretion	Larval	*Eriborus terebrans*	*Ostrinia nubilalis*
Mudd, & Corbet (1982). *Journal of Chemical Ecology*, *8*, 843–850. Mudd, et al. (1984). *Journal of Chemical Ecology*, *10*, 1597–1601.	2-Acylcyclohexane-1-3-diones	Mandibular glands	Larval	*Nemeritis canescens*	*Ephestia kuehniella*
Nemoto, et al. (1987). *Agriculture and Biological Chemistry*, *51*, 1805–1810.	2-Palmitoyl- 2-stearoylcyclohexane-1,3-dione	Frass	Larval	*V. canescens*	*Cadra cautella Plodia interpunctella*
Nettles, & Burks (1975). *Journal of Insect Physiology*, *21*, 965–978.	Protein (30 kD)	Frass, haemolymph	Entire larvae, pupae, emerged adults	*Archytas marmoratus*	*H. virescens*
Nordlund, & Lewis (1985). *Entomologia Experimentalis et Applicata*, *38*, 109–112.	13-Methylhentriacontane	Larval frass	Larval	*Microplitis demolitor*	*H. zea*
Obonyo, et al. (2010). *Biological Control*, *54*, 270–275.	Protein (enzyme?)	Larval body extract Larval frass regurgitants	Larval	*Cotesia flavipes* *Cotesia sesamiae*	*Chilo partellus* *Busseola fusca*
Ramachandran, et al. (1991). *Journal of Agricultural Food Chemistry*, *39*, 2310–2317.	3-Octanone and guaiacol	Larval frass	Larval	*M. demolitor*	*Pseudoplusia inchudens*

Rani, et al. (2007). *Journal of Chemical Ecology*, *33*, 59–73.	Long chain alkanes and alkenes like docosane, tetracosane, pentacosane, and eicosane	Adult extracts	Egg	*Trichogramma japonicum*	*Scripophaga incertulas*
Reddy, et al. (2002). *Journal of Chemical Ecology*, *28*, 131–143.	*Z*11-16:Ald, *Z*11-16:Ac, and *Z*11-16:OH in a 1: 1: 0.01 ratio	Sex pheromone	Egg	*T. chilonis*	*Plutella xylostella*
	*Z*11-16:Ac alone, 1:1 blend of *Z*11-16:Ac and *Z*11-16:Ald allyl isothiocyanate	Larval frass	Egg	*T. chilonis*	*Plutella xylostella*
Renou, et al. (1992). *Entomologia Experimentalis et Applicata*, *63*, 291–303.	(Heneicosane, tricosane) pentacosane, heptacosane and nonacosane), ethyl and palmitic acid palmitate	Egg extract	Egg	*T. brassicae*	*O. nubilalis* *Mamestra brassicae*
Roux, et al. (2007). *Chemoecology*, 17, 13–18.	Lipids	Larval cuticle	Larval	*Cotesia plutellae*	*Plutella xylostella*
Shu, et al. (1990). *Journal of Chemical Ecology*, *16*, 521–529.	Mixture of 11,15-, 13,17- and 15,19-dimethylno natriacontanes	Moth scale	Egg	*Trichogramma nubilale*	*O. nubilalis*
Silva, et al. (2006). *Pesquisa Agropecuaria Brasileira*, *41*, 1093–1098.	Methyl 2,6,10-trimethyltridecanoate	Male sexual pheromone	Egg	*Telenomus podisi*	*Euschistus heros*
Steidle, & Ruther (2000). *Journal of Chemical Ecology*, *26*, 2665–2675.	Alpha-tocopherol, beta-tocopherol beta-tocotrienol, cholesterol, ergostenol, and beta-sitosterol	Feces	Larval	*Lariophagus distinguendus*	*Sitophilus granaries*

(*Continued*)

Table 1 Diversity of Kairomones Involved in Host Recognition and Acceptance by Parasitoids and Their Sources—cont'd

References	Compound	Source	Category	Parasitoid Species	Host Species
Strand, et al. (1989). *Journal of Chemical Ecology*, *15*, 1491–1500.	2-Acylcyclohexane-1-3-diones	Mandibular glands	Larval	*Bracon hebetor*	*E. kuehniella*
Takabayashi, & Takahashi (1989). *Entomologia Experimentalis et Applicata*, *52*, 221–227.	2, 5-dialkyltetrahydrofuran	Frass	Larval	*Apanteles kariyai*	*Pseudaletia separata*
Thompson, et al. (1983). *Environmental Entomology*, *12*, 1312–1314.	Phenols, alcohols	Frass	Larval	*Lixophaga diatraeae*	*Diatraea saccharalis*
Vinson, et al. (1975). *Entomologia Experimentalis et Applicata*, *18*, 443–450.	Mixture of three long chain hydrocarbons (11-methyl-hentriacontane, 16-methyl-dotriacontane and 13-methyl-hentriacontane)	Mandibular glands	Pupal	*Cardiochiles nigriceps*	*H. virescens*
Weseloh (1977). *Journal of Chemical Ecology*, *3*, 723–735.	Sericin or fibrinogen likeprotein	Silk producing glands	Larval	*Apanteles melanoscelus*	*Limantria dispar*

Updated table of Afsheen, S., Wang, X., Li, R., Zhu, C.-S., & Lou, Y.-G. (2008). Differential attraction of parasitoids in relation to specificity of kairomones from herbivores and their by-products. *Insect Science*, *15*, 381–397.

Most of the identified kairomones have been associated with egg or larval parasitoids (Table 1). In the case of egg parasitoids, kairomones come from the insects' host itself (i.e., sex pheromones, adult host products such as scales or egg coating), whereas in the case of larval parasitoids, the kairomones are coming from the host feeding activities (i.e., frass, oral secretions). In the latter, kairomones may have plant origin or be derived from plant products and may thus be also classified as synomones. In addition, the wax layer of the host plants where the herbivores are located can also play an important role in the detection of host's kairomones for both egg (e.g., Colazza, Salerno, & Wajnberg, 1999; Conti, Salerno, Bin, Williams, & Vinson, 2003) and larval parasitoids (e.g., Rostas, Ruf, Zabka, & Hildebrandt, 2008). Parasitoids are sensitive to chemical footprints left by herbivores as they walk over the substrate. Furthermore, the chemical composition of the plant wax layer modulates the detectability of these footprints. However, the chemistry of both wax layer and footprints remains to be elucidated.

2.3 Learning and Memory Involved in Host Searching

Odour learning has been mostly studied in parasitic wasps. As detailed above, plants damaged by herbivore feeding often produce particular VOCs, which are attractive to parasitoid insects (Kaplan, 2012a; McCormick, Unsicker, & Gershenzon, 2012; Vet & Dicke, 1992). While some species show innate plant odour preferences (Kaplan, 2012b), others rely on experience, having to learn the host-associated VOCs (e.g., Canale, Geri, & Benelli, 2014; Simpson et al., 2011; Steidle & Schöller, 1997; Vet & Van Opzeeland, 1984). Even in species with innate preferences, their response can be strengthened by rewarding odour experiences, as has been recently reviewed by Giunti et al. (2015). From the plant's perspective, producing odours that attract the enemies of its herbivores has often been considered advantageous (Hare, 2011; Turlings et al., 1995). In general, plant-produced odours, such as those that are attractive from relatively large distances (Braasch & Kaplan, 2012; De Boer & Dicke, 2006; Geervliet, Ariens, Dicke, & Vet, 1998) are more easily detectable for foraging wasps than odours directly emitted by a potential host, which have evolved to be cryptic. Plant-derived odours are, on the other hand, less reliable and, as hosts may feed from more than one plant species, more variable than cues produced directly by the host. The ability to recognize reliable odours through learning from previous experience therefore has a great potential to increase a wasps' foraging success and most species studied in this regard are able to learn a plant odour in association with successful oviposition experience (Steidle & Van Loon, 2003; Turlings, Wäckers, Vet, Lewis, & Tumlinson, 1993).

Once a parasitoid can respond to the odour of a rewarding plant species, future host location is greatly facilitated. So how do unexperienced wasps locate their first host? Some studies have shown that naive wasps acquire the ability to orient to their developmental plant odour, either as immature (Barron, 2001; Gandolfi, Mattiacci, & Dorn, 2003), or at adult eclosion, during antennal exploration of plant tissues or host remains, a process termed 'early adult learning' (e.g., Van Emden, Sponagl, Baker, Ganguly, & Douloumpaka, 1996; Kester & Barbosa, 1991) (Fig. 2). Studies have also shown that preferences can be reversed when wasps have rewarding experiences with initially nonpreferred odours (e.g., Kester & Barbosa, 1991), or unrewarding experiences (Fig. 2) with initially preferred or neutral odours (Papaj, Snellen, Swaans, & Vet, 1994), rendering the odours attractive and repellent, respectively. In addition, odours that a wasp will not encounter in a natural habitat, e.g., vanilla or banana, can become highly attractive

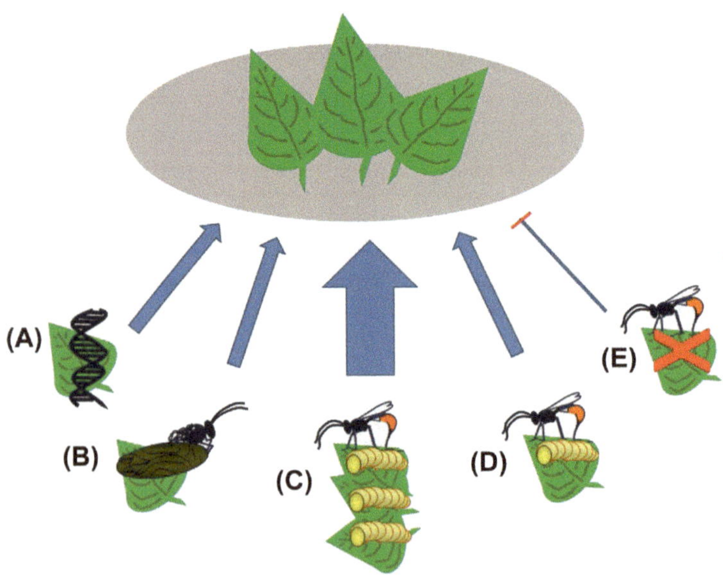

Figure 2 A parasitic wasps responsiveness towards specific plant volatiles can be genetically fixed (A). Very often however, relevant plant odours are learned early in life, e.g., during eclosion (B) or during experience with a suitable host (D). Existing preferences might then be strengthened by further rewarding experiences (C) or weakened by unrewarding experiences (E), allowing the wasp to develop a chemosensory search-pattern that matches the current host availability.

once a host has been encountered together with such an odour plume (e.g., Kaiser, Perez-Maluf, Sandoz, & Pham-Delegue, 2003; Takasu & Lewis, 2003). However, some wasp species restricted to plant-specialized host don't have this ability (e.g., Geervliet, Vreugdenhil, Dicke, & Vet, 1998).

Interestingly, laboratory experiments have revealed striking differences in learning speed and memory retention within (Koppik, Hoffmeister, Brunkhorst, Kiess, & Thiel, 2015; Thiel, Schlake, & Kosior, 2013) and between (Patt, Hamilton, & Lashomb, 1999; Tamo, Ricard, Held, Davison, & Turlings, 2006) parasitoid wasp species, and gene expression accounting for such differentiation is being explored (Van Vugt et al., 2015). Cognitive ability and the 'preparedness to learn' are most likely related to the specific environmental settings experienced by individual species, and the cost of learning and memory formation (Mery, 2013; Smid & Vet, 2016). Learning and memory dynamics is also plastic. The nature of the reward experienced during learning trials might also influence the strength of the learned association: with more valuable rewards, fewer repetitions are necessary until the information is memorized and/or until it enters long-term memory (Koppik et al., 2015; Kruidhof et al., 2012; Luo, Michaud, Li, Liu, & Zhang, 2013). Plasticity may be also linked to symbionts, as recently demonstrated with the tiny egg-parasitoid *Trichogramma brassicae* Bezdenko 1968, which has lower memory retention when infected with *Wolbachia* bacteria, possibly due to the costs of carrying *Wolbachia* (Farahani et al., 2016).

Experiments in field cages or under open field conditions have shown that wasps that had the opportunity to associate one of the plant odours present with the presence of hosts had a significantly higher probability of finding their hosts in semi-natural conditions as well as finding it faster (Kruidhof et al., 2015; Papaj & Vet, 1990). It has recently been suggested that training mass-reared wasps before releasing them for biocontrol purposes should be considered as a means of increasing wasp efficacy (Giunti et al., 2015; Kruidhof, Smid, Thiel, Hoffmeister, & Vet, 2014).

While olfactory orientation by female parasitoids towards host-related cues is probably the best studied learning response in parasitic wasps, these insects are also able to associate odours with food (nectar or honeydew) rewards (e.g., Canale et al., 2014; Ngumbi, Jordan, & Fadamiro, 2012; Patt et al., 1999; Takasu & Lewis, 1996). This ability has been studied in male wasps as well (Takasu, Rains, & Lewis, 2007). Finally, foraging parasitoids can learn to associate cues other than chemicals, such as colours or shapes with hosts, food or mates (e.g., Baeder & King, 2004; Benelli & Canale, 2012; Lucchetta, Bernstein, Thery, Lazzari, & Desouhant, 2008;

Oliai & King, 2000; Wäckers & Lewis, 1999). In some species, resources (prospective mates within the host's pupa or host egg masses for oviposition) are present in the wasp's environment for longer periods, but are suitable for exploitation only at a very specific time (i.e., at eclosion of the female or perfect age of the host egg). In these cases, wasps regularly revisit the resources, using learned landmark cues for orientation (Danci, Hrabar, Ikoma, Schaefer, & Gries, 2013; Van Nouhuys & Kaartinen, 2008).

3. THE PLANT, PLACE OF DEVELOPMENT

Physical and chemical plant traits can have both direct and indirect effects on entomophagous insects (Bottrell et al., 1998; Hare, 2002). Indirect effects are those that alter some aspect of herbivore abundance or quality, which in turn influences development/survivorship of insect natural enemies. Indirect effects can be further divided into 'density-mediated effects' that influence herbivore density (and ultimately natural enemy abundance) and 'trait-mediated indirect effects' that influence natural enemies through altering herbivore quality as a prey or host resource (Mooney & Singer, 2012). In general, but by no means exclusively, most studies have explored either the direct effects of plant physical and chemical traits on the foraging success of entomophagous insects or the indirect effects of plant chemical traits on the developmental success and survivorship of entomophagous insects.

3.1 Effects of Plant Physical Traits on Entomophagous Insect Development

While plant physical characteristics are well known to influence the foraging behaviour of both predators and parasitoids (Bottrell et al., 1998; Casas & Djemai, 2002; Price et al., 1980), far fewer studies have examined the effects of plant physical traits on the development of immature insect predators and parasitoids. In part, this is likely because relatively few studies have carefully distinguished between plant trait effects on oviposition decisions and trait effects on the development success and survivorship of immature entomophagous insects. Plant physical traits such as waxiness, trichomes and leaf thickness are expected to influence predators more than parasitoids, as most predators directly interact with the surface of the plant as they forage for prey. Several studies have demonstrated that predatory coccinellid beetle larvae develop more quickly and have lower mortality when developing on certain plant species or cultivars (e.g., Al-Zyoud, Tort, & Sengonca, 2005; Inbar & Gerling, 2008); however, in many studies, it is difficult to determine whether a physical aspect of the plant influences predator performance or if these effects are due to differences in defensive chemistry in the host plant. Yet, some plant physical characters, notably leaf domatia – small chambers produced by the plant that house predatory insects and mites, clearly have positive effects on some predators (Agrawal, Karban, & Colfer, 2000; O'Dowd & Willson, 1991). Such domatia appear to be important in reducing parasitism and predation of the predators' eggs and nymphs, which in turn results in higher predation of herbivores (Agrawal et al., 2000; Romero & Benson, 2005).

While most studies of the impact of plant physical traits on entomophagous insect development have focused on predators, a few studies suggest that plant physical traits can also influence parasitoid development. A study of *Encarsia pergandiella* Howard 1907, a parasitoid of the silverleaf whitefly *Bemisia argentifolii* Bellows & Perring, found that the number of parasitoids that successfully develop was substantially greater on a glossy (low-wax) variety of collard (*Brassica oleracea* L.) compared to a normal wax variety, even though the two varieties differ by only one gene for waxiness (McAuslane, Simmons, & Jackson, 2000). The difference between glossy versus normal wax plants could have been due to differential oviposition by adult females, differential mortality of developing parasitoids or a combination. In a study of two cultivars of poinsettia (*Euphorbia pulcherrima* Willd. ex Klotzsch) that differed by 15%, in trichome density, successful development of several species of *Encarsia* on *B. argentifolii* was greater on the cultivar with higher trichome densities, despite the fact that parasitism rates were greater on the poinsettia cultivar with lower trichome densities (Heinz & Parrella, 1994). This difference in the correlation between trichome density and attack rate versus trichome density and parasitoid emergence (survivorship) is suggestive of an effect of trichomes on parasitoid development.

3.2 Effects of Plant Defensive Chemistry on Entomophagous Insect Development

The effects of plant chemistry on parasitoids and predators have been well studied (Dicke & Baldwin, 2010; Harvey, 2005; Hunter, 2003; Ode, 2006, 2013; Price et al., 1980). Plant antiherbivore toxins can negatively affect the development and survivorship of natural enemies of herbivores. Studies of plant toxin effects on the third trophic level have taken a variety of approaches including (1) comparisons of plant populations or cultivars (which differ in their secondary plant chemistry profiles) in terms of their effects on herbivores and their natural enemies (e.g., Gols, Van Dam,

Raaijmakers, Dicke, & Harvey, 2009; Harvey & Gols, 2011; Ode, Berenbaum, Zangerl, & Hardy, 2004; Ode, Harvey, Reichelt, Gershenzon, & Gols, 2016), (2) artificial diet studies where known quantities of specific plant toxins are added to a herbivore's diet (e.g., Barbosa, Gross, & Kemper, 1991; Barbosa et al., 1986; Campbell & Duffey, 1979; Lampert, Zangerl, Berenbaum, & Ode, 2008, 2011) and (3) manipulation of the JA signalling pathway to alter the production of plant toxins through the use of exogenous applications of JA mimics (e.g., Thaler, 1999). The use of transformed lines, where genes in the JA pathway have been silenced, has been used in wild tobacco where it has been shown alter susceptibility to herbivores and attractiveness to natural enemies (e.g., Kessler, Halitschke, & Baldwin, 2004); however, we are unaware of application of this technique to explore how parasitoids and predators are affected by plant toxins.

Plant defensive chemistry can influence the development and survivorship of an entomophagous insect by either reducing the quality of its host/prey (e.g., smaller size, reduced nutritive quality) or by direct exposure to plant toxins that it encounters in the body of its host/prey (either unmetabolized or metabolic byproducts of ingested plant toxins) (Kaplan, Carrillo, Garvey, & Ode, 2016). Yet, relatively few studies have attempted to distinguish between these two possibilities. One approach has been to determine whether plant toxins pass unmetabolized into the haemolymph of a herbivore where it would be directly encountered by developing parasitoids. This has been demonstrated in several systems including the parasitoid *Hyposoter exiguae* (Viereck, 1912), which encounters the glycoalkaloid α-tomatine in its host *Heliothis zea* (Boddie, 1850) feeding on tomato (Campbell & Duffey, 1979), *Cotesia congregata* (Say, 1836) exposed to nicotine in its host *Manduca sexta* feeding on tobacco (Barbosa et al., 1986) and *Copidosoma sosares* Walker, 1837 encountering unmetabolized xanthotoxin in its host *Depressaria pastinacella* (Duponchel, 1838) when feeding on wild parsnip (Lampert et al., 2008, 2011; McGovern, Zangerl, Ode, & Berenbaum, 2006). In each of these cases, parasitoids experienced decreased survivorship, decreased body sizes and sometimes morphological deformities apart from any effects of reduced host quality. Whether parasitoids are able to metabolize plant toxins that they encounter in their hosts is largely unknown. In the case of *C. sosares*, neither larvae nor embyros show any capacity to metabolize the furanocoumarin xanthotoxin (Lampert et al., 2008, 2011; McGovern et al., 2006). At least in the cases of parasitoids of specialist herbivores, efficient metabolism and/or excretion of plant toxins by these herbivores may reduce significant exposure to plant toxins.

Some of the more spectacular examples of how plant toxins can negatively affect parasitoids are found in herbivores that sequester plant chemicals as defence against their natural enemies. Sequestration — the selective uptake, transport and storage of plant toxins — can result in 'enemy-free space' for herbivores, a process well documented for predators (Dyer, 1995; Nishida, 2002) and to a lesser extent for parasitoids (Gauld, Gaston, & Janzen, 1992). Sequestration is a widespread phenomenon, documented in over 250 insect species feeding on plants from over 40 families (Opitz & Müller, 2009). Although it is well established that specialist herbivores are more efficient than generalist ones at metabolizing and/or excreting plant toxins (Lampert et al., 2011; Ratzka, Vogel, Kliebenstein, Mitchell-Olds, & Kroymann, 2002; Wittstock et al., 2004), recent evidence suggests that there are also specialist herbivores among sequestering herbivores, and their efficiency at sequestering plant toxins thereby confers greater protection against parasitoids (Lampert, Dyer, & Bowers, 2014; Züst & Agrawal, 2016). The question of whether narrow host plant range leads to an ability to sequester or, if parasitism/predation pressure selects for this ability are best approached using phylogenetic studies and by gaining a better understanding of the physiological mechanisms of sequestration (Erb & Robert, 2016; Petschenka & Agrawal, 2016). Despite years of study of the ecological significance of sequestration, little is known about how insect herbivores transport, store and release bioactive plant origin defensive compounds (Erb & Robert, 2016).

A few studies have examined the role of sequestration of plant toxins on immune defence against parasitoids (e.g., Lampert, Dyer, & Bowers, 2010; Sime, 2002; Singer, Mace, & Bernays, 2009). For instance, the effects of sequestered chemicals have been assessed by measuring the encapsulation rates (host immune response) of inert silica beads, which have been widely used as proxies for parasitoid eggs. While such studies have provided valuable insight into how plant chemistry is related to herbivore's immune system (e.g., Quintero, Lampert, & Bowers, 2014; Smilanich, Dyer, Chambers, & Bowers, 2009; Vogelweith, Moreau, Thiéry, & Moret, 2015), parasitoids are much more than inert glass beads (Kaplan et al., 2016). Most parasitoids inject venoms along with their eggs during oviposition, and braconid and ichneumonid parasitoids also inject polydnaviruses — both of which are involved in the suppression of host immune responses (Drezen, Chevignon, & Huguet, 2014; Poirié, Colinet, & Gatti, 2014). Therefore, it is important to realize that parasitoids can be expected to evolve in response to changing selective pressures from host plant chemistry and host immune responses.

4. EFFECTS OF NATURAL ENEMIES ON PLANT DEFENCE TRAITS

The preceding sections of this chapter demonstrate the diversity of ways that plant traits, especially those involving antiherbivore chemistry, can influence the fitness of entomophagous insects. Such strong effects of plant traits on parasitoids, in particular, are expected as parasitoids are wholly dependent on their herbivorous hosts, which in turn generally feed on a single host plant. The vast majority of research involving multitrophic interactions has been from a 'bottom-up' perspective, focusing on the impact of plant traits on behavioural (e.g., foraging and acceptance decisions) and physiological (e.g., digestion, development) adaptations of carnivorous insects. Far fewer studies have considered top-down influences of entomophagous insects on plant traits. Therefore, the question of whether the presence of natural enemies of insect herbivores enhances plant fitness remains largely unanswered. Hare (2002) put forth several criteria necessary to demonstrate evolution of plant traits in response to entomophagous insects, including the presence of additive genetic variance for a plant trait that increases natural enemy success as well as enhances plant fitness.

Evidence for the potential selective impact of parasitoids on plant fitness remains scarce. On the one hand, the existence of many successful biological control programs against insect herbivores suggests that plant biomass and fruit/seed production is higher in the presence of natural enemies (DeBach & Rosen, 1991; Van den Bosch, 1971; Van Driesche et al., 2008; Williams, Arrendondo-Bernal, & Rodríguez-del-Bosque, 2013). However, it is important to realize that the effects of parasitoids and predators on individual plant fitness are rarely, if ever, measured in biological control, preventing an assessment of whether natural enemies can select for plant traits in such situations (Van der Meijden & Klinkhamer, 2000). A handful of field studies using exclusion cages to prevent parasitism have experimentally demonstrated that the presence of parasitoids can reduce rates of herbivory and increase plant fitness (Gómez & Zamora, 1994; Stiling & Moon, 2005). Similarly, a greenhouse study of *Arabidopsis thaliana* (L.) showed that herbivory by unparasitized *Pieris rapae* (Linnaeus) resulted in decreased lifetime seed production compared to plants attacked by *P. rapae* parasitized by the solitary *Cotesia rubecula* (Marshall, 1885) (Van Loon, De Boer, & Dicke, 2000).

In general, parasitism by solitary parasitoids (a single parasitoid offspring developing in or on a host) results in reduced feeding damage to an individual plant by the herbivore (e.g., Van Loon et al., 2000). However, parasitism by gregarious parasitoids (multiple parasitoid offspring per host) can result in either decreased or increased herbivory and seed production. This is because in some cases large broods of gregarious parasitoids increase the amount the herbivore eats (Karowe & Schoonhoven, 1992; Smallegange, Van Loon, Blatt, Harvey, & Dicke, 2008). In the case of polyembryonic parasitoids, where brood sizes can exceed 1000 clone mates, feeding damage by parasitized herbivores can be dramatically higher than that by unparasitized herbivores. For example, *Trichoplusia ni* (Hübner, 1803) parasitized by the polyembyronic *Copidosoma floridanum* (Ashmead, 1900) feeds approximately 50% more than unparasitized *T. ni* and induces higher production of the indole glucosinolates when the herbivores feed on wild cabbage *Brassica oleracea* (L.) (Ode et al., 2016). While the effects on plant fitness were not measured by Ode et al. (2016), it is likely that plants attacked by parasitized herbivores suffered decreased fitness as seen in Smallegange et al. (2008). While parasitoids that induce increased consumption by herbivores may, in the short term, harm plant fitness, it is possible that in the long-term plants in populations whose herbivores experience chronic parasitism pressure may benefit. Such an argument has been made in another system involving a polyembyronic species, *C. sosares*, which attacks the parsnip webworm — a specialist herbivore of the wild parsnip *Pastinaca sativa* L. Webworms parasitized by *C. sosares* feed 55% more than unparasitized larvae (McGovern et al., 2006). Nevertheless, parsnip plants in western Europe (where this system is native) experience reduced levels of chronic herbivory and, consequently, invest less in costly furanocoumarin defences compared to plants in eastern North America where herbivory is higher and the parasitoid is absent (Ode et al., 2004).

5. THE PLANT AS FOOD SOURCE

Many members of the third trophic level are not exclusively carnivorous but also consume plant products such as pollen, nectar or honeydew and plant tissue. Use of these plant-produced or plant-derived food sources can be mandatory or optional and occur occasionally during certain phases of insect development, or regularly throughout their life. Among the optional consumers are, for example, species of predatory mites, spiders, bugs or ants and a number of parasitoid species that feed primarily on their hosts (host feeding). Among the mandatory consumers are hoverflies and many species of parasitoids that do not feed on hosts as adult and thus solely depend on food resources from plants. Plants provide essential food for their survival and ability to forage as well as their ability to mature eggs (see Jervis & Kidd, 1996; or Wäckers, van Rijn, & Bruin, 2005 for reviews). Flowers' visual and olfactory signals can be used by entomophagous insects. Thus, unfed females of the parasitoid *C. rubecula* are particularly attracted to yellow, the most common flower colour (Weevers, 1952); they are also able to recognize the smell of the flowers (Wäckers & Lewis, 1999). The different sources of food derived from plants have characteristics that determine their use by entomophagous insects.

5.1 Characteristics and Use of Pollen

Pollen is mainly composed of proteins and free amino acids, but also contains some sterols, lipids and starch (Wäckers, 2001). Some predators such as ladybirds, adult hoverflies, predatory bugs and mites supplement their diet by consuming pollen (Wäckers et al., 2005). Although this consumption is less prevalent among parasitoids, some species like *Edovum puttleri* Grissel 1981 and *Pediobius foveolatus* (Crawford, 1912) (Hymenoptera, Eulophidae), parasitoids of beetles, also feed on pollen (Patt, Hamilton, & Lashomb, 1997). The relative proportion of these elements and their nature differ according to the species, and they influence the development and reproduction of entomophagous insects consuming pollen (Wäckers et al., 2005). As pollen is produced by insect-pollinated or wind-pollinated plant species, its availability and accessibility varies. In the case of insect-pollinated plants, which actively recruit pollinators, pollen is signalled by the colour or odour of the flower. The pollen itself can also emit odours, which are attractive at short distances (Dobson & Bergström, 2000).

5.2 Characteristics and Use of Floral Nectar

The nectar produced by flowers, usually in order to attract pollinators, can also be consumed by predators and parasitoids. Like pollen of entomophilous plants, nectar is easily detectable by entomophagous insects. Nectar is a major source of carbohydrates and can also contain amino acids, proteins, lipids and even vitamins. Nectar carbohydrates are mainly fructose, glucose and sucrose. Its consumption has been shown to increase longevity and fecundity of parasitoids (Araj & Wratten, 2015) and predators (Robinson, Jonsson, Wratten, Wade, & Buckley, 2008), but some sugars such as raffinose, galactose, mannose or xylose, present even in small quantities, can have negative effects on different life history traits of entomophagous insects (Wäckers, 2001). Depending on the floral architecture, accessibility of nectaries often limits the use of this resource, because the mouthparts of many entomophagous insects are short or their head is relatively large. Thus, the parasitoids *E. puttleri* and *P. foveolatu* differ in their ability to use the nectar: the first can only access exposed nectaries, while the second can also access nectaries hidden under petals and stamens (Patt et al., 1997). Their slight difference in size (1.5–2.5 mm vs 2–3.5 mm long, respectively) is sufficient to result in these contrasted abilities. Neither species can, however, access nectaries located at the bottom of tubular flowers.

The availability of floral nectar varies greatly in time and space. It depends particularly on the plant community composition, the phenology of species in these communities and climatic conditions (Brandenburg, Dell'Olivo, Bshary, & Kuhlemeier, 2009). Visiting flowers, whether to search for pollen or nectar, is not without risk. Flowers are indeed hunting sites for other entomophagous arthropods such as crab spiders (Thomisidae).

5.3 Characteristics and Use of Extrafloral Nectar

Flowers are not the only plant organs to carry nectaries; extrafloral nectaries can be found on the leaves, stalks, bracts and fruits of some plants. These nectaries are usually visited by predators and parasitoids, but not by pollinators. Predators using these food resources include ants, mites, lacewings, wasps, beetles and even some species of spiders (Heil, 2015; Wäckers et al., 2005). Extrafloral nectar has several advantages over the floral nectar: it is easy to access (nectaries are apparent and are found on the surface of plant organs that carry them), it is abundant (often much more than floral nectar) and it is available for a very large part of the vegetative growth of plants and not only at the time of flowering. The discovery of extrafloral nectar is however more difficult because nectaries are generally not associated with obvious visual or olfactory cues like in flowers (Belz, Kölliker, & Balmer, 2013). Some nectaries nevertheless emit odours detected at a short distance by parasitoids (Stapel, Cortesero, De Moraes, Tumlinson, & Lewis, 1997) or have special coloration that could help locate them on the plant. The composition of the floral nectar and that of extrafloral nectar is fairly similar: the latter contains sugars (mainly fructose, glucose and sucrose, and sometimes raffinose, galactose, arabinose or xylose), amino acids, lipids and vitamins. Thanks to these nutrients, the oophagous parasitoid *Gryon pennsylvanicum* (Ashmead, 1893) (Hymenoptera, Scelionidae) can live more than 17 days if it consumes the extrafloral nectar of zucchini (*Cucurbita pepo* L. (Cucurbitaceae)), whereas it survives on average 3 days without food (Olson & Nechols, 1995). Furthermore, parasitoids feeding on zucchini nectar produced three times the eggs as parasitoids that did not feed on nectar. Despite these interesting nutrients, extrafloral nectar does not contain all the elements essential to the diet of entomophagous insects. The low content of essential amino acids can be seen as an adaptation of the plants with extrafloral nectaries, since it encourages predators to seek other sources of protein including herbivorous prey (Wäckers et al., 2005). For example, the production of extrafloral nectar greatly increases levels of predation and parasitism of phytophagous arthropods (Jamont, Crépellière, & Jaloux, 2013; Rezende, Venzon, Perez, Cardoso, & Janssen, 2014) and is considered as an indirect defence strategy, as well as the emission of volatile compounds attracting entomophagous insects. Moreover, like these volatile compounds, nectar production can be seen as an induced defence, because the attack of the plant by a herbivore increases the secretion of nectar (Fig. 3) (Mathur et al., 2013).

5.4 Characteristics and Use of Honeydew

Even if it is not produced directly by plants, honeydew excreted by the Sternorrhyncha hemipterans, such as aphids, can play an important role in supplying entomophagous insects with sugars produced by plants and normally circulating in the phloem. This diet can be particularly useful outside the flowering periods. In a study on the parasitoids *C. glomerata* and *Microplitis mediator* (Haliday, 1834) (Hymenoptera, Braconidae), 80% and 50% (respectively) of individuals caught in cabbage fields had fed on honeydew (Wäckers & Steppuhn, 2003). The main constraint to feeding on honeydew is its detectability, because producers are generally less visible and fragrant than flowers (Wäckers et al., 2005). In addition, the nutritional quality of this resource seems generally lower than that of nectar (Wäckers, Van Rijn, & Heimpel, 2008). The composition and relative proportions of the various sugars can be modified by the sap-sucking insect. Some honeydews, for example, do not contain any glucose or sucrose (Wäckers, 2001). Honeydew also contains generally more amino acids than does nectar, but they are often nonessential amino acids, essential ones being assimilated by the sap-sucking insects. These differences in nutritional quality can have a significant impact on life history traits of entomophagous insects. For example, the parasitoid *Diadegma insulare* (Cresson, 1865) (Hymenoptera, Ichneumonidae) lives only 6–7 days if fed exclusively honeydew from *Aphis* (*Aphis*) *glycines* Matsumura, 1917 (Hemiptera, Aphididae) against more than

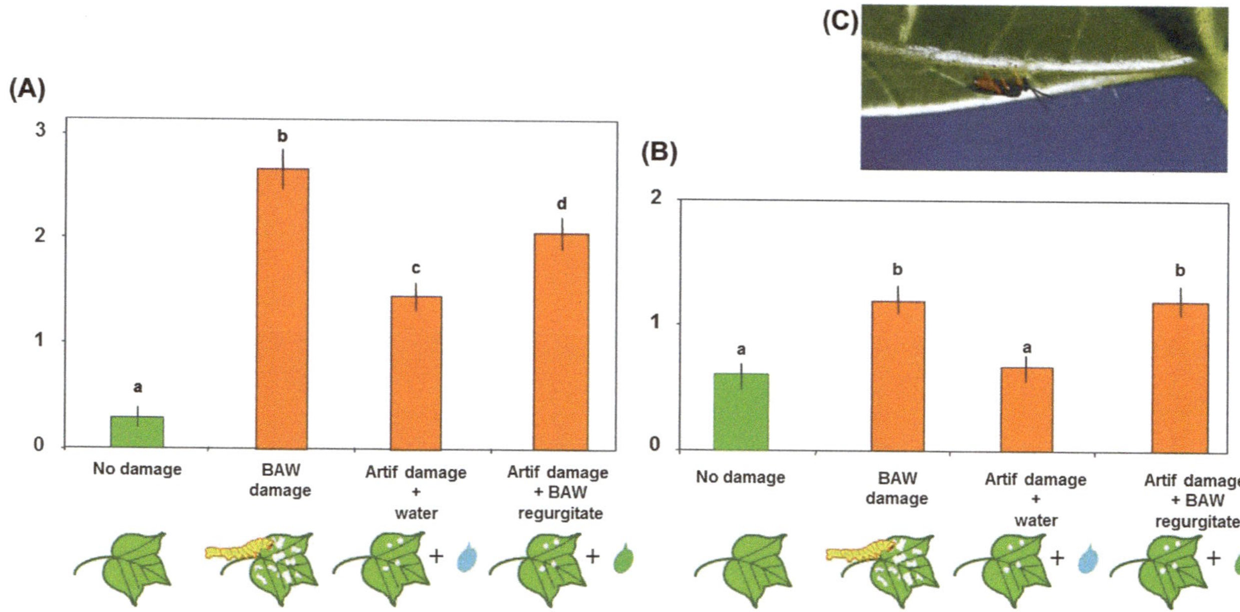

Figure 3 Average amount of extrafloral nectar per plant (+ESM) collected from damaged leaves (A) and undamaged leaves (B) of cotton plants that received *Spodoptera exigua* (BAW) damage, artificial damage + BAW regurgitate, artificial damage + water, and no damage for 7 days. Different letters in bars indicate significant difference in nectar production between treatments (Tukey's Multiple Comparison Test, $P < .05$). (C) The parasitic wasp *Microplitis demolitor* feeding on extrafloral nectar located under a cotton leaf. *(A and B) Unpublished data from A.M. Cortesero, J.O. Stapel, F.L. Wäckers, & W.J. Lewis. (C) Photo: A.M. Cortesero.*

15 days if fed buckwheat floral nectar (*Fagopyrum esculentum* Moench (Polygonaceae)) (Lee, Heimpel, & Leibee, 2004).

6. THE ROLES OF LOCAL PLANT COMPOSITION AND LANDSCAPE COMPLEXITY ON DIVERSITY, ABUNDANCE AND THERMOTOLERANCE OF ENTOMOPHAGOUS INSECTS

Beyond the influence of characteristics of plant individuals on the third trophic level, entomophagous insect behaviour as well as population and community dynamics is also influenced by the composition of the plant community. At a large scale, plant biomes (such as tropical forest, savanna or boreal forest) harbour different insect communities. Within a biome, the distribution of plants and habitat patches across a landscape is important. Finally, at the smallest scale is the heterogeneity of the foraging environment experienced by an individual foraging insect (Van Nouhuys & Hanski, 2002).

6.1 Insect Communities and Biomes

As for other species, the diversity of natural enemies has been found to be greatest in the tropics. Though the explanation for the latitudinal gradient in species diversity is still unresolved (Brown, 2014), it has been especially well illustrated over the last decade by large-scale barcoding projects revealing the great diversity parasitoids in tropical forests (Smith et al., 2008). These contrast strongly with what is being found in the much less structured high arctic insect community (Wirta et al., 2015). Within a latitudinal zone, there is also variation of natural enemy diversity associated with vegetation type. Hawkins (1994) conducted an extensive literature review to identify the determinants of species diversity and abundance of parasitoids. He found that while herbivore—host feeding niche (e.g., leaf eaters, leaf miners, gall-making insects) was very important, the type of plant (herbs, shrubs, trees) was not. However, larger plants, such as trees, have greater diversity than smaller plants simply because of their size and because their physical architecture provides a variety of habitats for a diversity of herbivores, and consequently a diversity of natural enemies.

6.2 Landscape-Scale Habitat Complexity and Fragmentation

The effects of plants on entomophagous insects on a landscape scale have been primarily approached in two ways: through the roles of landscape complexity and through the connectivity of suitable habitats in the landscape. These aspects will be more developed in chapter: Plant-Insect Interactions in a Changing World by Pincebourde et al. (2016). The study of how landscape complexity affects entomophagous insects had mostly been driven by the interest in controlling insect pests of agricultural crops (Tscharntke et al., 2008). A series of studies in which habitat complexity is quantified as the percentage of land devoted to agriculture in a landscape have demonstrated that abundance of herbivorous pests of agricultural crops is generally independent of landscape complexity or increase of landscape simplification (a larger fraction of the land that is agricultural) (reviewed in Bianchi, Booij, & Tscharntke, 2006; Chaplin-Kramer, O'Rourke, Blitzer, & Kremen, 2011). Entomophagous insects, on the other hand, tend to increase in abundance or diversity (Gardiner et al., 2009) or rate of parasitism (e.g., Marino & Landis, 1996; Thies, Steffan-Dewenter, & Tscharntke, 2003) with increasing habitat complexity (a smaller fraction of the land agricultural). This association is stronger for generalist natural enemies than for natural enemies with narrow host ranges (Chaplin-Kramer et al., 2011). In some cases, the increase in natural enemy diversity with habitat complexity is associated with subsequent increased pest control (Rusch et al., 2016). The simplest explanation for the overall pattern is that the crop provides a large resource for growth of the herbivore population (Rand, Waters, Blodgett, Knodel, & Harris, 2014), but that the entomophagous insects benefit from alternate food (both alternate host or prey species and nectar for adult parasitoids) and shelter, so they thrive in a more diverse landscape. There are other more subtle aspects of this pattern that have been explored (see Marino, Landis, & Hawkins, 2006; Tscharntke et al., 2012).

The role of habitat connectivity in determining the community structure and population dynamics of entomophagous insects has mostly been explored in the context of conservation and habitat degradation due to fragmentation. As suitable habitat becomes fragmented, there is less of it, it becomes inaccessible to species that don't disperse easily between fragments, and the habitat quality tends to decrease as the amount of edge increases. All of these things lead to decreased abundance of individual species and thus decreased species diversity (Fahrig, 2003; Hanski, 2015). Entomophagous insects experience a more fragmented landscape than do their hosts because the hosts are present in only a fraction of the available habitat. Thus we expect that, like other higher trophic level species, they suffer more from habitat fragmentation than do species at lower trophic levels (Holt, Lawton, Polis, & Martinez, 1999; Komonen, Penttila, Lindgren, & Hanski, 2000). This has proven to be the case for some parasitoids (Cronin & Reeve, 2005) and predators (Langellotto & Denno, 2004), but it is not universal because generalist natural enemies may use more than one habitat type (Van Nouhuys, 2005). Furthermore, some specialized natural enemies are well adapted to using a host that is spread out in the landscape (reviewed, for parasitoids, in Nair & Van Nouhuys, 2016).

6.3 Local-Scale Plant Heterogeneity

Most entomophagous insects forage at a scale larger than an individual plant, especially if we consider both their juvenile and adult movements. Thus, they must locate prey or hosts on a plant, but they also have to narrow their search to that plant. Heterogeneity at the scale of a patch of plants has generally been found to reduce foraging efficiency of predators (Kareiva, 1987) and parasitoids (Gols et al., 2005; Kruidhof et al., 2015). This can be explained by olfactory cues and associative learning being obscured by surrounding plants and the nonhost herbivores on those plants. Alternatively, the high potential resource concentration of a homogeneous environment may attract entomophagous insects and cause them to stay in the area rather than leaving (Hambäck & Englund, 2005).

Nonetheless, local plant heterogeneity can also be advantageous to entomophagous insects that use more than one type of plant, or type of vegetation. For instance, parasitoids benefit from plants that provide nectar even if no hosts are present (Bianchi & Wäckers, 2008). Both predators and parasitoids might prey on insects that live on a variety of plants, at different times, so a heterogeneous environment may be a more stable resource than a homogeneous one (Tylianakis, Tscharntke, & Klein, 2006). Finally, predators and some parasitoids need an overwintering environment that might be different than their feeding environment. This has been demonstrated in agricultural setting where hedgerows provide overwintering sites for predators (e.g., Gareau, Letourneau, & Shennan, 2013).

6.4 Plant-Made Microclimate and Consequences on the Third Trophic Level

Climate changes such as variations of precipitation and temperature patterns, variation of CO_2, O_3, methane in the atmosphere, can affect the third trophic level by their impact on plant quality, which in turn affect herbivores and their parasitoids. For example, Yarnes and Boecklen (2006) showed that in warmer, dryer sites, trees exhibit greater seasonal variation in foliar nitrogen which was found to be positively correlated to the parasitism rate on the leaf miners *Phyllonorycter* spp. Calatayud, Polonia, Seligmann, and Bellotti (2002) showed that reducing water supply to cassava plants favoured the growth and development of the mealybug *Phenacoccus herreni* Cox & Williams, 1981 (Sternorrhyncha, Pseudococcidae) and impaired the development of its parasitoids (increased host immune resistance, reduction in the size of the adult parasitoids). The emission of plant volatiles is influenced by air pollution and it can modify their attractiveness for the third trophic level (Blande, Holopainen, & Niinemets, 2014).

In addition to the effects of climate on plant quality and consequences on herbivore parasitism or predation, plants themselves create a microclimate at the local habitat level or at the landscape level, which conditions thermal responses of insects. Air temperature can be several Celsius degrees lower than temperature measured on the leaf surface and inside structures like leaf miner gallery, due to gas exchanges (Pincebourde & Casas, 2006) and light reflection/absorption balance. The temperature pattern on the plant also depends on the interaction with herbivorous insect development, as shown for the Lepidoptera *M. sexta* (Linnaeus, 1763) which increases leaf temperature relative to air depending on larval instars (Woods, 2013). Vegetation cover offers a microclimate which can buffer atmospheric variations depending on pant height, density, architecture and colour, and benefits the plant dwelling insects, whatever their trophic level (Suh, Orr, Van Duyn, & Borcher, 2002; see Pumarino et al., 2015 for a review in agroforestry). Thermal refuges created by plants could play a growing role in increasing the resilience of the ecosystem to climate changes (increase of extreme climatic events such as heat or cold waves, Easterling et al., 2000, 1997; IPCC, 2013; Karl, Knight, & Baker, 2000, Walther et al., 2002), and allowing some individuals to escape thermal stresses.

The structure of the vegetation can create microclimates at the landscape scale that can affect the thermotolerance of phytophagous insects and their natural enemies (Tougeron, Van Baaren, Burel, & Alford, 2016). This latter case is detailed below.

Several studies have shown that vegetation structure can affect microclimates at the landscape level. For example, the role of hedges as windbreaks and in reducing freezing has been widely confirmed in agricultural landscapes. There is also evidence that the mean temperature and extreme temperatures in wooded and open areas differ over both daily and season scales (Argent, 1992; Chen et al., 1999; Quénol & Beltrando, 2006). Few studies have investigated the effects of landscape structure on thermal tolerance of ectotherms such as entomophagous insects, which are highly dependent on the external temperature to ensure their biological functions. Their maintenance in the environment and their resistance to unfavourable temperatures are ensured by their physiological capacity of thermal tolerance. In an agricultural environment in Britain, Tougeron et al. (2016) investigated the effects of the landscape structure and composition on thermal tolerance of three groups of insects: aphids, their natural enemies (braconid parasitic wasps) and carabid beetles. Individuals were sampled along a gradient of landscape complexity and their physiological thermotolerance was measured according to the experimental design of Powell and Bale (2006). For the collected braconids, insects originating from open landscapes were more thermotolerant than those collected in woody landscapes, regardless of their species (Fig. 4). This landscape effect was however not found in aphids or carabids, for which the behavioural thermoregulation was invoked: carabids can indeed buffer the effects of thermal stress using their mobility to find shelters in different parts of their habitats, and aphids can resist to cold temperatures thanks to their ability to stay on their host plant (Alford, Andrade, Georges, Burel, & Van Baaren, 2014). These studies showed that landscape structure could impact the

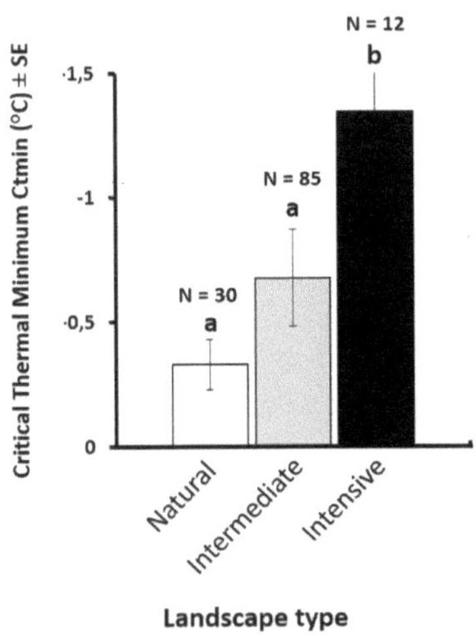

Figure 4 Average critical thermal minimum of braconid parasitoids of aphids in winter, depending on the type of landscape, (Ctmin $\pm$ standard error). The number of individuals tested is given for each category. The letters show significant differences between landscapes ($\alpha = 0.05$).

thermotolerance of some ectotherm species but not all, depending on their behavioural thermoregulation capacities. It suggests that landscape management could be used to favour some species to increase ecosystem services.

7. CONCLUSION

This chapter has given an overview of the multiple ways, biological processes and scales at which entomophagous insects interact with the plants of their herbivorous hosts or prey, and more widely with the vegetal community they live in. In conclusion, we would like to point out recent focuses and emerging approaches that will contribute to innovative management of natural enemies.

There is a complex food web centred on the plant and current researches investigate biochemical and population functioning of trophic networks between plants, herbivorous, primary and secondary entomophagous insects and pathogens as well as symbionts microorganisms. Identification of plant quality changes in response to interacting plant-dwelling organisms is important for understanding the outcome on fitness components of natural enemies like odour-guided host selection and development (Pashalidou et al., 2015). Studies of multitrophic interactions are called for in the context of sustainable agriculture. They can use high-throughput technologies for exhaustive identification of molecular changes at play during these trophic interactions. Such knowledge encompassing ecological and molecular aspects facilitates the possibility of our breeding plants for traits beneficial to entomophagous insects.

Increasing researches consider how plant communities, their spatio-temporal landscape and surrounding micro- and macro-climates condition the abundance and diversity of natural enemies' communities. These studies benefit from improving tools and methods to acquire and manage big data sets and from current progress in mathematical modelling. Ability to develop scenarios that depend on multiple environmental factors will help manage agricultural landscapes and forest habitats for efficient conservation of biological control and maintenance of ecosystem services. Including species' genetic diversity data in the models will help to simulate spatially explicit eco-evolutionary dynamics in complex food webs (Moya-Larano, 2011).

ACKNOWLEDGEMENTS

We are grateful to Alice Arnauld de Sartre for cross-checking and editing the references listed below. We thank Johnnie Van der Berg for his picture of *Iphiaulax* and Claude-Julie Parisot and Galaxie Presse-Arte France for their pictures of *Cotesia sesamiae*.

REFERENCES

Afsheen, S., Wang, X., Li, R., Zhu, C.-S., & Lou, Y.-G. (2008). Differential attraction of parasitoids in relation to specificity of kairomones from herbivores and their by-products. *Insect Science, 15*, 381–397.

Agrawal, A. (2000a). Mechanisms, ecological consequences and agricultural implications of tri-trophic interactions. *Current Opinion in Plant Biology, 3*, 329–335.

Agrawal, A. A. (2000b). Benefits and costs of induced plant defense for Lepidium virginicum (Brassicaceae). *Ecology, 81*, 1804–1813.

Agrawal, A. A., Karban, R., & Colfer, R. G. (2000). How leaf domatia and induced plant resistance affect herbivores, natural enemies and plant performance. *Oikos, 89*, 70–80.

Alford, L., Andrade, T. O., Georges, R., Burel, F., & Van Baaren, J. (2014). Could behaviour and not physiological thermal tolerance determine winter survival of aphids in cereal fields? *PLoS One, 9*, e114982.

Al-Zyoud, F., Tort, N., & Sengonca, C. (2005). Influence of host plant species on *Bemisia tabaci* (Genn.) (Hom., Aleyrodidae) on some of the biological and ecological characteristics of the entomophagous *Serangium parcesetosum* Sicard (Col., Coccinellidae). *Journal of Pest Science, 78*, 25–30.

Appel, H. M., Maqbool, S. B., Raina, S., Jagadeeswaran, G., Acharya, B. R., Hanley, J. C., Jr., … Schultz, J. C. (2014). Transcriptional and metabolic signatures of *Arabidopsis* responses to chewing damage by an insect herbivore and bacterial infection and the consequences of their interaction. *Frontiers in Plant Science, 5*, 441.

Araj, S. E., & Wratten, S. D. (2015). Comparing existing weeds and commonly used insectary plants as floral resources for a parasitoid. *Biological Control, 81*, 15–20.

Argent, R. M. (1992). The influence of a plant canopy on shelter effect. *Journal of Wind Engineering and Industrial Aerodynamics, 44*, 2643–2653.

Baeder, J. M., & King, B. H. (2004). Associative learning of color by males of the parasitoid wasp *Nasonia vitripennis* (Hymenoptera: Pteromalidae). *Journal of Insect Behavior, 17*, 201–213.

Barbosa, P., Gross, P., & Kemper, J. (1991). Influence of plant allelochemicals on the tobacco hornworm and its parasitoid *Cotesia congregata*. *Ecology, 72*, 1567–1575.

Barbosa, P., Saunders, J. A., Kemper, J., Trumbule, R., Olechno, J., & Martinat, P. (1986). Plant allelochemicals and insect parasitoids: Effects of nicotine on *Cotesia congregata* (Say) (Hymenoptera: Braconidae) and *Hyposoter annulipes* (Cresson) (Hymenoptera: Ichneumonidae). *Journal of Chemical Ecology, 12*, 1319–1328.

Barron, A. B. (2001). The life and death of Hopkins' host-selection principle. *Journal of Insect Behavior, 14*, 725–737.

Belz, E., Kölliker, M., & Balmer, O. (2013). Olfactory attractiveness of flowering plants to the parasitoid *Microplitis mediator*: potential implications for biological control. *BioControl, 58*, 163–173.

Benelli, G., & Canale, A. (2012). Learning of visual cues in the fruit fly parasitoid *Psyttalia concolor* (Szepligeti) (Hymenoptera: Braconidae). *BioControl, 57*, 767–777.

Bianchi, F. J. J. A., Booij, C. J. H., & Tscharntke, T. (2006). Sustainable pest regulation in agricultural landscapes: a review on landscape composition, biodiversity and natural pest control. *Proceedings of the Royal Society B: Biological Sciences, 273*, 1715–1727.

Bianchi, F. J. J. A., & Wäckers, F. L. (2008). Effects of flower attractiveness and nectar availability in field margins on biological control by parasitoids. *Biological Control, 46*, 400–408.

Blande, J. D., Holopainen, J. K., & Niinemets, U. (2014). Plant volatiles in polluted atmospheres: Stress responses and signal degradation. *Plant Cell and Environment, 37*, 1892–1904.

Bottrell, D. G., Barbosa, P., & Gould, F. (1998). Manipulating natural enemies by plant variety selection and modification: A realistic strategy? *Annual Review of Entomology, 43*, 347–367.

Braasch, J., & Kaplan, I. (2012). Over what distance are plant volatiles bioactive? Estimating the spatial dimensions of attraction in an arthropod assemblage. *Entomologia Experimentalis et Applicata, 145*, 115–123.

Brandenburg, A., Dell'Olivo, A., Bshary, R., & Kuhlemeier, C. (2009). The sweetest thing: advances in nectar research. *Current Opinion in Plant Biology, 12*, 486–490.

Brown, J. H. (2014). Why are there so many species in the tropics? *Journal of Biogeography, 41*, 8–22.

Calatayud, P.-A., Polonia, M. A., Seligmann, C. D., & Bellotti, A. C. (2002). Influences of water-stressed cassava on *Phenacoccus herenii* and their associated parasitoids. *Entomologia Experimentalis et Applicata, 102*, 163–175.

Campbell, B. C., & Duffey, S. S. (1979). Tomatine and parasitic wasps: Potential incompatibility of plant antibiosis with biological control. *Science, 205*, 700–702.

Canale, A., Geri, S., & Benelli, G. (2014). Associative learning for host-induced fruit volatiles in *Psyttalia concolor* (Hymenoptera: Braconidae), a koinobiont parasitoid of tephritid flies. *Bulletin of Entomological Research, 104*, 774–780.

Carroll, M. J., Schmelz, E. A., Meagher, R. L., & Teal, P. E. (2006). Attraction of *Spodoptera frugiperda* larvae to volatiles from herbivore-damaged maize seedlings. *Journal of Chemical Ecology, 32*, 1911–1924.

Casas, J., & Djemai, I. (2002). Plant canopy architecture and multitrophic interactions. In T. Tscharntke, & B. A. Hawkins (Eds.), *Multitrophic level interactions*. Cambridge, UK: Cambridge University Press.

Chaplin-Kramer, R., O'Rourke, M. E., Blitzer, E. J., & Kremen, C. (2011). A meta-analysis of crop pest and natural enemy response to landscape complexity. *Ecology Letters, 14*, 922–932.

Chen, J. Q., Saunders, S. C., Crow, T. R., Naiman, R. J., Brosofske, K. D., Mroz, G. D., ... Franklin, J. F. (1999). Microclimate in forest ecosystem and landscape ecology – Variations in local climate can be used to monitor and compare the effects of different management regimes. *Bioscience, 49*, 288–297.

Colazza, S., Fucarino, A., Peri, E., Salerno, G., Conti, E., & Bin, F. (2004). Insect oviposition induces volatile emission in herbaceous plants that attracts egg parasitoids. *Journal of Experimental Biology, 207*, 47–53.

Colazza, S., Salemo, G., & Wajnberg, E. (1999). Volatile and contact chemicals released by *Nezara viridula* (Heteroptera: Pentatomidae) have a kairomonal effect on the egg parasitoid *Trissolcus basalis* (Hymenoptera: Scelionidae). *Biological Control, 16*, 310–317.

Conti, E., Salemo, G., Bin, F., Williams, H. J., & Vinson, S. B. (2003). Chemical cues from *Murgantia histrionica* eliciting host location and recognition in the egg parasitoid Trissolcus brochymenae. *Journal of Chemical Ecology, 29*, 115–130.

Cronin, J. T., & Reeve, J. D. (2005). Host parasitoid spatial ecology: A plea for a landscape-level synthesis. *Proceedings of the Royal Society B: Biological Sciences, 272*, 2225–2235.

Cusumano, A., Weldegergis, B. T., Colazza, S., Dicke, M., & Fatouros, N. E. (2015). Attraction of egg-killing parasitoids toward induced plant volatiles in a multi-herbivore context. *Oecologia, 179*, 163–174.

Danci, A., Hrabar, M., Ikoma, S., Schaefer, P. W., & Gries, G. (2013). Learning provides mating opportunities for males of a parasitoid wasp. *Entomologia Experimentalis et Applicata, 149*, 229–240.

De Boer, J. G., & Dicke, M. (2006). Olfactory learning by predatory arthropods. *Animal Biology, 56*, 143–155.

De Rijk, M., Dicke, M., & Poelman, E. H. (2013). Foraging behaviour by parasitoids in multiherbivore communities. *Animal Behaviour, 85*, 1517–1528.

DeBach, P., & Rosen, D. (1991). *Biological control by natural enemies* (2nd ed.). Cambridge, UK: Cambridge University Press, 456 pp.

Desurmont, G. A., Xu, H., & Turlings, T. C. (2016). Powdery mildew suppresses herbivore-induced plant volatiles and interferes with parasitoid attraction in *Brassica rapa*. *New Phytologist, 210*, 1344–1356.

Dicke, M. (2016). Induced plant volatiles: Plant body odours structuring ecological networks. *New Phytologist, 210*, 10–12.

Dicke, M., & Baldwin, I. T. (2010). The evolutionary context for herbivore-induced plant volatiles: beyond the "cry for help". *Trends in Plant Science, 15*, 167–175.

Dicke, M., & Sabelis, M. W. (1988). Infochemicals terminology: based on cost-benefit analysis rather than origin of compounds? *Functional Ecology, 2*, 131–139.

Dicke, M., Sabelis, M. W., & De Jong, M. (1988). Analysis of prey preference in phytoseiid mites by using an olfactometer, predation models and electrophoresis. *Experimental & Applied Acarology, 5*, 225–241.

Dobson, H. E. M., & Bergström, G. (2000). Ecology and evolution of pollen odors. In A. Dafni, M. Hesse, & E. Pacini (Eds.), *Pollen and pollination*. Vienna, Autriche: Springer.

Drezen, J.-M., Chevignon, G., & Huguet, L. F. (2014). Origin and evolution of symbiotic viruses associated with parasitoid wasps. *Current Opinion in Insect Science, 6*, 35–43.

Dyer, L. A. (1995). Tasty generalists and nasty specialists? Antipredator mechanisms in tropical lepidopteran larvae. *Ecology, 76*, 1483–1496.

Easterling, D. R., Evans, J. L., Groisman, P. Y., Karl, T. R., Kunkel, K. E., & Ambenje, P. (2000). Observed variability and trends in extreme climate events: A brief review. *Bulletin of the American Meteorological Society, 81*, 417–425.

Easterling, D. R., Horton, B., Jones, P. D., Peterson, T. C., Karl, T. R., Parker, D. E., … Folland, C. K. (1997). Maximum and minimum temperature trends for the globe. *Science, 277*, 364–367.

Erb, M., & Robert, C. A. M. (2016). Sequestration of plant secondary metabolites by insect herbivores: molecular mechanisms and ecological consequences. *Current Opinion in Insect Science, 14*, 8–11.

Erb, M., Robert, C. A. M., Hibbard, B. E., & Turlings, T. C. J. (2011). Sequence of arrival determines plant-mediated interactions between herbivores. *Journal of Ecology, 99*, 7–15.

Fahrig, L. (2003). Effects of habitat fragmentation on biodiversity. *Annual Review of Ecology, Evolution and Systematics, 34*, 487–515.

Farahani, H. K., Ashouri, A., Goldansaz, S. H., Shapiro, M. S., Pierre, J. S., & Van Baaren, J. (2016). Decrease of memory retention in a parasitic wasp: an effect of host manipulation by *Wolbachia*? *Insect Science* (in press).

Gandolfi, M., Mattiacci, L., & Dorn, S. (2003). Preimaginal learning determines adult response to chemical stimuli in a parasitic wasp. *Proceedings of the Royal Society B: Biological Sciences, 270*, 2623–2629.

Gardiner, M. M., Landis, D. A., Gratton, C., DiFonzo, C. D., O'Neal, M., Chacon, J. M., … Heimpel, G. E. (2009). Landscape diversity enhances biological control of an introduced crop pest in the north-central USA. *Ecological Applications, 19*, 143–154.

Gareau, T. L., Letourneau, D. K., & Shennan, C. (2013). Relative densities of natural enemy and pest insects within California hedgerows. *Environmental Entomology, 42*, 688–702.

Gauld, I. D., & Bolton, B. (1988). *The Hymenoptera*. Oxford, GB: Oxford University Press, 332 pp.
Gauld, I. D., Gaston, K. J., & Janzen, D. H. (1992). Plant allelochemicals, tritrophic interactions and the anomalous diversity of tropical parasitoids: the "nasty" host hypothesis. *Oikos, 65*, 353–357.
Geervliet, J. B. F., Ariens, S., Dicke, M., & Vet, L. E. M. (1998). Long-distance assessment of patch profitability through volatile infochemicals by the parasitoids *Cotesia glomerata* and *C. rubecula* (Hymenoptera: Braconidae). *Biological Control, 11*, 113–121.
Geervliet, J. B. F., Vreugdenhil, A. I., Dicke, M., & Vet, L. E. M. (1998). Learning to discriminate between infochemicals from different plant-host complexes by the parasitoids *Cotesia glomerata* and *C. rubecula*. *Entomologia Experimentalis et Applicata, 86*, 241–252.
Gish, M., De Moraes, C. M., & Mescher, M. C. (2015). Herbivore-induced plant volatiles in natural and agricultural ecosystems: open questions and future prospects. *Current Opinion in Insect Science, 9*, 1–6.
Giunti, G., Canale, A., Messing, R. H., Donati, E., Stefanini, C., Michaud, J. P., & Benelli, G. (2015). Parasitoid learning: current knowledge and implications for biological control. *Biological Control, 90*, 208–219.
Godfray, H. C. J. (1994). In J. R. Krebs, & T. Clutton-Brock (Eds.), *Parasitoids: Behavioral and evolutionary ecology*. Princeton, USA: Princeton University Press, 473 pp.
Gols, R., Bukovinszky, T., Hemerik, L., Harvey, J. A., Van Lenteren, J. C., & Vet, L. E. M. (2005). Reduced foraging efficiency of a parasitoid under habitat complexity: Implications for population stability and species coexistence. *Journal of Animal Ecology, 74*, 1059–1068.
Gols, R., Van Dam, N. M., Raaijmakers, C. E., Dicke, M., & Harvey, J. A. (2009). Are population differences in plant quality reflected in the preference and performance of two endoparasitoid wasps? *Oikos, 118*, 733–743.
Gómez, J. M., & Zamora, R. (1994). Top-down effects in a tritrophic system: Parasitoids enhance plant fitness. *Ecology, 75*, 1023–1030.
Hambäck, P. A., & Englund, G. (2005). Patch area, population density and the scaling of migration rates: the resource concentration hypothesis revisited. *Ecology Letters, 8*, 1057–1065.
Hanski, I. (2015). Habitat fragmentation and species richness. *Journal of Biogeography, 42*, 989–993.
Hare, J. D. (2002). Plant genetic variation in tritrophic interactions. In T. Tscharntke, & B. A. Hawkins (Eds.), *Multitrophic level interactions* (pp. 8–43). Cambridge, UK: Cambridge University Press.
Hare, J. D. (2011). Ecological role of volatiles produced by plants in response to damage by herbivorous insects. In M. R. Berenbaum, R. T. Carde, & G. E. Robinson (Eds.), *Annual review of entomology* (Vol. 56, pp. 161–180).
Harvey, J. A. (2005). Factors affecting the evolution of development strategies in parasitoid wasps: The importance of functional constraints and incorporating complexity. *Entomologia Experimentalis et Applicata, 117*, 1–13.
Harvey, J. A., & Gols, R. (2011). Population-related variation in plant defense more strongly affects survival of an herbivore than its solitary parasitoid wasp. *Journal of Chemical Ecology, 37*, 1081–1090.
Hawkins, B. A. (1994). *Pattern and process in host-parasitoid interactions*. Cambridge: Cambridge University Press, 190 pp.
Heil, M. (2008). Indirect defence via tritrophic interactions. *New Phytologist, 178*, 41–61.
Heil, M. (2015). Extrafloral nectar at the plant-insect interface: A spotlight on chemical ecology, phenotypic plasticity, and food webs. *Annual Review of Entomology, 60*, 213–232.

Heil, M., & Karban, R. (2010). Explaining evolution of plant communication by airborne signals. *Trends in Ecology & Evolution, 25*, 137–144.

Heinz, K. M., & Parrella, M. P. (1994). Poinsettia (*Euphorbia pulcherrima* Willd. Ex Koltz.) cultivar-mediated differences in performance of five natural enemies of *Bemisia argentifolii* Bellows and Perring, n. sp. (Homoptera: Aleyrodidae). *Biological Control, 4*, 305–318.

Hilker, M., & Fatouros, N. E. (2015). Plant responses to insect egg deposition. *Annual Review of Entomology, 60*, 493–515.

Holt, R. D., Lawton, J. H., Polis, G. A., & Martinez, N. D. (1999). Trophic rank and the species-area relationship. *Ecology, 80*, 1495–1504.

Hunter, M. D. (2003). Effects of plant quality on the population ecology of parasitoids. *Agricultural and Forest Entomology, 5*, 1–8.

Inbar, M., & Gerling, D. (2008). Plant-mediated interactions between whiteflies, herbivores, and natural enemies. *Annual Review of Entomology, 53*, 431–448.

IPCC. (2013). In T. F. Stocker, D. Qin, G.-K. Plattner, M. Tignor, S. K. Allen, J. Boschung, et al. (Eds.), *Climate change 2013: The physical science basis: Final draft underlying scientific-technical assessment: Working group I contribution to the IPCC fifth assessment report*. Cambridge, UK and New York: Cambridge University Press, 1535 pp. (Summary for Policymakers, 28 pp.).

Jaloux, B. (2004). *La discrimination interspécifique par Eupelmus vuilleti (Hymenoptera: Eupelmidae) des hôtes parasites par Dinarmus basalis (Hymenoptera: Pteromalidae)* (PhD thesis). *Univ. François Rabelais, Tours*. https://tel.archives-ouvertes.fr/tel-001108221 (HAL Id: tel-00108221).

Jamont, M., Crépellière, S., & Jaloux, B. (2013). Effect of extrafloral nectar provisioning on the performance of the adult parasitoid *Diaeretiella rapae*. *Biological Control, 65*, 271–277.

Jervis, M., & Kidd, N. (1996). Phytophagy. In M. Jervis, & N. Kidd (Eds.), *Insect natural enemies. Practical approaches to their study and evaluation*. London, UK: Chapman & Hall.

Kaiser, L., Couty, A., & Cortesero, A.-M. (2013). La plante, un biotope de choix pour les entomophages (QUAE-IRD. Chap. 31). In N. Sauvion, P.-A. Calatayud, D. Thiery, & F. Marion-Poll (Eds.), *Des insectes et des plantes* (pp. 513–536).

Kaiser, L., Perez-Maluf, R., Sandoz, J. C., & Pham-Delegue, M. H. (2003). Dynamics of odour learning in *Leptopilina boulardi*, a hymenopterous parasitoid. *Animal Behaviour, 66*, 1077–1084.

Kaplan, I. (2012a). Attracting carnivorous arthropods with plant volatiles: the future of biocontrol or playing with fire? *Biological Control, 60*, 77–89.

Kaplan, I. (2012b). Trophic complexity and the adaptive value of damage-induced plant volatiles. *PLoS Biology, 10*(11), e1001437.

Kaplan, I., Carrillo, J., Garvey, M., & Ode, P. J. (2016). Indirect plant-parasitoid interactions mediated by changes in herbivore physiology. *Current Opinion in Insect Science, 14*, 112–119.

Kareiva, P. (1987). Habitat fragmentation and the stability of predator-prey interactions. *Nature, 326*, 388–390.

Karl, T. R., Knight, R. W., & Baker, B. (2000). The record breaking global temperatures of 1997 and 1998: Evidence for an increase in the rate of global warming? *Geophysical Research Letters, 27*, 719–722.

Karowe, D. N., & Schoonhoven, L. M. (1992). Interactions among three trophic levels: the influence of host plant on performance of *Pieris brassicae* and its parasitoid, *Cotesia glomerata*. *Entomologia Experimentalis et Applicata, 62*, 241–251.

Kessler, A., Halitschke, R., & Baldwin, I. T. (2004). Silencing the jasmonate cascade: induced plant defenses and insect populations. *Science, 305*, 665–668.

Kester, K. M., & Barbosa, P. (1991). Postemergence learning in the insect parasitoid, *Cotesia congregata* (Say) (Hymenoptera: Braconidae). *Journal of Insect Behavior, 4*, 727–742.

Kimani-Njogu, S. W., & Overholt, W. A. (1997). Biosystematics of the *Cotesia flavipes* species complex (Hymenoptera: Braconidae), parasitoids of the gramineous stemborers. *Insect Science and Its Application, 17*, 119–130.
Komonen, A., Penttila, R., Lindgren, M., & Hanski, I. (2000). Forest fragmentation truncates a food chain based on an old-growth forest bracket fungus. *Oikos, 90*, 119–126.
Koppik, M., Hoffmeister, T. S., Brunkhorst, S., Kiess, M., & Thiel, A. (2015). Intraspecific variability in associative learning in the parasitic wasp *Nasonia vitripennis*. *Animal Cognition, 18*, 593–604.
Kruidhof, H. M., Pashalidou, F. G., Fatouros, N. E., Figueroa, I. A., Vet, L. E. M., Smid, H. M., & Huigens, M. E. (2012). Reward value determines memory consolidation in parasitic wasps. *PLoS One,* 7(8), e39615.
Kruidhof, H. M., Roberts, A. L., Magdaraog, P., Munoz, D., Gols, R., Vet, L. E. M., ... Harvey, J. A. (2015). Habitat complexity reduces parasitoid foraging efficiency, but does not prevent orientation towards learned host plant odours. *Oecologia, 179*, 353–361.
Kruidhof, H. M., Smid, H., Thiel, A., Hoffmeister, T. S., & Vet, L. E. M. (2014). Olfactory conditioning of natural enemies: potential benefits for pest control in greenhouse crops. *IOBC-WPRS Bulletin, 102*, 121–126.
Lampert, E. C., Dyer, L. A., & Bowers, M. D. (2010). Caterpillar chemical defense and parasitoid success: *Cotesia congregata* parasitism of *Ceratomia catalpae*. *Journal of Chemical Ecology, 36*, 992–998.
Lampert, E. C., Dyer, L. A., & Bowers, M. D. (2014). Dietary specialization and the effects of plant species on potential multitrophic interactions of three species of nymphaline caterpillars. *Entomologia Experimentalis et Applicata, 153*, 207–216.
Lampert, E. C., Zangerl, A. R., Berenbaum, M. R., & Ode, P. J. (2008). Tritrophic effects of xanthotoxin on the polyembryonic parasitoid *Copidosoma sosares* (Hymenoptera: Encyrtidae). *Journal of Chemical Ecology, 34*, 783–790.
Lampert, E. C., Zangerl, A. R., Berenbaum, M. R., & Ode, P. J. (2011). Generalist and specialist host-parasitoid associations respond differently to wild parsnip (*Pastinaca sativa*) defensive chemistry. *Ecological Entomology, 36*, 52–61.
Langellotto, G. A., & Denno, R. F. (2004). Responses of invertebrate natural enemies to complex-structured habitats: A meta-analytical synthesis. *Oecologia, 139*, 1–10.
Lee, J. C., Heimpel, G. E., & Leibee, G. L. (2004). Comparing floral nectar and aphid honeydew diets on the longevity and nutrient levels of a parasitoid wasp. *Entomologia Experimentalis et Applicata, 111*, 189–199.
Lucchetta, P., Bernstein, C., Thery, M., Lazzari, C., & Desouhant, E. (2008). Foraging and associative learning of visual signals in a parasitic wasp. *Animal Cognition, 11*, 525–533.
Luo, S. P., Michaud, J. P., Li, J. C., Liu, X. X., & Zhang, Q. W. (2013). Odor learning in *Microplitis mediator* (Hymenoptera: Braconidae) is mediated by sugar type and physiological state. *Biological Control, 65*, 207–211.
Marino, P. C., & Landis, D. A. (1996). Effect of landscape structure on parasitoid diversity and parasitism in agroecosystems. *Ecological Applications, 6*, 276–284.
Marino, P. C., Landis, D. A., & Hawkins, B. A. (2006). Conserving parasitoid assemblages of North American pest Lepidoptera: Does biological control by native parasitoids depend on landscape complexity? *Biological Control, 37*, 173–185.
Mathur, V., Wagenaar, R., Caissard, J. C., Reddy, A. S., Vet, L. E., Cortesero, A. M., & Van Dam, N. M. (2013). A novel indirect defense in Brassicaceae: Structure and function of extrafloral nectaries in *Brassica juncea*. *Plant, Cell and Environment, 36*, 528–541.
McAuslane, H. J., Simmons, A. M., & Jackson, D. M. (2000). Parasitism of *Bemisia argentifolii* on collard with reduced or normal leaf wax. *Florida Entomologist, 83*, 428–437.

McCormick, A. C., Unsicker, S. B., & Gershenzon, J. (2012). The specificity of herbivore-induced plant volatiles in attracting herbivore enemies. *Trends in Plant Science, 17*, 303–310.

McGovern, J. L., Zangerl, A. R., Ode, P. J., & Berenbaum, M. R. (2006). Furanocoumarins and their detoxification in a tri-trophic interaction. *Chemoecology, 16*, 45–50.

Meiners, T., & Hilker, M. (2000). Induction of plant synomones by oviposition of a phytophagous insect. *Journal of Chemical Ecology, 26*, 221–232.

Meiners, T., & Peri, E. (2013). Chemical ecology of insect parasitoids: Essential elements for developing effective biological control programmes. In E. Wajnberg, & S. Colazza (Eds.), *Chemical ecology of insect parasitoids* (pp. 191–224). Chichester, UK: John Wiley & Sons, Ltd.

Mery, F. (2013). Natural variation in learning and memory. *Current Opinion in Neurobiology, 23*, 52–56.

Mooney, K. A., & Singer, M. S. (2012). Plant effects on herbivore—enemy interactions in natural systems. In T. Ohgushi, O. Schmitz, & R. D. Holt (Eds.), *Trait-mediated indirect interactions: Ecological and evolutionary perspectives* (pp. 107–130). Cambridge, UK: Cambridge University Press.

Moujahed, R., Frati, F., Cusumano, A., Salerno, G., Conti, E., Peri, E., & Colazza, S. (2014). Egg parasitoid attraction toward induced plant volatiles is disrupted by a non-host herbivore attacking above or belowground plant organs. *Frontiers in Plant Science, 5*, 601.

Moya-Larano, J. (2011). Genetic variation, predator-prey interactions and food web structure. *Philosophical Transactions of the Royal Society of London, Series B, 366*, 1425–1437.

Mumm, R., & Dicke, M. (2010). Variation in natural plant products and the attraction of bodyguards involved in indirect plant defense. *Canadian Journal of Zoology, 88*, 628–667.

Nair, A., & Van Nouhuys, S. (2016). Spatial and temporal genetic structure at the fourth trophic level in a fragmented landscape. *Proceedings of the Royal Society B: Biological Sciences, 283*.

Ngumbi, E., Jordan, M., & Fadamiro, H. (2012). Comparison of associative learning of host-related plant volatiles in two parasitoids with different degrees of host specificity, *Cotesia marginiventris* and *Microplitis croceipes*. *Chemoecology, 22*, 207–215.

Nishida, R. (2002). Sequestration of defensive substances from plants by *Lepidoptera*. *Annual Review of Entomology, 47*, 57–92.

Nordlund, D. A., Jones, R. L., & Lewis, W. J. (1981). *Semiochemicals: Their role in pest control.* New York, USA: John Whiley.

Nordlund, D. A., & Lewis, W. J. (1976). Terminology of chemical releasing stimuli in intraspecific and interspecific interactions. *Journal of Chemical Ecology, 2*, 211–220.

Ode, P. J. (2006). Plant chemistry and natural enemy fitness: Effects on herbivore and natural enemy interactions. *Annual Review of Entomology, 51*, 163–185.

Ode, P. J. (2013). Plant defences and parasitoid chemical ecology. In E. Wajnberg, & S. Colazza (Eds.), *Chemical ecology of insect parasitoids* (pp. 11–36). West Sussex, UK: Wiley-Blackwell.

Ode, P. J., Berenbaum, M. R., Zangerl, A. R., & Hardy, I. C. W. (2004). Host plant, host plant chemistry and the polyembryonic parasitoid *Copidosoma sosares*: Indirect effects in a tritrophic interaction. *Oikos, 104*, 388–400.

Ode, P. J., Harvey, J. A., Reichelt, M., Gershenzon, J., & Gols, R. (2016). Differential induction of plant chemical defenses by parasitized and unparasitized herbivores: consequences for reciprocal, multitrophic interactions. *Oikos, 125*, 1398–1407.

O'Dowd, D. J., & Willson, M. F. (1991). Associations between mites and leaf domatia. *Trends in Ecology & Evolution, 6*, 179–182.

Oliai, S. E., & King, B. H. (2000). Associative learning in response to color in the parasitoid wasp *Nasonia vitripennis* (*Hymenoptera* : Pteromalidae). *Journal of Insect Behavior, 13*, 55–69.

Olson, D. L., & Nechols, J. R. (1995). Effect of squash leaf trichome exudate and honey on adult feeding, survival, and fecundity of the squash bug (Heteroptera: Coreidae) egg parasitoid *Gryon pennsylvanicum* (Hymenoptera: Scelionidae). *Biological Control, 24*, 454–458.

Opitz, S. W., & Müller, C. (2009). Plant chemistry and insect sequestration. *Chemoecology, 19*, 117–154.

Papaj, D. R., Snellen, H., Swaans, K., & Vet, L. E. M. (1994). Unrewarding experiences and their effect on foraging in the parasitic wasp *Leptopilina heterotoma* (Hymenoptera: Eucoilidae). *Journal of Insect Behavior, 7*, 465–481.

Papaj, D. R., & Vet, L. E. M. (1990). Odor learning and foraging success in the parasitoid, *Leptopilina heterotoma*. *Journal of Chemical Ecology, 16*, 3137–3150.

Pashalidou, F. G., Frago, E., Griese, E., Poelman, E. H., Van Loon, J. J. A., Dicke, M., & Fatouros, N. E. (2015). Early herbivore alert matters: Plant-mediated effects of egg deposition on higher trophic levels benefit plant fitness. *Ecology Letters, 18*, 927–936.

Patt, J. M., Hamilton, G. C., & Lashomb, J. H. (1997). Foraging success of parasitoid wasps on flowers: Interplay of insect morphology, floral architecture and searching behavior. *Entomologia Experimentalis et Applicata, 83*, 21–30.

Patt, J. M., Hamilton, G. C., & Lashomb, J. H. (1999). Responses of two parasitoid wasps to nectar odors as a function of experience. *Entomologia Experimentalis et Applicata, 90*, 1–8.

Petschenka, G., & Agrawal, A. A. (2016). How herbivores coopt plant defenses: Natural selection, specialization, and sequestration. *Current Opinion in Insect Science, 14*, 17–24.

Pieterse, C. M. J., Van der Does, D., Zamioudis, C., Leon-Reyes, A., & Van Wees, S. C. M. (2012). Hormonal modulation of plant immunity. *Annual Review of Cell and Developmental Biology, 28*, 489–521.

Pincebourde, S., & Casas, J. (2006). Leaf miner-induced changes in leaf transmittance cause variations in insect respiration rates. *Journal of Insect Physiology, 52*, 194–201.

Pincebourde, S., van Baaren, J., Rasmann, S., Rasmont, P., Rodet, G., Martinet, B., & Calatayud, P.-A. (2016). Plant–insect interactions in a changing world. In N. Sauvion, D. Thiery, & P.-A. Calatayud (Eds.), *Insect-Plant Interactions in a Crop Protection Perspective* (Vol. 81, pp. 289–332).

Poelman, E. H., Bruinsma, M., Zhu, F., Weldegergis, B. T., Boursault, A. E., Jongema, Y., … Dicke, M. (2012). Hyperparasitoids use herbivore-induced plant volatiles to locate their parasitoid host. *PLoS Biology, 10*, e1001435.

Poelman, E. H., & Dicke, M. (2014). Plant-mediated interactions among insects within a community ecological perspective. In C. Voelckel, & G. Jander (Eds.), *Annual plant reviews* (Vol. 47). Chichester, UK: John Wiley & Sons, Ltd.

Poelman, E. H., Zheng, S.-J., Zhang, Z., Heemskerk, N. M., & Cortesero, A.-M. (2011). Parasitoid-specific induction of plant responses to parasitized herbivores affects colonization by subsequent herbivores. *Proceedings of the National Academy of Sciences B: Biological Sciences, 108*, 19647–19652.

Poirié, M., Colinet, D., & Gatti, J.-L. (2014). Insights into function and evolution of parasitoid wasp venoms. *Current Opinion in Insect Science, 6*, 52–60.

Ponzio, C., Gols, R., Weldegergis, B. T., & Dicke, M. (2014). Caterpillar-induced plant volatiles remain a reliable signal for foraging wasps during dual attack with a plant pathogen or non-host insect herbivore. *Plant, Cell and Environment, 37*, 1924–1935.

Powell, S. J., & Bale, J. S. (2006). Effect of long-term and rapid cold hardening on the cold torpor temperature of an aphid. *Physiological Entomology, 31*, 348–352.

Price, P. W., Bouton, C. E., Gross, P., McPheron, B. A., Thompson, J. N., & Weis, A. E. (1980). Interactions among three trophic levels: Influence of plants on interactions between insect herbivores and natural enemies. *Annual Review of Ecology, Evolution and Systematics, 11*, 41–65.

Pumarino, L., Sileshi, G. W., Gripenberg, S., Kaartinen, R., Barrios, E., Muchame, M. N., … Jonsson, M. (2015). Effects of agroforestry on pest, disease and weed control: a meta-analysis. *Basic and Applied Ecology, 16*, 573–582.

Quénol, H., & Beltrando, G. (2006). Microclimate in forest ecosystem and landscape ecology. *Climatologie, 3*, 9–23.

Quicke, D. L. J. (1988). Host relationships in the Braconinae – How little we know! *Queensland Entomological Society News Bulletin, 16*, 85–92.

Quicke, D. L. J. (1997). *Parasitic wasps*. London, UK: Chapman and Hall, 488 pp.

Quicke, D. L. J. (2015). *The braconid and ichneumonid parasitoid wasps: Biology, systematics, evolution and ecology*. Chichester, UK: John Wiley & Sons, Ltd, 704 pp.

Quintero, C., Lampert, E. C., & Bowers, M. D. (2014). Time is of the essence: direct and indirect effects of plant ontogenetic trajectories on higher trophic levels. *Ecology, 95*, 2589–2602.

Rand, T. A., Waters, D. K., Blodgett, S. L., Knodel, J. J., & Harris, M. O. (2014). Increased area of a highly suitable host crop increases herbivore pressure in intensified agricultural landscapes. *Agriculture, Ecosystems & Environment, 186*, 135–143.

Rani, P. U., Kumari, S. I., Sriramakrishna, T., & Sudhakar, T. R. (2007). Kairomones extracted from rice yellow stem borer and their influence on egg parasitization by *Trichogramma japonicum* Ashmead. *Journal of Chemical Ecology, 33*, 59–73.

Ratzka, A., Vogel, H., Kliebenstein, D. J., Mitchell-Olds, T., & Kroymann, J. (2002). Disarming the mustard oil bomb. *Proceedings of the National Academy of Sciences of the United States of America (PNAS), 99*, 11223–11228.

Rezende, M. Q., Venzon, M., Perez, A. L., Cardoso, I. M., & Janssen, A. (2014). Extrafloral nectaries of associated trees can enhance natural pest control. *Agriculture, Ecosystems & Environment, 188*, 198–203.

Robinson, K. A., Jonsson, M., Wratten, S. D., Wade, M. R., & Buckley, H. L. (2008). Implications of floral resources for predation by an omnivorous lacewing. *Basic and Applied Ecology, 9*, 172–181.

Romero, G. Q., & Benson, W. W. (2005). Biotic interactions of mites, plants and leaf domatia. *Current Opinion in Plant Biology, 8*, 436–440.

Rostas, M., Ruf, D., Zabka, V., & Hildebrandt, U. (2008). Plant surface wax affects parasitoid's response to host footprints. *The Science of Nature – Naturwissenschaften, 95*, 997–1002.

Rusch, A., Chaplin-Kramer, R., Gardiner, M. M., Hawro, V., Holland, J., Landis, D., … Winqvist, C. (2016). Agricultural landscape simplification reduces natural pest control: A quantitative synthesis. *Agriculture, Ecosystems & Environment, 221*, 198–204.

Schuman, M. C., & Baldwin, I. T. (2016). The layers of plant responses to insect herbivores. *Annual Review of Entomology, 61*, 373–394.

Sime, K. (2002). Chemical defense of *Battus philenor* larvae against attack by the parasitoid *Trogus pennator*. *Ecological Entomology, 27*, 237–345.

Simpson, M., Gurr, G. M., Simmons, A. T., Wratten, S. D., James, D. G., Leeson, G., … Orre-Gordon, G. U. S. (2011). Attract and reward: Combining chemical ecology and habitat manipulation to enhance biological control in field crops. *Journal of Applied Ecology, 48*, 580–590.

Singer, M. S., Mace, K. C., & Bernays, E. A. (2009). Self-medication as adaptive plasticity: Increased ingestion of plant toxins by parasitized caterpillars. *PLoS One, 4*, e4796.

Smallegange, R. C., Van Loon, J. J. A., Blatt, S. E., Harvey, J. A., & Dicke, M. (2008). Parasitoid load affects plant fitness in a tritrophic system. *Entomologia Experimentalis et Applicata, 128*, 172–183.

Smid, H., & Vet, L. E. M. (2016). The complexity of learning, memory and neural processes in an evolutionary ecological context. *Current Opinion in Insect Science, 15*, 61–69.

Smilanich, A. M., Dyer, L. A., Chambers, J. Q., & Bowers, M. D. (2009). Immunological cost of chemical defence and the evolution of herbivore diet breadth. *Ecology Letters, 12*, 612–621.

Smith, M. A., Rodriguez, J. J., Whitfield, J. B., Deans, A. R., Janzen, D. H., Hallwachs, W., & Hebert, P. D. N. (2008). Extreme diversity of tropical parasitoid wasps exposed by iterative integration of natural history, DNA barcoding, morphology, and collections. *Proceedings of the National Academy of Sciences of the United States of America, 105*, 12359–12364.

Stapel, J. O., Cortesero, A.-M., De Moraes, C. M., Tumlinson, J. H., & Lewis, W. J. (1997). Extrafloral nectar, honeydew and sucrose effects on searching behavior and efficiency of *Microplitis croceipes* (Hymenoptera: Braconidae) in cotton. *Environmental Entomology, 26*, 617–623.

Steidle, J. L. M., & Schöller, M. (1997). Olfactory host location and learning in the granary weevil parasitoid *Lariophagus distinguendus* (Hymenoptera: Pteromalidae). *Journal of Insect Behavior, 10*, 331–342.

Steidle, J. L. M., & Van Loon, J. J. A. (2003). Dietary specialization and infochemical use in carnivorous arthropods: testing a concept. *Entomologia Experimentalis et Applicata, 108*, 133–148.

Stiling, P., & Moon, D. C. (2005). Quality or quantity: The direct and indirect effects of host plants on herbivores and their natural enemies. *Oecologia, 142*, 413–420.

Suh, C. P. C., Orr, D. B., Van Duyn, J. W., & Borcher, M. D. (2002). Influence of cotton microhabitat on temperature and survival of *Trichogramma* (Hymenoptera: Trichogrammatidae) within cardboard capsules. *Environmental Entomology, 31*, 361–366.

Takasu, K., & Lewis, W. J. (1996). The role of learning in adult food location by the larval parasitoid, *Microplitis croceipes* (Hymenoptera: Braconidae). *Journal of Insect Behavior, 9*, 265–281.

Takasu, K., & Lewis, W. J. (2003). Learning of host searching cues by the larval parasitoid *Microplitis croceipes*. *Entomologia Experimentalis et Applicata, 108*, 77–86.

Takasu, K., Rains, G. C., & Lewis, W. J. (2007). Comparison of detection ability of learned odors between males and females in the larval parasitoid *Microplitis croceipes*. *Entomologia Experimentalis et Applicata, 122*, 247–251.

Tamo, C., Ricard, I., Held, M., Davison, A. C., & Turlings, T. C. J. (2006). A comparison of naive and conditioned responses of three generalist *endoparasitoids* of *lepidopteran larvae* to host-induced plant odours. *Animal Biology, 56*, 205–220.

Thaler, J. S. (1999). Jasmonate-inducible plant defences cause increased parasitism of herbivores. *Nature, 399*, 686–688.

Thiel, A., Schlake, S., & Kosior, D. (2013). *Omnia tempus habent*: habitat-specific differences in olfactory learning and decision making in parasitic wasps. *Animal Cognition, 16*, 223–232.

Thies, C., Steffan-Dewenter, I., & Tscharntke, T. (2003). Effects of landscape context on herbivore and parasitism at different spatial scales. *Oikos, 18*, 18–25.

Tougeron, K., Van Baaren, J., Burel, F., & Alford, L. (2016). Comparing thermal tolerance across contrasting landscapes: first steps towards understanding how landscape management could modify ectotherm thermal tolerance. *Insect Conservation and Diversity, 9*, 171–180.

Trapero, C., Wilson, I. W., Stiller, W. N., & Wilson, L. J. (2016). Enhancing integrated pest management in GM cotton systems using host plant resistance. *Frontiers in Plant Science, 7*, 500.

Tscharntke, T., Bommarco, R., Clough, Y., Crist, T. O., Kleijn, D., Rand, T. A., … Vidal, S. (2008). Conservation biological control and enemy diversity on a landscape scale. *Biological Control, 45*, 238–253.

Tscharntke, T., Tylianakis, J. M., Rand, T. A., Didham, R. K., Fahrig, L., Batáry, P., … Westphal, C. (2012). Landscape moderation of biodiversity patterns and processes – eight hypotheses. *Biological Reviews of the Cambridge Philosophical Society, 87*, 661–685.

Turlings, T. C. J., Loughrin, J. H., McCall, P. J., Rose, U. S. R., Lewis, W. J., & Tumlinson, J. H. (1995). How caterpillar damaged plants protect themselves by attracting parasitic wasps. *Proceedings of the National Academy of Sciences of the United States of America, 92*, 4169–4174.

Turlings, T. C. J., Tumlinson, J. H., & Lewis, W. J. (1990). Exploitation of herbivore-induced plant odors by host-seeking parasitic wasps. *Science, 250*, 1251–1253.

Turlings, T. C. J., & Wäckers, F. L. (2004). Recruitment of predators and parasitoids by herbivore-damaged plants. In R. T. Cardé, & J. G. Millar (Eds.), *Advances in insect chemical ecology* (pp. 21–75). Cambridge, UK: Cambridge University Press.

Turlings, T. C. J., Wäckers, F. L., Vet, L. E. M., Lewis, W. J., & Tumlinson, J. H. (1993). Learning of host-finding cues by hymenopterous parasitoids. In D. R. Papaj, & A. C. Lewis (Eds.), *Insect learning: Ecology and evolutionary perspectives* (pp. 51–78). New York: Chapman & Hall.

Tylianakis, J. M., Tscharntke, T., & Klein, A.-M. (2006). Diversity, ecosystem function, and stability of parasitoid host interactions across a tropical habitat gradient. *Ecology, 87*, 3047–3057.

Van den Bosch, R. (1971). Biological control of insects. *Annual Review of Ecology, Evolution and Systematics, 2*, 45–66.

Van der Meijden, E., & Klinkhamer, P. G. L. (2000). Conflicting interests of plants and the natural enemies of herbivores. *Oikos, 89*, 202–208.

Van Driesche, R., Hoddle, M., & Center, T. (2008). *Control of pests and weeds by natural enemies: An introduction to biological control.* Oxford, UK: Blackwell Publishing, 473 pp.

Van Emden, H. F., Sponagl, B., Baker, T., Ganguly, S., & Douloumpaka, S. (1996). Hopkins' "host selection principle", another nail in its coffin. *Physiological Entomology, 21*, 325–328.

Van Loon, J. J. A., De Boer, J. G., & Dicke, M. (2000). Parasitoid-plant mutualism: Parasitoid attack of herbivore increases plant reproduction. *Entomologia Experimentalis et Applicata, 97*, 219–227.

Van Nouhuys, S. (2005). Effects of habitat fragmentation at different trophic levels in insect communities. *Annales Zoologici Fennici, 42*, 433–447.

Van Nouhuys, S., & Hanski, I. (2002). Multitrophic interactions in space: Metacommunity dynamics in fragmented landscapes. In T. Tscharntke, & B. A. Hawkins (Eds.), *Multitrophic level interactions* (pp. 124–147). Cambridge: Cambridge University Press.

Van Nouhuys, S., & Kaartinen, R. (2008). A parasitoid wasp uses landmarks while monitoring potential resources. *Proceedings of the Royal Society B: Biological Sciences, 275*, 377–385.

Van Vugt, J., Hoedjes, K. M., Van de Geest, H. C., Schijlen, E., Vet, L. E. M., & Smid, H. M. (2015). Differentially expressed genes linked to natural variation in long-term memory formation in *Cotesia* parasitic wasps. *Frontiers in Behavioral Neuroscience, 9*.

Vet, L. E. M., & Dicke, M. (1992). Ecology of infochemical use by natural enemies in a tritrophic context. *Annual Review of Entomology, 37*, 141–172.

Vet, L. E. M., & Van Opzeeland, K. (1984). The influence of conditioning on olfactory microhabitat and host location in *Asobara tabida* (Nees) and *A. rufescens* (Foerster) (Braconidae: Alysiinae) larval parasitoids of *Drosophilidae*. *Oecologia, 63*, 171–177.

Vogelweith, F., Moreau, J., Thiéry, D., & Moret, Y. (2015). Food-mediated modulation of immunity in a phytophagous insect: an effect of nutrition rather than parasitic contamination. *Journal of Insect Physiology, 77*, 55–61.

Wäckers, F. L. (2001). A comparison of nectar- and honeydew sugars with respect to their utilization by the hymenopteran parasitoid Cotesia glomerata. *Journal of Insect Physiology, 47*, 1077–1084.

Wäckers, F. L., & Lewis, W. J. (1999). A comparison of color-, shape- and pattern-learning by the hymenopteran parasitoid *Microplitis croceipes*. *Journal of Comparative Physiology A, 184*, 387–393.

Wäckers, F. L., & Steppuhn, A. (2003). Characterizing nutritional state and food source use of parasitoids collected in fields with high and low nectar availability. *IOBC-WPRS Bulletin, 26*, 203–208.

Wäckers, F. L., van Rijn, P. C. J., & Bruin, J. (2005). *Plant-provided food for carnivorous insects: A protective mutualism and its applications*. Cambridge, UK: Cambridge University Press.

Wäckers, F. L., van Rijn, P. C., & Heimpel, G. E. (2008). Honeydew as a food source for natural enemies: Making the best of a bad meal? *Biological Control, 45*(2), 176–184.

Wajnberg, E., Bernstein, C., & Van Alphen, J. (2007). *Behavioural ecology of insect parasitoids: From theoretical approaches to field applications*. Dordrecht, Pays-Bas: Blackwell publishing.

Wajnberg, E., & Colazza, S. (2013). *Chemical ecology of insect parasitoids*. New York, United States: John Wiley & Sons, 328 pp.

Walther, G. R., Post, E., Convey, P., Menzel, A., Parmesan, C., Beebee, T. J. C., … Bairlein, F. (2002). Ecological responses to recent climate change. *Nature, 416*, 389–395.

Weevers, T. (1952). Flower color and their frequency. *Acta Botanica Neerlandica, 1*, 81–92.

Williams, T., Arrendondo-Bernal, H. C., & Rodríguez-del-Bosque, L. A. (2013). Biological pest control in Mexico. *Annual Review of Entomology, 58*, 119–140.

Wirta, H. K., Vesterinen, E. J., Hambäck, P. A., Weingartner, E., Rasmussen, C., Reneerkens, J., … Roslin, T. (2015). Exposing the structure of an Arctic food web. *Ecology and Evolution, 5*, 3842–3856.

Wittstock, U., Agerbirk, N., Stauber, E. J., Olsen, C. E., Hippler, M., Mitchell-Olds, T., … Vogel, H. (2004). Successful herbivore attack due to metabolic diversion of a plant chemical defense. *Proceedings of the National Academy of Sciences of the United States of America, 101*, 4859–4864.

Woods, H. A. (2013). Ontogenetic changes in the body temperature of an insect herbivore. *Functional Ecology, 27*, 1322–1331.

Yarnes, C. T., & Boecklen, W. J. (2006). Abiotic mosaics affect seasonal variation of plant resources and influence the performance and mortality of a leaf-miner in Gambel's oak (*Quercus gambelii*, Nutt.). *Ecological Research, 21*, 157–163.

Züst, T., & Agrawal, A. A. (2016). Population growth and sequestration of plant toxins along a gradient of specialization in four aphid species on the common milkweed, *Asclepias syriaca*. *Functional Ecology, 30*, 547–555.

CHAPTER SEVEN

Influence of Microbial Symbionts on Plant–Insect Interactions

D. Giron*[,1], F. Dedeine*, G. Dubreuil*, E. Huguet*, L. Mouton[§], Y. Outreman[¶], F. Vavre[§], J.-C. Simon[||]

*Université François-Rabelais de Tours, Centre National de la Recherche Scientifique, Tours, France
[§]Université de Lyon 1, Centre National de la Recherche Scientifique, Villeurbanne, France
[¶]Agrocampus Ouest, Institut National de la Recherche Scientifique, Université de Rennes 1, Université Bretagne-Loire, Rennes, France
[||]INRA, UMR 1349 IGEPP INRA-Agrocampus Ouest Rennes-Université Rennes 1, Le Rheu, France
[1]Corresponding author: E-mail: david.giron@univ-tours.fr

Contents

Advances in Botanical Research, Volume 81
ISSN 0065-2296
http://dx.doi.org/10.1016/bs.abr.2016.09.007

Abstract

There is growing evidence that microorganisms are important 'hidden players' in insect–plant interactions. Insect symbionts can directly affect these interactions by providing insects with key nutrients or by interfering with the plant to modulate food provisioning to insects and plant defences. Insect symbionts can also have indirect cascading ecological consequences at the community level through insect- and plant-mediated effects that include their impact on insect reproduction, on natural enemies of herbivores or on plant-associated microorganisms. Identification of symbiotic communities associated with insects, characterization of transmission and acquisition patterns as well as understanding of molecular mechanisms underlying these plant–insect–microbe interactions have important ecological and evolutionary consequences. This review highlights the excitement that surrounds these investigations and the promise they hold for a better understanding of the functional, ecological and evolutionary impacts of symbionts on plant–insect interactions, with implications and relevance for both applied and fundamental researches.

1. INTRODUCTION

Nutrition is the cornerstone of most interactions between organisms. With more than 4 million estimated species, insects are among the most significant evolutionary successes on Earth (Novotny et al., 2002). The origin of this success can be directly linked to the diversity of their feeding strategies, of which herbivory is the most common (Schoonhoven, van Loon, & Dicke, 2005; Slansky & Rodriguez, 1987). However, plant tissues are typically suboptimal nutritionally, due to unbalanced ratios and/or low levels of key nutrients and frequent requirement to detoxify plant-defensive allelochemicals (Schoonhoven et al., 2005). The ability of phytophagous arthropods to exploit plant resources requires them to employ a suite of pre- and postingestive mechanisms to address the nutritional mismatch between what plants provide and what insects require (Raubenheimer, Simpson, & Mayntz, 2009). These strategies include specific behavioural and physiological adaptations (Behmer, 2009), intricate interactions that involve insect reprogramming of host plant development (Giron, Huguet, Stone, & Body, 2016; see also chapter: From Plant Exploitation to Mutualism by Lieutier et al., 2017), symbioses in which plants have evolved food rewards specifically for insects (e.g., Heil & McKey, 2003) and also associations with one or more symbiotic partners (Sugio, Dubreuil, Giron, & Simon, 2015).

Microorganisms have been shown to be important 'hidden players' in insect–plant interactions (Biere & Bennett, 2013; Frago, Dicke, & Godfray, 2012; Sugio et al., 2015) and can affect, among other traits, insect host plant range (Chu, Spencer, Curzi, Zavala, & Seufferheld, 2013; Hosokawa, Kikuchi, Shimada, & Fukatsu, 2007), feeding efficiency of the insect (Brune & Dietrich, 2015), insect metabolism (Douglas, 2013), ability of the insect to manipulate the plant physiology for their own benefit (Giron et al., 2016; Kaiser, Huguet, Casas, Commin, & Giron, 2010) and more generally insect diversification and speciation (Vavre & Kremer, 2014). Insect symbionts can indeed directly or indirectly affect the plant by interfering with plant signal transduction pathways, repressing or counteracting the expression of plant defence–related genes or altering plant primary and secondary metabolisms (Body, Kaiser, Dubreuil, Casas, & Giron, 2013; Giron, Frago, Glevarec, Pieterse, & Dicke, 2013; Sugio et al., 2015; Zhu, Poelman, & Dicke, 2014). Insect symbionts can also affect plant–insect interactions through their direct or indirect effects on their insect host by providing new metabolic pathways (Douglas, 2013; Moran, McCutcheon, & Nakabachi, 2008) and/or by altering insect reproduction (Engelstädter & Hurst, 2009; Ferrari & Vavre, 2011) or insect immunity with consequences on plant exploitation (Dubreuil, Deleury, Crochard, Simon, & Coustau, 2014). Finally, they can also modulate insect interactions with natural enemies or plant-associated organisms such as other herbivores, plant symbionts or plant pathogens (Biere & Bennett, 2013; Chuche, Danet, Salar, Foissac, & Thiéry, 2016; Frago et al., 2012; Sugio et al., 2015).

This chapter focuses on recent studies on symbionts associated with herbivorous insects that directly or indirectly influence insect–plant interactions. Although plant-associated symbionts are another important and active research field, it is not addressed here. This chapter details (1) the diversity of microbial communities and the ecological dynamics of insect host–microbe interactions, (2) the direct and (3) indirect effects of symbionts on plant–insect interactions, and (4) the roles symbionts may play on insect diversification and specialization on host plants. This review more particularly aims at highlighting the excitement that surrounds investigations on plant–insect–symbionts interactions and the promise they hold for a global understanding of plant–insect interactions.

2. DIVERSITY OF INSECT MICROBIAL COMMUNITIES AND ECOLOGICAL DYNAMICS OF INSECT HOST–MICROBE INTERACTIONS

2.1 Diversity of the Symbionts Associated With Herbivorous Insects

The term 'symbiont' generally refers to microorganisms that live in intimate interaction with a host permanently or at least during a substantial part of the host's life cycle. Symbiotic associations are extremely diverse in herbivorous insects not only due to the taxonomic diversity of the microbial partners engaged (i.e., fungi, protists, bacteria, archae or viruses) but also due to other attributes such as: (1) the location of the symbionts relative to the host body, (2) the transmission mode of the symbionts through host generations, (3) the number of distinct microbial taxa coexisting within host individuals, (4) the nature of the host–symbiont interactions along the parasitism–mutualism continuum or (5) the degree of dependence between symbiotic partners for survival and reproduction. Symbionts associated with herbivorous insects can therefore be classified in many ways. However, the location of the symbionts is a relevant categorization criterion to characterize symbionts that possibly influence insect–plant interactions. The location of the symbionts in insect tissues may indeed restrict the nature and intensity of actions on host plants (Hansen & Moran, 2014).

Many herbivorous insects harbour intracellular symbionts that are inherited maternally through the germ line. The most studied intracellular symbionts are undoubtedly bacteria living in specialized host cells (i.e., bacteriocytes) that are required for the host's nutrition and survival (Baumann, 2005). Such obligate symbionts, also called 'primary symbionts', have been described in a variety of herbivorous insect taxa, although they have been particularly well studied in sap-feeding hemipterans in which they complement the unbalanced diet (i.e., phloem, xylem) of their hosts by providing nitrogen, essential amino acids and vitamins (Moran et al., 2008). Recent studies revealed that intracellular symbionts also include a variety of bacteria that are facultative for host survival and reproduction. These 'secondary symbionts' can nevertheless deeply influence their hosts' biology and ecology in a variety of ways along the parasitism–mutualism continuum. Some secondary symbionts such as *Wolbachia* impact host reproduction by inducing various phenotypic effects (Engelstädter & Hurst, 2009). However, intracellular secondary symbionts can also have beneficial effects for their hosts, conferring protection to natural enemies (Oliver, Smith, & Russell, 2014; see also Section 4.2) or mediating interactions between their hosts and the plants they consume (Kaiser et al., 2010; see also Sections 3.3 and 3.4). They can act in conjunction with primary symbionts and even replace them (see Sections 3.1 and 3.4).

Herbivorous insects, as most other animals, also harbour microorganisms in their intestinal tracts. Most of these gut microorganisms are bacteria, but insects feeding on wood or plant litter can also harbour fungi, protists or methanogenic archaea (Brune & Dietrich, 2015). The diversity and composition of gut microbial communities strongly vary among insects from very simple to highly complex microbial assemblages composed by hundreds of taxa representative of the three domains of life (i.e., bacteria, archaea and eukaryotic microbes) (Engel & Moran, 2013). Recent investigations also support the general view that both diet and evolutionary history of the hosts shape gut communities (Colman, Toolson, & Takacs-Vesbach, 2012; Jones, Sanchez, & Fierer, 2013). The consequences of intestinal symbionts in insects have been relatively less investigated than those induced by intracellular symbionts. Nevertheless, it has become clear that gut symbionts can have beneficial effects on their hosts, contributing, for example, to nutrition, protection from parasites and pathogens, modulation of immune responses and communication (Engel & Moran, 2013; Lizé, McKay, & Lewis, 2013).

Some insects have domesticated external symbionts, mostly fungi, which live outside their body (Aylward et al., 2014). The most studied ectosymbionts are known to help their hosts to feed directly or indirectly on fresh or decaying plant materials. Fungus-farming ants and termites cultivate these symbionts in their nests (Mueller, Gerardo, Aanen, Six, & Schultz, 2005; Poulsen et al., 2014), while ambrosia beetles cultivate them in their galleries (Kostovcik et al., 2015). In other cases, such as stink bugs, symbionts can be acquired by feeding on a capsule deposited by the mothers nearby their eggs (Hosokawa, Kikuchi, Meng, & Fukatsu, 2005) or directly from the soil at every generation (Kikuchi, Hosokawa, & Fukatsu, 2007). In these mutualistic associations, the symbionts contribute to the nutrition of their hosts, whereas the symbionts benefit from the association for food provisioning and dispersion. It is likely that ectosymbionts are much more common in herbivorous insects than described so far and provide benefits not only for nutrition but also possibly for modulating host plant recognition or detoxifying secondary plant compounds (Hansen & Moran, 2014).

2.2 Characterization of Microbial Diversity in Insects: Identification and Localization

Because most symbionts cannot be cultivated outside their hosts, characterization and identification of microbes associated with insects rely primarily on molecular techniques. Before the development of next-generation sequencing technologies, assessment of symbiont diversity and composition was mainly achieved by cloning and sequencing 16S or 18S rRNA partial or complete genes for prokaryotes and eukaryotes, respectively. These genomic regions generally contain sufficient molecular variations to discriminate microbial taxa and to infer symbiont species from DNA sequences. Specific primers can then be designed to selectively amplify the symbiont(s) of interest in order to study, for example, the prevalence and transmission patterns of associated microbes. Multilocus sequence typing is also a widespread technique to discriminate between strains of microbes that can infect the same or different insect hosts (e.g., Henry et al., 2013).

In addition to qualitative data, these specific primers can be used to measure the abundance of symbionts within the host by quantitative polymerase chain reaction. Metagenomic, genomic and transcriptomic data obtained by high-throughput sequencing is now replacing classical polymerase chain reaction amplification and cloning techniques to detect microbial partners. These genomic data sets are then processed with adapted bioinformatics tools allowing to identify the full diversity of microbial communities associated with insects (i.e., the microbiome) and to tackle the way the microbiota influence the host's phenotype. Complete genomes of insect symbionts are now available for many systems and have been decisive to elucidate evolutionary patterns of insect symbiosis and to reveal symbiotic functions through metabolic networks inference.

Besides the characterization of symbiont diversity and functions, localizing microbes inside their hosts is also important to understand host–symbiont or symbiont–symbiont interactions as well as transmission patterns of the symbionts. Fluorescence in situ hybridization method is generally employed to visualize symbionts and symbiotic organs inside the host. Transmission electron microscopic observations allow a much deeper analysis of symbiont ultrastructure and can allow to distinguish between different symbiont taxa inhabiting the same host tissue by using specific immunegold labelling (Tsuchida, Koga, Fujiwara, & Fukatsu, 2014).

2.3 Symbiont Transmission and Acquisition at the Intergenerational Level

Symbiont transmission maintains symbiotic associations through host generations and represents a pivotal factor in their evolutionary stability and diversification (Salem, Florez, Gerardo, & Kaltenpoth, 2015). Although the transmission mechanisms are diverse, three principal modes of symbiont transmission can be distinguished: vertical, horizontal and mixed.

In herbivorous insects, most vertically transmitted symbionts are transferred from the mother to the offspring (maternal inheritance). This is the case in the widespread intracellular symbionts, which are, in host females, translocated from bacteriocytes to the germ line where they are internalized in maturing oocytes. Maternal inheritance is nevertheless not restricted to intracellular symbionts and it also occurs in intestinal and external symbioses (Salem et al., 2015). In the European firebug, for instance, the transmission of beneficial gut symbionts relies on secretions that are smeared over the egg surface following oviposition (Kaltenpoth, Winter, & Kleinhammer, 2009). It is worth noting that, in rare instances, the transmission of symbionts to the offspring is ensured by both parents (bi-parental inheritance). For example, the gut symbionts of termites are transferred from the royal couple founding a new colony to their first hatched larvae, which lick and ingurgitate symbiont-rich fluids excreted by their parents (i.e., proctodeal trophallaxis) (Brune & Dietrich, 2015).

Horizontally transmitted symbionts can be acquired by the hosts either from conspecific or hetero-specific host individuals, or directly from the environment. In the former situation, coprophagy might play a major role in some bugs, cockroaches and termites (Salem et al., 2015). In these cases, symbiont acquisition by symbiont-free individuals requires direct contact with faeces after excretion. In other insects such as in the bean bug, *Riptortus pedestris* (Fabricius, 1775) (Hemiptera: Coreoidea), laboratory studies revealed that some beneficial gut symbionts are acquired directly from the environment (Kikuchi et al., 2007).

In many instances, symbionts can be transmitted both vertically and horizontally. Many facultative maternally transmitted intracellular symbionts such as *Wolbachia* are known to be occasionally transferred between unrelated insect species through a range of mechanisms (Koehncke, Telschow, & Kondoh, 2012). An interesting example is the intracellular *Rickettsia* bacteria infecting the whitefly, *Bemisia tabaci* (Gennadius, 1889) (Hemiptera: Aleyrodidae). Whereas this facultative symbiont is primarily transmitted maternally via the eggs, the bacteria can also be transferred among *B. tabaci* host lineages via the host plant, the symbiont being found in the phloem of several plant species following feeding by an infected whitefly (Caspi-Fluger et al., 2012).

3. DIRECT EFFECTS OF SYMBIONTS IN PLANT—INSECT INTERACTIONS

3.1 Influence on Insect Nutrition and Metabolism

Most insects that feed exclusively on unbalanced diet such as plant sap have developed symbiosis with microorganisms that provide essential amino acids and vitamins that are present in short supply in their food and that insects cannot synthetize on their own (Baumann, 2005; Buchner, 1965). These obligate symbioses have been keys in the ability of some insects to colonize new ecological niches. Most often these interactions involve intracellular bacteria that are maternally inherited and located in dedicated organs. The most documented example is probably the association between *Buchnera aphidicola* and aphids (Buchner, 1965), but association between *Sulcia* and Auchenorrhyncha (a sap-feeding insect group including plant hoppers, cicadas, spittlebugs, leafhoppers – Cryan & Urban, 2012) is the oldest symbiosis reported in insects (~270 millions of years; Moran, Degnan, Santos, Dunbar, & Ochman, 2005). Usually, all nutrient biosynthetic pathways are present and complete in these obligate symbionts despite extremely reduced genomes (Moran et al., 2008). However, some of them have lost some key metabolic genes, involved in the production of essential amino acids, for example, relying on the host and/or another symbiont to compensate for the loss of essential metabolic traits by the obligatory symbiont. For example, *Sulcia muelleri*, the obligate symbiont of sharpshooters and other Auchenorrhyncha, can produce 8 out of the 10 essential amino acids, the 2 missing amino acids being provided by partnering symbionts, *Baumannia cicadellinicola* and *Hodgkinia cicadicola*, respectively (McCutcheon, McDonald, & Moran, 2009). Metabolic interdependency can also occur when obligatory symbionts have incomplete biosynthetic pathways. The primary symbiont *Portiera aleyrodidarum* of the whitefly *B. tabaci* has lost three genes involved in lysine synthesis which are present in the genome of the common facultative symbiont *Hamiltonella defensa* (Rao et al., 2015; Rollat-Farnier et al., 2015). In this intricate interaction, the genome of *B. tabaci* also contributes to multiple metabolic reactions through genes of insect origin but also thanks to other genes that were horizontally acquired from other bacteria (Luan et al., 2015).

3.2 Influence on Insect Immunity and Plant Exploitation

The invertebrate's innate immune system was reported to show some forms of adaptive features including highly diversified recognition systems, complex regulatory processes and specific effectors (e.g., Baeza Garcia et al., 2010; Hoffmann & Reichhart, 2002; Schulenburg, Boehnisch, & Michiels, 2007). In aphids, adaptation to feeding on phloem is largely ensured by their association with *Buchnera*. Aphids also interact with bacterial secondary endosymbionts that are facultative (Oliver, Degnan, Burke, & Moran, 2010) and can influence the immune response of their insect hosts (Laughton, Fan, & Gerardo, 2014; Schmitz et al., 2012). In pea aphids, a negative effect of symbionts on expression of immune gene members of the macrophage migration inhibitory factor (MIF) family was reported

(Dubreuil et al., 2014). MIFs are known as important pro-inflammatory cytokines regulating immune responses of vertebrates (Calandra & Roger, 2003). The presence of facultative symbionts correlated with a decreased expression of the MIF genes in aphids and an increased expression of MIF genes was observed in aphids without secondary symbionts after injection with the gram-negative bacteria *Escherichia coli*. Additionally, among the five members of MIF multigene family, ApMIF1 is the unique member encoding MIF protein that is secreted during aphid feeding (Naessens et al., 2015). The RNA interference targeting the ApMIF1 resulted in a significant decrease in aphid success in phloem feeding and functional analysis showed that MIF interferes with the plant immune system, suggesting that MIF secretion mimics or antagonizes plant proteins to repress plant immune responses. Presence of symbionts could thus modulate the ability of the pea aphid to exploit its host possibly by repressing the plant immune response.

3.3 Influence on Plant Nutritional Status and Morphology

Phytohormones lay at the very core of molecular mechanisms controlling plant growth, defence and/or nutritional status (Erb, Meldau, & Howe, 2012; Giron et al., 2013). The ability to control the plant phytohormonal balance is a well-characterized mechanism used by several plant-associated microorganisms to colonize and exploit the plant (Giron & Glevarec, 2014). Indeed, plant-associated microorganisms potentially influence the levels of phytohormones by inducing plant genes involved in phytohormone biosynthesis, metabolism, degradation or response, but they can also produce and secrete relevant phytohormones themselves (Giron et al., 2013). There is also growing evidence that insect-associated microbes are active players in plant manipulation to the benefit of the insect host (Body et al., 2013; Kaiser et al., 2010; Sugio et al., 2015).

Curing the apple tree leaf-miner, *Phyllonorycter blancardella* (Fabricius, 1781) (Lepidoptera: Gracillariidae) of its endosymbiotic *Wolbachia* bacteria resulted in the loss of the cytokinin (CK)-induced green-island phenotype on apple tree leaves and in the absence of detectable CKs in larvae compared to nontreated controls (Body et al., 2013; Kaiser et al., 2010). These results suggest that these insects have the ability to modify the phytohormonal profile in mined leaf tissues and to deliver CKs to the plant via their association with symbiotic bacteria (Giron & Glevarec, 2014; Zhang et al., 2016). This allows insects to 'hijack' plant metabolism, thus enabling them to successfully invade the plant by inhibiting plant defences and withdrawing plant resources for their own benefit (Zhang et al., 2016). The first survey of bacteria associated with the gut of a plant-manipulating insect, the Hessian fly, has recently revealed a predominance of *Pseudomonas* species (Bansal et al., 2014), the genomes of which were identified in whole-genome sequencing of the Hessian fly, *Mayetiola destructor* (Say, 1817) (Diptera: Cecidomyiidae) (Zhao et al., 2015). It remains to be seen whether these bacteria, or other microbes associated with the insect, modify host plant nutrition and development, leading to gall induction.

Besides bacteria, other symbionts may also help insects to manipulate their host plant. Some gall midges have a symbiotic association with biotrophic fungi that are essential for invasion of plant stems and access to vascular tissue, for providing larvae with highly nutritious food and for gall development (Rohfritsch, 2008). The molecular mechanisms underlying such tripartite interactions involving fungi still need to be uncovered. Cornell (1983) suggested that viruses or viral proteins could be involved in the delivery of stimuli in gall-inducing cynipids.

3.4 Impact on Plant Secondary Metabolism/Plant Immunity

Insect symbionts can also be involved in the suppression of phytohormone-mediated plant defence signalling (Giron et al., 2013, 2016; Sugio et al., 2015). A striking example is the Colorado potato beetle, *Leptinotarsa decemlineata* (Say, 1824) (Coleoptera: Chrysomelidae), that releases bacteria in its oral secretions, resulting in the activation of a plant microbial defence response through the induction of the salicylic acid–signalling pathway. This leads in turn, by negative cross-talk, to downregulation of the jasmonic acid (JA)–responsive antiherbivore resulting in improved larval growth (Chung et al., 2013). These results show that the herbivore disrupts plant perception and evades antiherbivore defences by exploiting symbiotic bacteria. However, symbionts do not always benefit their host as shown in the aphid-*Buchnera* system, where a symbiont protein delivered in the insect saliva is recognized by the plant and elicits reaction defences (Chaudhary, Atamiana, Shenc, Briggsc, & Kaloshian, 2014).

Plant defence suppression involving insect-associated bacteria was also suggested in the maize–corn rootworm [*Diabrotica virgifera virgifera* (Le Conte, 1868) (Coleoptera: Chrysomelidae)] interaction, in which *Wolbachia* infection was positively correlated with the ability of the larvae to inhibit defence gene expression in the maize (Barr, Hearne, Briesacher, Clark, & Davis, 2010). However, further work showed that endosymbiont-free insects do not elicit different maize defence responses in comparison to

Wolbachia-infected insects (Robert et al., 2013), suggesting that symbiont effects can be context dependent.

In the whitefly *B. tabaci*, saliva of individuals harbouring the facultative symbiont *H. defensa* is able to suppress JA-related defences in tomato compared to saliva from noninfected controls (Su et al., 2015). Putative non-proteinaceous effectors were identified in the saliva, but it remains to know their origin and exactly how *H. defensa* mediates the suppression of plant defences in this system. *H. defensa* also serves as a nutrient provider in whiteflies (Luan et al., 2015; Rao et al., 2015; Rollat-Farnier et al., 2015; see Section 3.1), illustrating the multiple ways in which a symbiont can impact overall insect fitness. Feeding by the silverleaf whitefly has been shown to induce SA defences and to suppress JA responses in *Arabidopsis* (Zarate, Kempema, & Walling, 2007). Whether this ability is endogenous to the insect or is symbiont-associated awaits validation. Recently, leaf-mining larvae of *Scaptomyza flava* (Fallén, 1823) (Diptera: Drosophilidae) have been shown to vector *Pseudomonas syringae* bacteria to and from feeding sites and that the larvae perform better on plants infected with *P. syringae*. Here, the suggested mechanism is that *P. syringae* acts by suppressing antiherbivore defences mediated by reactive oxygen species (Groen et al., 2016).

Rather than interfering with plant defence signalling, insect symbionts can also inhibit or counteract plant defences as suggested in the cigarette beetles (Dowd & Shen, 2011) and in the gypsy moth (Broderick, Raffa, Goodman, & Handelsman, 2004). This could be achieved through the direct or indirect production of enzymes targeting plant-defensive compounds. The microbial community of the mountain pine beetle, *Dendroctonus ponderosae* (Hopkins, 1902) (Coleoptera: Curculionoidea), seems to contribute to overcome the plant's terpenoid-based defences by degrading them (Boone et al., 2013). *Dendroctonus ponderosae* is strongly associated with microbial communities that are enriched with genes involved in terpene degradation compared with other plant biomass—processing microbial communities (Adams et al., 2013). Furthermore, the bacteria associated with *D. ponderosae* were shown to metabolize monoterpenes and diterpene acids that are toxic to beetles (Boone et al., 2013). Similarly, gut bacteria isolated from the velvet bean caterpillar, *Anticarsia gemmatalis* (Hübner, 1818) (Lepidoptera: Noctuidae), a soybean pest, are involved in serine proteinase production. Higher production of proteases induced or produced by the bacteria might contribute to the adaptation of the caterpillar to the soybean plant, which is rich in protease inhibitors (Visotto, Oliveira,

Guedes, Ribon, & Good-God, 2009). A comparison of gut bacterial microbiota of two different variants of *D. virgifera virgifera* showed that the gut bacterial communities of 'rotation-resistant' populations were different from those of wild-type populations. The 'rotation-resistant' variant microbiota contributes to the proteolysis and survival of *D. virgifera virgifera* on nonhost soybeans, suggesting that this adaptation of the western corn rootworm to a new host plant is directly linked with a modification of the gut bacteria adapted to tolerate the antiherbivory defences expressed in soybean foliage (Chu et al., 2013).

3.5 Genes Acquired by Horizontal Gene Transfer That Influence Plant–Insect Interactions

Mechanisms and actual microbial genes involved in symbionts-associated interference with plant defence signalling and plant physiological status still await characterization. Candidate microbial genes most likely playing important roles in plant–herbivore interactions, have however been identified within genomes of insect pests following acquisition via horizontal gene transfer (HGT) events (Boto, 2014). Indeed, there is increasing evidence that HGTs (i.e., transmission of genetic material between organisms other than by descent) play a role in eukaryotic evolution leading to acquisition of novel traits (Boto, 2014; Soucy, Huang, & Gogarten, 2015). Several examples of HGT described in insects concern microbial genes involved in the adaptation of insects to plants, either because these genes encode specific enzymes allowing degradation and metabolism of plant products or because they may enable detoxification of potentially harmful plant components.

Several studies have reported the presence of genes encoding Plant Cell Wall Degrading Enzymes in different herbivorous insects. Some of these genes are likely endogenous insect genes (Calderon-Cortes, Quesada, Watanabe, Cano-Camacho, & Oyama, 2012), while others most likely derived from HGTs from different microbial sources (Kirsch et al., 2014). These horizontally acquired genes related to plant feeding have been particularly well investigated in beetles and weevils (e.g., Acuña et al., 2012; Kirsch, Heckel, & Pauchet, 2016). For example, the coffee berry beetle harbours in its genome a functional mannanase gene, phylogenetically related to *Bacillus* genes, that is presumed to facilitate feeding within the coffee berry by hydrolyzing galactomannan, the major storage polysaccharide in this plant (Acuña et al., 2012). Studies combining molecular evolution and enzymatic assays show that many of these genes have undergone duplication and diversification events since their acquisition suggesting their important role in beetle adaptation to plants (Kirsch et al., 2016). Similarly, evidence for multiple HGT of genes involved in sugar and amino acid metabolism followed by duplications and diversification were reported in lepidopteran genomes (Sun et al., 2003) and 30 candidate HGT events involving mainly carbohydrate metabolic enzymes in Hessian fly (Zhao et al., 2015) suggest that the acquisition of these genes could be recurrent in herbivorous insect species, allowing better utilization of plant carbohydrates by these insects.

Acquisition of genes by phytophagous insects via horizontal transfer from microorganisms could also be involved in detoxification of plant products. A gene of bacterial origin encoding β-cyanoalanine synthase was shown to allow mites and Lepidoptera to feed on plants releasing toxic hydrogen cyanide upon tissue disruption (Wybouw et al., 2014). In the silkworm, β-fructofuranosidase genes of probable bacterial origin have been proposed to play a critical role in this caterpillar's ability to avoid the toxic effects of sugar mimic alkaloids present in mulberry latex, that are highly toxic to nonmulberry specialist insects (Daimon et al., 2008). In the same vein, carotenoid biosynthesis genes of fungal origin have been identified in different herbivorous arthropods (Cobbs, Heath, Stireman, & Abbot, 2013; Grbić et al., 2011; Moran & Jarvik, 2010) and have been suggested to contribute to the herbivorous lifestyle by playing a role as antioxidants (Cobbs et al., 2013). Finally, in hemipteran species known to host obligate endosymbiotic bacteria, HGT events could be identified in the insect genomes that correspond to genes phylogenetically distinct from those of the endosymbionts and that facilitate the mutualistic associations. These HGT genes contribute to the association with the host plant by facilitating the nutritional symbiosis (Luan et al., 2015; Nikoh et al., 2010).

The fact that HGT events have been identified in several other plant parasite species, such as mites and nematodes, suggests that the acquisition of genes by HGT may play an important role in transitions to plant parasitic lifestyles or to herbivory on specific host plants or tissues (Grbić et al., 2011). It is predictable that growing genomic data on insects will unveil new evidence of HGT. Furthermore, new functional approaches allowing targeted inactivation of genes may help to formally link the acquisition of these genes with insect adaptive traits.

4. INDIRECT EFFECT OF SYMBIONTS IN PLANT–INSECT INTERACTIONS: INSECT- AND PLANT-MEDIATED INDIRECT EFFECTS

4.1 Impact on Insect Reproduction

One strategy employed by symbionts to maintain and invade insect populations is the induction of reproductive manipulations, which have been most extensively studied in the case of vertically transmitted endosymbionts. Indeed, this uniparental transmission favours strategies that increase daughter production at the expense of son production. These effects include the feminization of genetic males, the induction of thelytokous parthenogenesis, the killing of infected sons and the induction of cytoplasmic incompatibility (CI), a form of postzygotic reproductive sterility occurring when infected males mate with uninfected females or females infected with another strain of the symbiont (Werren, Baldo, & Clark, 2008). While the most famous reproductive manipulator is the bacterium *Wolbachia*, other symbionts, such as *Arsenophonus*, *Cardinium*, *Rickettsia* or *Spiroplasma*, are also able to manipulate their host reproduction (Engelstädter & Hurst, 2009). These symbionts infect a large number of insect species (see, for example, Duron et al., 2008). Interestingly, relatively few cases of reproductive manipulation by symbionts have been described in Hemiptera. In aphids, one potential reason for this is that many species exhibit complete or cyclical parthenogenesis, which may impair the efficiency of reproductive manipulation. Simon et al. (2011) however demonstrated that *Spiroplasma* induces male killing in the pea aphid. Male-killing bacteria can be maintained in their host populations when killing males procure an indirect advantage to infected females either through the limitation of competition with their brothers or through avoidance of inbreeding depression. This latter point makes a lot of sense in the case of the pea aphid that exhibits cyclical parthenogenesis with maintenance of the genotypic composition during subsequent clonal generations. Indeed, if inbreeding depression is important, limitation of inbreeding due to male killing during the annual event of sexual reproduction may provide an advantage not only at that moment but also during subsequent clonal generations, thus benefiting to the females harbouring the male killer all along the year. This hypothesis would be worth testing either theoretically or by field surveys. Alternatively, *Spiroplasma* may provide other benefits to its hosts that would explain its maintenance in pea aphid populations (Łukasik, van Asch, Guo, Ferrari, & Godfray, 2013). In whiteflies, the recent invasion of *Rickettsia* in the

United States has also been associated with increased female bias in the progeny of infected females (Himler et al., 2011), which may in turn affect the population dynamics of whiteflies.

While not directly influencing the insect—plant interactions, reproductive manipulators can modulate the population dynamics (either positively or negatively) and tend to reduce the genetic diversity and/or recombination rates in infected species (Engelstädter & Hurst, 2009). Through cascading effects, this may ultimately impact the functioning of ecological networks and their co-evolutionary dynamics (Ferrari & Vavre, 2011). These effects could impact not only plant—insect interactions, prey—predator or host—parasitoid interactions but also competitive interactions at a given trophic level (e.g., between herbivores). As an example, the virus LbFv, which manipulates the reproductive behaviour of the parasitoid *Leptopilina boulardi* (Barbotin & Carton et Keiner-Pillault, 1979) (Hymenoptera: Eucoilidae) decreases the competitive ability of this species against *Leptopilina heterotoma* (Thomson, 1862) (Patot, Allemand, Fleury, & Varaldi, 2012). Importantly, the effects of reproductive manipulators on ecological networks, and notably on insect—plant interactions, remain mostly unstudied despite their potential importance.

Reproductive manipulators may also provide interesting candidate for the development of alternative control strategies. In particular, CI-inducing *Wolbachia* could be used to develop insect incompatible techniques (IITs). The proof of principle has been obtained in the fruit fly *Ceratitis capitata* (Wiedemann, 1824) (Diptera: Tephritidae) (Zabalou et al., 2004). This species is naturally not infected by any reproductive manipulator. Through artificial transfection, a *Wolbachia* strain from *Rhagoletis cerasi* (Linnaeus, 1758) has been introduced in *C. capitata*, where it induces 100% CI. Population cage experiments have shown that releasing infected males can indeed be used to control the host populations, in a way very similar to the sterile insect techniques (SIT). One advantage of IIT over SIT is that infected males are generally much more competitive than irradiated males. One drawback is however that only males should be released, which requires perfect sexing strategies, even though alternative strategies have recently been proposed that combine IIT and SIT (Bourtzis, Lees, Hendrichs, & Vreysen, 2016).

4.2 Interactions With Natural Enemies of Herbivores: Protection Versus Attraction

For their maintenance in host populations, some heritable facultative symbionts have adopted strategies that have direct beneficial effects on host fitness. These effects include host plant adaptation (Tsuchida, Koga, & Fukatsu, 2004) and resistance to adverse abiotic (i.e., heat tolerance, Montllor, Maxmen, & Purcell, 2002) but also biotic stresses (Oliver et al., 2010). Microbial symbionts have become increasingly recognized to mediate interactions between herbivorous insects and their natural enemies and, as such, they are important players in the effectiveness of natural enemies to regulate herbivore populations. In herbivorous insect species, symbiont-mediated protection has been first demonstrated in aphids and subsequently observed in other host species such as *Drosophila* spp. Many more examples of protective symbioses in insects will surely be discovered (Oliver et al., 2014). These protective phenotypes may of course impact directly the population dynamics of the host insect, but can also greatly affect the entire community of phytophagous insects through cascading effects (Sanders et al., 2016).

In aphids, facultative bacterial symbionts confer protection against various natural enemies. Different strains of *H. defensa* have been shown to protect the pea aphid *Acyrthosiphon pisum* (Harris, 1776) (Hemiptera: Aphididae), the black bean aphid *Aphis fabae* (Scopoli, 1763) and the cowpea aphid *Aphis craccivora* (Koch, 1854) against insect parasitoids (Asplen et al., 2014; Oliver, Russell, Moran, & Hunter, 2003; Schmid, Sieber, Zimmermann, & Vorburger, 2012). *Hamiltonella defensa* would protect against a large range of parasitoid species in *A. pisum*, although this facultative symbiont may not protect all its host aphid species [e.g., the grain aphid *Sitobion avenae* (Fabricius, 1775), Łukasik, Dawid, Ferrari, & Godfray, 2013]. Protection of aphids against parasitoids could also be associated with other bacterial symbionts: *Regiella insecticola* in the green peach aphid, *Myzus (Nectarosiphon)* persicae (Sulzer, 1776) (Vorburger, Gehrer, & Rodriguez, 2010) and both *Serratia symbiotica* and the Pea Aphid X-type Symbiont in the pea aphid (Guay, Boudreault, Michaud, & Cloutier, 2009; Oliver et al., 2003). Symbionts can also protect aphids against predators and pathogens: *Rickettsiella viridis* provides protection of the pea aphid against ladybirds (Polin, Gallic, Simon, Tsuchida, & Outreman, 2015) and this aphid species may also be protected against fungal pathogens when infected with *R. insecticola*, *R. viridis* or *Spiroplasma* (Ferrari, Darby, Daniell, Godfray, & Douglas, 2004; Łukasik, van Asch, et al., 2013). As protection against natural enemies can be favoured by natural selection, protective symbionts are expected to be fixed within natural aphid populations. Interestingly, these symbionts are found at intermediate frequencies within populations, suggesting that some forces limit their prevalence such as infection costs (Simon et al., 2011).

In *Drosophila*, examples of symbiont-mediated protection are accumulating, and two lineages of bacteria that infect the genus are known to be protective: *Wolbachia* and *Spiroplasma*. Xie, Vilchez, and Mateos (2010) showed that *Spiroplasma* enhances larva-to-adult survival of the fruit fly *Drosophila hydei* (Sturtevant, 1921) (Diptera: Drosophilidae) when parasitized by *L. heterotoma* parasitoid, which attacks the larvae of many *Drosophila* spp. *Wolbachia* has been shown to defend the fruit flies *Drosophila melanogaster* (Meigen, 1830), *Drosophila simulans* (Sturtevant, 1919), *Drosophila innubila* (Spencer, 1943) and *Drosophila suzukii* (Matsumura, 1931) against multiple RNA viruses (e.g., Cattel, Martinez, Jiggins, Mouton, & Gibert, 2016; Hedges, Brownlie, O'Neill, & Johnson, 2008). So far, *Wolbachia* are not known to defend *Drosophila* against other enemies such as parasitoids or predators (Hamilton & Perlman, 2013).

Parasitoid protection provided by *H. defensa* in the pea aphid is associated with the presence of a bacteriophage (APSE), which encodes toxins responsible for prematurely arresting the development of parasitoid immatures (Degnan & Moran, 2008). Underlying mechanisms of parasitoid protection in other systems are often unknown and may involve toxins or diverse biologically active compounds produced by symbionts and targeting insects' natural enemies. Microbial symbionts may also confer protection through indirect mechanisms by competing with natural enemies for limited host resources, by priming the herbivore immune system against subsequent infections (Oliver et al., 2014).

Microbial symbionts may also modulate parasitoid attraction through indirect plant-mediated effects. In response to herbivore attack, plants release distinct bouquets of volatile organic compounds (VOCs) that increase the plant's attractiveness to natural enemies of herbivores (Dicke, van Loon, & Soler, 2009). Given emerging evidences that insect symbionts can modulate the plant signalling pathways (Sections 3.3 and 3.4), it is most likely that volatile emission is influenced by insect-associated symbionts. Alteration of plant VOCs may also be linked to complex interactions with soil-borne microbes such as mycorrhizal fungi also known to influence VOC emissions (Fontana, Reichelt, Hempel, Gershenzon, & Unsicker, 2009) with possible interference between plant and insect symbionts (Hackett, Karley, & Bennett, 2013).

4.3 Interactions With Plant Pathogens: Influence of Insect Symbionts on Plant Pathogen Transmission

Many plant pathogens are transmitted by insect vectors through different types of mechanisms. Plant viruses and phytoplasma are by far those pathogens that rely on insect vectors for their transmission. In some conditions, interests of virus and insect vector may be aligned so that mutualistic associations can evolve. This is, for example, the case of some viruses which reduce plant defences upon their insect vectors' attacks, favouring both the transmission rate of the virus and the growth rate of the insect population (e.g., Luan et al., 2013). Viruses can also increase plant quality with regard to their insect vectors, with mutual benefits for the virus–vector association (Belliure, Janssen, Maris, Peters, & Sabelis, 2005). In other cases, viruses are thought to 'manipulate' their insect vectors' behaviour and physiology in order to favour their acquisition and/or transmission from plant to plant (Ingwell, Eigenbrode, & Bosque-Pérez, 2012; Stafford, Walker, & Ullman, 2011; Su et al., 2013). However, the relationships between plant pathogens, insect vectors and host plants should also consider insect symbionts as a fourth player. They can modulate plant–pathogen transmission by influencing directly or indirectly insect–plant–pathogen interactions (Pinheiro, Kliot, Ghanim, & Cilia, 2015). In hemipterans such as aphids and whiteflies, the circulative transmission of a luteovirus and geminivirus, respectively, was suggested to depend on the presence of a GroEL protein produced by endosymbiotic bacteria associated with these insects (Gottlieb et al., 2010; Hogenhout, van der Wilk, Verbeek, Goldbach, & van den Heuvel, 1998; Morin et al., 1999). In the whitefly *B. tabaci*, vectors of tomato yellow leaf curl virus (TYLCV), different biotypes can be distinguished based on their associations with different symbiotic bacteria (Gueguen et al., 2010). In this system, efficient transmission of TYLCV could be correlated with the presence of a specific symbiotic bacterium, *H. defensa* (Gottlieb et al., 2010; Su et al., 2013). However, the molecular mechanisms involved in plant pathogenic virus–bacterial endosymbiont interactions need refining. A study on an aphid–luteovirus barley yellow dwarf virus interaction suggests that GroEL proteins do not co-localize with virus particles in vivo (Bouvaine, Boonham, & Douglas, 2011). A more recent study showed that *Rickettsia*, another facultative symbiont of *B. tabaci*, improves TYLCV acquisition, transmission and retention in the insects with no involvement of GroEL (Kliot, Cilia, Czosnek, & Ghanim, 2014). These results are in sharp contrast with the protective phenotypes provided by symbionts discussed earlier. In addition, infection by the endosymbiont *Wolbachia* is currently used in mosquitoes to limit transmission of pathogenic viruses such as dengue to humans (Bourtzis et al., 2016). The role of insect symbionts on insect-vectored plant pathogens is a whole research field requiring further investigation. While its potential for application is important, there is a crucial need to determine the circumstances in which positive or negative interactions have to be expected, which probably relies on the direct and indirect costs and benefits associated to co-infections.

5. ECOLOGICAL DIVERSIFICATION AND INSECT DIVERSIFICATION AND SPECIALIZATION

5.1 Ecological Diversification and Plant Specialization

As described earlier, phytophagous insects have established a wide range of symbiotic associations with an array of microorganisms. These symbionts can bring key innovations to their hosts allowing them to colonize new habitats or to extend their ecological niches. Symbionts play a major role in the adaptation of phytophagous insects to different lifestyles and feeding regimens (Janson, Stireman, Singer, & Abbot, 2008). Because sap represents an unbalanced diet for insects, being deprived in some essential amino acids and vitamins, the development of bacterial partnerships has been instrumental in the ecological and evolutionary success of sap-feeding insects (Hansen & Moran, 2014). Each hemipteran group has established an obligate symbiosis with a specific bacterial lineage: *Buchnera* for aphids, *Carsonella* for psyllids, *Portiera* for whiteflies and *Tremblaya* for mealy bugs. Diversification of host and symbiont associations during the course of evolution (180 My for the aphid and whitefly symbioses, 120 My for the psyllid symbiosis and 40 My for the mealy bug symbiosis, Moran et al., 2008) allowed these hemipterans to exploit virtually all plant species on earth and to generate a substantial amount of biodiversity (e.g., 4000 aphid species, more than 1500 whitefly species, 3000—3500 psyllid species, 7000 coccid species).

Acquisition by insects of facultative symbionts can be also very influential for ecological diversification. While these symbionts are largely maternally inherited, they can be horizontally transferred through occasional jumps within or between host species. These jumps may represent for their hosts an instantaneous acquisition of ecologically important traits (e.g., defences against parasitoids or fungi). Facultative symbionts may thus be viewed as a horizontal gene pool that provides the novel host with adaptive traits allowing the exploitation of new habitats and resources or a better success in the current ecological niche (Henry et al., 2013). In addition, a symbiotic association is a dynamic process which could have many types of evolutionary trajectories. For example, the obligate symbiont may be replaced by a facultative symbiont such as in some lineages of weevils (Toju, Tanabe, Notsu, Sota, & Fukatsu, 2013) or may evolve complementation with one or more other symbionts such as in conifer aphids (Manzano-Marín, Simon, & Latorre, 2016), whiteflies (Luan et al., 2015; Rao et al., 2015; Rollat-Farnier et al., 2015), leafhoppers (McCutcheon et al., 2009) or mealy bugs (Husnik et al., 2013).

In some instances, microbial partners may play a role in plant specialization of their host insects. This could be revealed indirectly by showing changes in symbiotic composition of host's populations according to plant species or genus. For example, populations of the chestnut curculio, *Curculio sikkimensis* (Heller, 1927) (Coleoptera: Curculionidae), differ in the prevalence of a range of bacterial symbionts when found on chestnuts or on acorns of wild oaks (Toju & Fukatsu, 2011). In the hemipteran *Phylloxera notabilis* (Pergande, 1904), the bacteria *Pantoea agglomerans* and *Serratia marcescens* are absent in the host populations feeding on water hickory, whereas they are abundant in insects feeding on pecan (Medina, Nachappa, & Tamborindeguy, 2011). Associations between aphids and their commonest facultative symbionts were found to be related more by host plant affiliation than by phylogenetic relationships (Henry et al., 2013). However, other factors than plants can influence symbiont composition of host populations. Therefore, experimental studies are required to support evidence for symbiont-mediated plant specialization in insects. Exchange of the symbiont 'Candidatus *Ishikawaella capsulata*' between two related stink bug species modifies their performances on crop legumes (Hosokawa et al., 2007). Similarly, the facultative symbiont *R. insecticola* was shown to increase fecundity of the aphid *A. pisum* and *Megoura crassicauda* (Mordvilko, 1919) specifically on clover (Tsuchida et al., 2004). However, note that these results have been contradicted by other studies (e.g., McLean, van Asch, Ferrari, & Godfray, 2011). In the polyphagous aphid, *A. craccivora*, the symbiont *Arsenophonus* promotes specialization on locust (*Robinia pseudoacacia* L.) and could expand dietary breath depending on host genotype (Wagner et al., 2015). In the same line, it was demonstrated that *B. tabaci* performed better on tomato when infected by the facultative symbiont *H. defensa*, the proposed underlying mechanisms being a suppression of induced plant defences by the symbiont (Su et al., 2015). The same seems to apply to the olive fly, *Bactrocera oleae* (Rossi, 1790) (Diptera: Tephritidae), whose gut symbiont *Erwinia dacicola* enables fly larvae to develop in unripe olives by overcoming plant defences that accumulate during this phenological stage (Ben-Yosef, Pasternak, Jurkevitch, & Yuval, 2015). In terms of ecological and evolutionary consequences, symbiont mediation of plant specialization may create the conditions for divergent selection among plant-adapted populations or races and eventually trigger speciation events (Wagner et al., 2015).

5.2 Reproductive Isolation and Speciation

Symbiosis has been suggested to promote speciation for a long time. However, the microevolutionary mechanisms by which this process could occur were not identified and made this hypothesis too speculative to be truly considered. This idea has however recently resurged (Brucker & Bordenstein, 2012; Vavre & Kremer, 2014), which distinguished three processes through which symbiosis could facilitate speciation:

- First, as mentioned in the preceding paragraph, acquisition of new symbionts can lead to the adaptation to new niches through biological innovations. Importantly, when populations are facing contrasted environments, different symbiotic communities may be selected for in response to divergent selection, resulting in the local adaptation of the symbiotic community. Once local adaptation is in place, additional mechanisms are however required to reach speciation. In particular, mechanisms limiting gene flow between populations are required. Interestingly, symbionts have been shown to potentially influence both pre- and postmating isolation.
- Symbiosis could influence premating isolation through different ways. The 'easiest' one is when symbionts are involved in habitat specialization and when mating occurs preferentially in the specialized habitat, such as in aphids, for example, that shows high phylopatry (Peccoud, Ollivier, Plantegenest, & Simon, 2009). For example, symbiont-mediated specialization of phytophagous insect populations to a new host plant may facilitate reproductive isolation when insect populations reproduce on the plant they are adapted to. Recent results have also provided thought-provoking examples on the ability of symbionts to modify their host behaviour (Lewis & Lizé, 2015), including modifications involved in premating isolation. These cases involve not only gut symbionts (Lizé, McKay, & Lewis, 2014; Sharon et al., 2010) but also intracellular bacteria of the genus *Wolbachia* such as in

the species complex of *Drosophila paulistorum* (Miller, Ehrman, & Schneider, 2010). In this latter case, symbionts could act on both sexes by affecting emission and perception of cues associated with mate choice.
- The effect of symbiosis on postmating isolation has been particularly studied in the case of the *Wolbachia*-induced CI. When two populations, infected by different strains of mutually incompatible CI-*Wolbachia*, come into contact, gene flow can be drastically reduced (Bordenstein, O'Hara, & Werren, 2001). Unidirectional CI can also favour reinforcement process as proposed in *Drosophila subquinaria* (Spencer, 1942) (Shoemaker, Katju, & Jaenike, 1999). Other results suggest that symbiosis could also participate to postmating reproductive isolation through more indirect effects. In a few cases, hybrid unviability or sterility has indeed been associated with proliferation of symbionts, notably in *D. paulistorum*, where symbionts are also involved in prezygotic isolation (Miller et al., 2010). In a similar way, hybrid mortality can be rescued in *Nasonia* through antibiotic treatments (Brucker & Bordenstein, 2013). In all these cases, the current hypothesis is that hybrids are not able to control the proliferation of symbiotic bacteria that then turn into pathogens, inducing postzygotic isolation. Reduction in hybrid fitness could thus be due to the rupture of co-adapted gene–symbiont interactions, a specific case of Dobzhansky–Bateson–Muller incompatibilities (Brucker & Bordenstein, 2012).

All the points mentioned earlier make symbionts potential actors of speciation. Importantly, symbionts, by potentially affecting both ecological specialization and reproductive isolation, may behave act as 'magic traits' promoting ecological speciation (Servedio, Doorn, van Kopp, Frame, & Nosil, 2011).

5.3 Evolutionary Dynamics of Plant–Insect–Microbe Interactions

As already highlighted in previous sections, acquisition of new microbes is a potent mechanism of adaptation for insects. One important question is thus, which are the mechanisms that facilitate or limit the acquisition of symbionts at an ecological and evolutionary time scale? Understanding these mechanisms first requires adopting a community ecology approach of insect–microbe interactions. Indeed, acquisition of a new symbiont first requires that the symbiont comes into contact with the new host species, and this will obviously occur preferentially between species sharing ecological networks (Henry, Maiden, Ferrari, & Godfray, 2015). For example, in the hymenopteran seed feeders of the genus *Megastigmus*, horizontal transmission of parthenogenesis-inducing *Wolbachia* has occurred repeatedly, but is constrained by the insect specialization on different plant families (Boivin et al., 2014). Similarly, while different *Curculio* species developing on oak acorns have distinct predominant symbionts, residual infection by different symbionts probably acquired through recurrent horizontal transmission are observed (Merville et al., 2013). In these examples, horizontal transmission may also be facilitated by the phylogenetic proximity of the hosts, which probably allows better evasion of the host immune system by the symbiont. This phenomenon has also been recently demonstrated using artificial transinfection of symbionts in aphids (Łukasik et al., 2015). However, horizontal transfer between highly divergent species may also occur occasionally (e.g., Vavre, Fleury, Lepetit, Fouillet, & Bouletreau, 1999). Another factor that may affect the establishment of a new symbiont is the presence of other symbionts.

Interestingly, recent results have demonstrated that some symbionts may be horizontally transmitted in herbivores through interactions with parasitoids (e.g., *Rickettsia* in whiteflies, Chiel et al., 2009; *H. defensa* and *R. insecticola* in aphids, Gehrer & Vorburger, 2012) or the host plant (e.g., *Rickettsia* in whiteflies, Caspi-Fluger et al., 2012; or *Cardinium* in Cicadellidae, Gonella et al., 2015). Importantly, in the case of *Rickettsia*, the symbiont was apparently metabolically active in the plant, suggesting that at least some symbionts may be able to also exploit the host plant. How these acquisitions may lead to transitions in the lifestyle of symbionts is an open question, but the genus *Arsenophonus* provides an interesting example of these potential transitions. Indeed, phenotypic effects associated with *Arsenophonus* are extremely diverse. Known first as a reproductive manipulator, *Arsenophonus* is also a mutualist in book lice (Perotti, Allen, Reed, & Braig, 2007), and provision of benefits is also suspected in aphids and whiteflies (Wagner et al., 2015). On the other hand, *Arsenophonus* is also a plant pathogen vectored by plant hoppers (reviewed in Bressan, 2014). This example highlights the blurred line between insect symbionts and plant pathogens, at least at an evolutionary scale.

6. CONCLUSION AND OUTLOOK

In the past few years, insect microbial symbionts have emerged as key players in plant–insect interactions with tremendous ecological and evolutionary implications highlighted in this review. They can exert their influence through direct effects on their insect host as well as through indirect plant-mediated effects. Indirect effects can impact the whole plant-associated ecological networks including plant-associated pathogens or mutualistic symbionts but also other herbivores that will exploit the same host plant. One of the forthcoming challenges will be now to connect environment, plant and insect microbiota to shed light on the evolution and functioning of complex multi-trophic interactions in which plants, herbivorous insects and microorganisms are inserted.

As mentioned in this review, higher trophic levels such as parasitoids can also be impacted through direct or indirect interactions with phytophagous insects' symbionts. The rising awareness of the important roles that microbial symbionts play in natural enemies ecology has led to a steep increase in the identification of ecologically important traits being attributed to symbiosis. Because insect microbial symbionts modulate the effectiveness of natural enemies and may be manipulated, they are potential targets for biological control programmes. Qualitative or quantitative alterations of the microbiome may largely modify the ability of insect pests to exploit their host plants but may also improve the ability of parasitoids to detect and control them. Reproductive manipulators may also provide interesting candidates for the development of alternative control strategies by altering the population dynamics of crop pests. Use of symbiotic microorganisms as potential biological agents for controlling insect pests now needs to be fully considered.

Finally, most of the focus has been so far on insect-associated bacteria but other insect symbionts can be instrumental and may help insects to exploit their host plant or to modulate their interactions with their whole ecological community. Extending our knowledge to all types of insect symbionts (e.g., viruses) is now required to gain a deeper understanding of the ecology and evolution of plant–insect interactions. Deciphering molecular mechanisms underlying such plant–insect–microbe interactions in model systems under controlled environments and in more natural ecological settings still also needs to be uncover. The revolution in our understanding of the role of symbionts has been made possible by the many advances in molecular biology. The next decade is likely to see major progress in unravelling the mechanisms underlying these interactions. How microbes associated with insect modify the host plant physiology and development leading to extravagant plant alterations such as galls or how HGT is mediated by plant–insect–microbe interactions and how this promotes the colonization of new ecological niches are questions that can now be fully addressed. Understanding mechanisms underlying plant—insect—microbe interactions will shed light on these exciting research topics and hold promise for a global understanding of plant biotic interactions.

ACKNOWLEDGEMENTS

We are grateful to Akiko Sugio for her constructive comments on an earlier version of this book chapter.

REFERENCES

Acuña, R., Padilla, B. E., Flórez-Ramos, C. P., Rubio, J. D., Herrera, J. C., Benavides, P., … Rose, J. K. (2012). Adaptive horizontal transfer of a bacterial gene to an invasive insect pest of coffee. *Proceedings of the National Academy of Sciences of the United States of America, 109*, 4197—4202.

Adams, A. S., Aylward, F. O., Adams, S. M., Erbilgin, N., Aukema, B. H., Currie, C. R., … Raffa, K. F. (2013). Mountain pine beetles colonizing historical and naive host trees are associated with a bacterial community highly enriched in genes contributing to terpene metabolism. *Applied and Environmental Microbiology, 79*, 3468—3475.

Asplen, M. K., Bano, N., Brady, C. M., Desneux, N., Hopper, K. R., Malouines, C., … Heimpel, G. E. (2014). Specialisation of bacterial endosymbionts that protect aphids from parasitoids. *Ecological Entomology, 39*, 736—739.

Aylward, F. O., Suen, G., Biedermann, P. H., Adams, A. S., Scott, J. J., Malfatti, S. A., … Klepzig, K. D. (2014). Convergent bacterial microbiotas in the fungal agricultural systems of insects. *mBio, 5*(6), e02077-14.

Baeza Garcia, A., Pierce, R. J., Gourbal, B., Werkmeister, E., Colinet, D., Reichhart, J. M., … Coustau, C. (2010). Involvement of the cytokine MIF in the snail host immune response to the parasite *Schistosoma mansoni*. *PLoS Pathogens, 6*(9), e1001115.

Bansal, R., Hulbert, S. H., Reese, J. C., Whitworth, R. J., Stuart, J. J., & Chen, M. S. (2014). Pyrosequencing reveals the predominance of Pseudomonadaceae in gut microbiome of a gall midge. *Pathogens, 3*, 459—472.

Barr, K. L., Hearne, L. B., Briesacher, S., Clark, T. L., & Davis, G. E. (2010). Microbial symbionts in insects influence down-regulation of defense genes in maize. *PLoS One, 5*, e11339.

Baumann, P. (2005). Biology bacteriocyte-associated endosymbionts of plant sap-sucking insects. *Annual Review of Microbiology, 59*, 155—189.

Behmer, S. T. (2009). Insect herbivore nutrient regulation. *Annual Review of Entomology, 54*, 165—187.

Belliure, B., Janssen, A., Maris, P. C., Peters, D., & Sabelis, M. W. (2005). Herbivore arthropods benefit from vectoring plant viruses. *Ecology Letters, 8*, 70—79.

Ben-Yosef, M., Pasternak, Z., Jurkevitch, E., & Yuval, B. (2015). Symbiotic bacteria enable olive fly larvae to overcome host defences. *Royal Society Open Science, 2*, 150170.

Biere, A., & Bennett, A. E. (2013). Three-way interactions between plants, microbes and insects. *Functional Ecology, 27*, 567—573.

Body, M., Kaiser, W., Dubreuil, G., Casas, J., & Giron, D. (2013). Leaf-miners co-opt microorganisms to enhance their nutritional environment. *Journal of Chemical Ecology, 39*, 969—977.

Boivin, T., Henri, H., Vavre, F., Gidoin, C., Veber, P., Candau, J. N., ... Auger-Rozenberg, M. A. (2014). Epidemiology of asexuality induced by the endosymbiotic *Wolbachia* across phytophagous wasp species: Host plant specialization matters. *Molecular Ecology, 23*, 2362–2375.

Boone, C. K., Keefover-Ring, K., Mapes, A. C., Adams, A. S., Bohlmann, J., & Raffa, K. F. (2013). Bacteria associated with a tree-killing insect reduce concentrations of plant defense compounds. *Journal of Chemical Ecology, 39*, 1003–1006.

Bordenstein, S. R., O'Hara, F. P., & Werren, J. H. (2001). *Wolbachia*-induced incompatibility precedes other hybrid incompatibilities in *Nasonia*. *Nature, 409*, 707–710.

Boto, L. (2014). Horizontal gene transfer in the acquisition of novel traits by metazoans. *Proceedings of the Royal Society B-Biological Sciences, 281*. http://dx.doi.org/10.1098/rspb.2013.2450.

Bourtzis, K., Lees, R. S., Hendrichs, J., & Vreysen, M. J. B. (2016). More than one rabbit out of the hat: Radiation, transgenic and symbiont-based approaches for sustainable management of mosquito and tsetse fly populations. *Acta Tropica, 157*, 115–130.

Bouvaine, S., Boonham, N., & Douglas, A. E. (2011). Interactions between a luteovirus and the GroEL chaperonin protein of the symbiotic bacterium *Buchnera aphidicola* of aphids. *Journal of General Virology, 92*, 1467–1474.

Bressan, A. (2014). Emergence and evolution of *Arsenophonus* bacteria as insect-vectored plant pathogens. *Infection, Genetics and Evolution, 22*, 81–90.

Broderick, N. A., Raffa, K. F., Goodman, R. M., & Handelsman, J. (2004). Census of the bacterial community of the gypsy moth larval midgut by using culturing and culture-independent methods. *Applied and Environmental Microbiology, 70*, 293–300.

Brucker, R. M., & Bordenstein, S. R. (2012). Speciation by symbiosis. *Trends in Ecology & Evolution, 27*, 443–451.

Brucker, R. M., & Bordenstein, S. R. (2013). The hologenomic basis of speciation: Gut bacteria cause hybrid lethality in the genus *Nasonia*. *Science, 341*, 667–669.

Brune, A., & Dietrich, C. (2015). The gut microbiota of termites: Digesting the diversity in the light of ecology and evolution. *Annual Review of Microbiology, 69*, 145–166.

Buchner, P. (1965). *Endosymbiosis of animals with plant microorganisms*. New York.

Calandra, T., & Roger, T. (2003). Macrophage migration inhibitory factor: A regulator of innate immunity. *Nature Reviews Immunology, 3*(10), 791–800.

Calderon-Cortes, N., Quesada, M., Watanabe, H., Cano-Camacho, H., & Oyama, K. (2012). Endogenous plant cell wall digestion: A key mechanism in insect evolution. *Annual Review of Ecology, Evolution and Systematics, 43*, 45–71.

Caspi-Fluger, A., Inbar, M., Mozes-Daube, N., Katzir, N., Portnoy, V., Belausov, E., ... Zchori-Fein, E. (2012). Horizontal transmission of the insect symbiont *Rickettsia* is plant-mediated. *Proceedings of the Royal Society B: Biological Sciences, 279*, 1791–1796.

Cattel, J., Martinez, J., Jiggins, F. M., Mouton, L., & Gibert, P. (2016). *Wolbachia*-mediated protection against viruses in the invasive pest *Drosophila suzukii*. *Insect Molecular Biology*. http://dx.doi.org/10.1111/imb.12245.

Chaudhary, R., Atamiana, H. S., Shenc, Z., Briggsc, S. P., & Kaloshian, I. (2014). GroEL from the endosymbiont *Buchnera aphidicola* betrays the aphid by triggering plant defense. *Proceedings of the National Academy of Sciences of the United States of America, 111*, 8919–8924.

Chiel, E., Zchori-Fein, E., Inbar, M., Gottlieb, Y., Adachi-Hagimori, T., Kelly, S. E., ... Hunter, M. S. (2009). Almost there: Transmission routes of bacterial symbionts between trophic levels. *PLoS One, 4*, e4767.

Chu, C. C., Spencer, J. L., Curzi, M. J., Zavala, J. A., & Seufferheld, M. J. (2013). Gut bacteria facilitate adaptation to crop rotation in the western corn rootworm. *Proceedings of the National Academy of Sciences of the United States of America, 110*, 11917–11922.

Chuche, J., Danet, J.-L., Salar, P., Foissac, X., & Thiéry, D. (2016). Transmission of 'Candidatus *Phytoplasma solani*' by *Reptalus quinquecostatus* (Hemiptera: Cixiidae). *Annals of Applied Biology, 169*, 214–223. http://dx.doi.org/10.1111/aab.12291.
Chung, S. H., Rosa, C., Scully, E. D., Peiffer, M., Tooker, J. F., Hoover, K., ... Felton, G. W. (2013). Herbivore exploits orally secreted bacteria to suppress plant defenses. *Proceedings of the National Academy of Sciences of the United States of America, 110*, 15728–15733.
Cobbs, C., Heath, J., Stireman, J. O., & Abbot, P. (2013). Carotenoids in unexpected places: Gall midges, lateral gene transfer, and carotenoid biosynthesis in animals. *Molecular Phylogenetics and Evolution, 68*, 221–228.
Colman, D. R., Toolson, E. C., & Takacs-Vesbach, C. D. (2012). Do diet and taxonomy influence insect gut bacterial communities? *Molecular Ecology, 21*, 5124–5137.
Cornell, H. V. (1983). The secondary chemistry and complex morphology of galls formed by the Cynipinae (Hymenoptera): Why and how? *The American Midland Naturalist Journal, 110*, 225–234.
Cryan, J. R., & Urban, J. M. (2012). Higher-level phylogeny of the insect order Hemiptera: Is Auchenorrhyncha really paraphyletic? *Systematic Entomology, 37*, 7–21.
Daimon, T., Taguchi, T., Meng, Y., Katsuma, S., Mita, K., & Shimada, T. (2008). Beta-fructofuranosidase genes of the silkworm, *Bombyx mori*: Insights into enzymatic adaptation of *B. mori* to toxic alkaloids in mulberry latex. *The Journal of Biological Chemistry, 283*, 15271–15279.
Degnan, P. H., & Moran, N. A. (2008). Diverse phage-encoded toxins in a protective insect endosymbiont. *Applied and Environmental Microbiology, 74*, 6782–6791.
Dicke, M., van Loon, J. J. A., & Soler, R. (2009). Chemical complexity of volatiles from plants induced by multiple attack. *Nature Chemical Biology, 5*, 317–324.
Douglas, A. E. (2013). Microbial brokers of insect-plant interactions revisited. *Journal of Chemical Ecology, 39*, 952–961.
Dowd, P. F., & Shen, S. K. (2011). The contribution of symbiotic yeast to toxin resistance of the cigarette beetle (*Lasioderma serricorne*). *Entomological Experimentalis et Applicata, 56*, 241–248.
Dubreuil, G., Deleury, E., Crochard, D., Simon, J.-C., & Coustau, C. (2014). Diversification of MIF immune regulators in aphids: Link with agonistic and antagonistic interactions. *BMC Genomics, 15*, 762.
Duron, O., Bouchon, D., Boutin, S., Bellamy, L., Zhou, L., Engelstädter, J., & Hurst, G. D. (2008). The diversity of reproductive parasites among arthropods: *Wolbachia* do not walk alone. *BMC Biology, 6*, 27.
Engel, P., & Moran, N. A. (2013). The gut microbiota of insects – Diversity in structure and function. *FEMS Microbiology Reviews, 37*, 699–735.
Engelstädter, J., & Hurst, G. D. D. (2009). The ecology and evolution of microbes that manipulate host reproduction. *Annual Review of Ecology, Evolution and Systematics, 40*, 127–149.
Erb, M., Meldau, S., & Howe, G. A. (2012). Role of phytohormones in insect-specific plant reactions. *Trends in Plant Science, 17*, 250–259.
Ferrari, J., Darby, A. C., Daniell, T. J., Godfray, H. C. J., & Douglas, A. E. (2004). Linking the bacteria community in pea aphids with host-plant use and natural enemy resistance. *Ecological Entomology, 29*, 60–65.
Ferrari, J., & Vavre, F. (2011). Bacterial symbionts in insects or the story of communities affecting communities. *Philosophical Transactions of the Royal Society B, 366*, 1389–1400.
Fontana, A., Reichelt, M., Hempel, S., Gershenzon, J., & Unsicker, S. B. (2009). The effects of arbuscular mycorrhizal fungi on direct and indirect defense metabolites of *Plantago lanceolata* L. *Journal of Chemical Ecology, 35*, 833–843.

Frago, E., Dicke, M., & Godfray, H. C. J. (2012). Insect symbionts as hidden players in insect-plant interactions. *Trends in Ecology & Evolution, 27*, 705–711.

Gehrer, L., & Vorburger, C. (2012). Parasitoids as vectors of facultative bacterial endosymbionts in aphids. *Biology Letters, 8*, 613–615.

Giron, D., Frago, E., Glevarec, G., Pieterse, C. M. J., & Dicke, M. (2013). Cytokinins as key regulators in plant-microbe-insect interactions: Connecting plant growth and defence. *Functional Ecology, 27*, 599–609.

Giron, D., & Glevarec, G. (2014). Cytokinin-induced phenotypes in plant–insect interactions: Learning from the bacterial world. *Journal of Chemical Ecology, 40*, 826–835.

Giron, D., Huguet, E., Stone, G. N., & Body, M. (2016). Insect-induced effects on plants and possible effectors used by galling and leaf-mining insects to manipulate their host-plant. *Journal of Insect Physiology, 84*, 70–89.

Gonella, E., Pajoro, M., Marzorati, M., Crotti, E., Mandrioli, M., Pontini, M., … Daffonchio, D. (2015). Plant-mediated interspecific horizontal transmission of an intracellular symbiont in insects. *Scientific Reports, 5*, 15811.

Gottlieb, Y., Zchori-Fein, E., Mozes-Daube, N., Kontsedalov, S., Skaljac, M., Brumin, M., … Ghanim, M. (2010). The transmission efficiency of tomato yellow leaf curl virus by the whitefly *Bemisia tabaci* is correlated with the presence of a specific symbiotic bacterium species. *Journal of Virology, 84*, 9310–9317.

Grbić, M., Van Leeuwen, T., Clark, R. M., Rombauts, S., Rouzé, P., Grbić, V., … Hernández-Crespo, P. (2011). The genome of *Tetranychus urticae* reveals herbivorous pest adaptations. *Nature, 479*, 487–492.

Groen, S. C., Humphrey, P. T., Chevasco, D., Ausubel, F. M., Pierce, N. E., & Whiteman, N. K. (2016). *Pseudomonas syringae* enhances herbivory by suppressing the reactive oxygen burst in *Arabidopsis*. *Journal of Insect Physiology, 84*, 90–102.

Guay, J. F., Boudreault, S., Michaud, D., & Cloutier, C. (2009). Impact of environmental stress on aphid clonal resistance to parasitoids: Role of *Hamiltonella defensa* bacterial symbiosis in association with a new facultative symbiont of the pea aphid. *Journal of Insect Physiology, 55*, 919–926.

Gueguen, G., Vavre, F., Gnankine, O., Peterschmitt, M., Charif, D., Chiel, E., … Fleury, F. (2010). Endosymbiont metacommunities, mtDNA diversity and the evolution of the *Bemisia tabaci* (Hemiptera: Aleyrodidae) species complex. *Molecular Ecology, 19*, 4365–4376.

Hackett, S. C., Karley, A. J., & Bennett, A. E. (2013). Unpredicted impacts of insect endosymbionts on interactions between soil organisms, plants and aphids. *Proceedings of the Royal Society B: Biological Sciences, 280*, 2131275. http://dx.doi.org/10.1098/rspb.2013.1275.

Hamilton, P. T., & Perlman, S. J. (2013). Host defense via symbiosis in *Drosophila*. *PLoS Pathogens, 9*(12), e1003808.

Hansen, A. K., & Moran, N. A. (2014). The impact of microbial symbionts on host plant utilization by herbivorous insects. *Molecular Ecology, 23*, 1473–1496.

Hedges, L. M., Brownlie, J. C., O'Neill, S. L., & Johnson, K. N. (2008). *Wolbachia* and virus protection in insects. *Science, 322*, 702.

Heil, M., & McKey, D. (2003). Protective ant-plant interactions as model systems in ecological and evolutionary research. *Annual Review of Ecology, Evolution and Systematics, 34*, 425–453.

Henry, L. M., Maiden, M. C. J., Ferrari, J., & Godfray, H. C. J. (2015). Insect life history and the evolution of bacterial mutualism. *Ecology Letters, 18*, 516–525.

Henry, L. M., Peccoud, J., Simon, J.-C., Hadfield, J. D., Maiden, M. J. C., Ferrari, J., & Godfray, H. C. J. (2013). Horizontally transmitted symbionts and host colonization of ecological niches. *Current Biology, 23*, 1713–1717.

Himler, A. G., Adachi-Hagimori, T., Bergen, J. E., Kozuch, A., Kelly, S. E., Tabashnik, B. E., … Hunter, M. S. (2011). Rapid spread of a bacterial symbiont in an invasive whitefly is driven by fitness benefits and female bias. *Science, 332*, 254–256.

Hoffmann, J. A., & Reichhart, J. M. (2002). *Drosophila* innate immunity: An evolutionary perspective. *Nature Immunology, 3*, 121–126.

Hogenhout, S. A., van der Wilk, F., Verbeek, M., Goldbach, R. W., & van den Heuvel, J. F. (1998). Potato leafroll virus binds to the equatorial domain of the aphid endosymbiotic GroEL homolog. *Journal of Virology, 72*, 358–365.

Hosokawa, T., Kikuchi, Y., Meng, X. Y., & Fukatsu, T. (2005). The making of symbiont capsule in the plataspid stinkbug *Megacopta punctatissima. FEMS Microbiology Ecology, 54*, 471–477.

Hosokawa, T., Kikuchi, Y., Shimada, M., & Fukatsu, T. (2007). Obligate symbiont involved in pest status of host insect. *Proceedings of the Royal Society B-Biological Sciences, 274*, 1979–1984.

Husnik, F., Nikoh, N., Koga, R., Ross, L., Duncan, R. P., Fujie, M., … Von Dohlen, C. D. (2013). Horizontal gene transfer from diverse bacteria to an insect genome enables a tripartite nested mealybug symbiosis. *Cell, 153*, 1567–1578.

Ingwell, L. L., Eigenbrode, S. D., & Bosque-Pérez, N. A. (2012). Plant viruses alter insect behavior to enhance their spread. *Scientific Reports, 2*, 578. http://dx.doi.org/10.1038/srep00578.

Janson, E. M., Stireman, J. O., Singer, M. S., & Abbot, P. (2008). Phytophagous insect-microbe mutualisms and adaptive evolutionary diversification. *Evolution, 62*, 997–1012.

Jones, R. T., Sanchez, L. G., & Fierer, N. (2013). A cross-taxon analysis of insect-associated bacterial diversity. *PLoS One, 8*(4), e61218.

Kaiser, W., Huguet, E., Casas, J., Commin, C., & Giron, D. (2010). Plant green-island phenotype induced by leaf-miners is mediated by bacterial symbionts. *Proceedings of the Royal Society B-Biological Sciences, 277*, 2311–2319.

Kaltenpoth, M., Winter, S. A., & Kleinhammer, A. (2009). Localization and transmission route of *Coriobacterium glomerans*, the endosymbiont of pyrrhocorid bugs. *FEMS Microbiology Ecology, 69*, 373–383.

Kikuchi, Y., Hosokawa, T., & Fukatsu, T. (2007). Insect-microbe mutualism without vertical transmission: A stinkbug acquires a beneficial gut symbiont from the environment every generation. *Applied and Environmental Microbiology, 73*, 4308–4316.

Kirsch, R., Gramzow, L., Theißen, G., Siegfried, B. D., Ffrench-Constant, R. H., Heckel, D. G., & Pauchet, Y. (2014). Horizontal gene transfer and functional diversification of plant cell wall degrading polygalacturonases: Key events in the evolution of herbivoryin beetles. *Insect Biochemistry and Molecular Biology, 52*, 33–50.

Kirsch, R., Heckel, D. G., & Pauchet, Y. (2016). How the rice weevil breaks down the pectin network: Enzymatic synergism and sub-functionalization. *Insect Biochemistry and Molecular Biology, 71*, 72–82.

Kliot, A., Cilia, M., Czosnek, H., & Ghanim, M. (2014). Implication of the bacterial endosymbiont *Rickettsia* spp. in interactions of the whitefly *Bemisia tabaci* with tomato yellow leaf curl virus. *Journal of Virology, 88*, 5652–5660.

Koehncke, A., Telschow, A., & Kondoh, M. (2012). Invasibility as an emergent property of native metapopulation structure. *Oikos, 122*, 332–340.

Kostovcik, M., Bateman, C. C., Kolarik, M., Stelinski, L. L., Jordal, B. H., & Hulcr, J. (2015). The ambrosia symbiosis is specific in some species and promiscuous in others: Evidence from community pyrosequencing. *ISME Journal, 9*, 126–138.

Laughton, A. L., Fan, M. H., & Gerardo, N. M. (2014). Combined effects of symbionts and ageing on life history traits in the pea aphid, *Acyrthosiphon pisum. Applied and Environmental Microbiology, 80*, 470–477.

Lewis, Z., & Lizé, A. (2015). Insect behaviour and the microbiome. *Current Opinion in Insect Science, 9*, 86–90.
Lieutier, F., Bermudez-Torres, K., Cook, J., Harris, M. O., Legal, L., Sallé, A., … Giron, D. (2017). From plant exploitation to mutualism. In N. Sauvion, D. Thiery, & P. A. Calatayud (Eds.), *Insect-Plant Interactions in a Crop Protection Perspective* (Vol. 81, pp. 55–109).
Lizé, A., McKay, R., & Lewis, Z. (2013). Gut microbiota and kin recognition. *Trends in Ecology & Evolution, 28*, 325–326.
Lizé, A., McKay, R., & Lewis, Z. (2014). Kin recognition in *Drosophila*: The importance of ecology and gut microbiota. *ISME Journal, 8*, 469–477.
Luan, J. B., Chen, W., Hasegawa, D. K., Simmons, A. M., Wintermantel, W. M., Ling, K. S., … Douglas, A. E. (2015). Metabolic coevolution in the bacterial symbiosis of whiteflies and related plant sap-feeding insects. *Genome Biology and Evolution, 7*, 2635–2647. http://dx.doi.org/10.1093/gbe/evv170.
Luan, J. B., Yao, D. M., Zhang, T., Walling, L. L., Yang, M., Wang, Y. J., & Liu, S. S. (2013). Suppression of terpenoid synthesis in plants by a virus promotes its mutualism with vectors. *Ecology Letters, 16*, 390–398.
Łukasik, P., van Asch, M., Guo, H., Ferrari, J., & Godfray, H. C. J. (2013). Unrelated facultative endosymbionts protect aphids against a fungal pathogen. *Ecology Letters, 16*, 214–218.
Łukasik, P., Dawid, M. A., Ferrari, J., & Godfray, H. C. J. (2013). The diversity and fitness effects of infection with facultative endosymbionts in the grain aphid, *Sitobion avenae*. *Oecologia, 173*, 985–996.
Łukasik, P., Guo, H., van Asch, M., Henry, L. M., Godfray, H. C. J., & Ferrar, J. (2015). Horizontal transfer of facultative endosymbionts is limited by host relatedness. *Evolution, 69*, 2757–2766.
Manzano-Marín, A., Simon, J.-C., & Latorre, A. (2016). Reinventing the wheel and making it round (but smaller) again: Evolutionary convergence in *Buchnera–Serratia* symbiotic consortia between the distantly related aphids *Tuberolachnus salignus* (Lachninae, Lachnini) and *Cinara cedri* (Lachninae, Eulachnini). *Genome Biology and Evolution, 8*, 1440–1458. http://dx.doi.org/10.1093/gbe/evw085.
McCutcheon, J. P., McDonald, B. R., & Moran, N. A. (2009). Convergent evolution of metabolic roles in bacterial co-symbionts of insects. *Proceedings of the National Academy of Sciences of the United States of America, 106*, 15394–15399.
McLean, A. H. C., van Asch, M., Ferrari, J., & Godfray, H. C. J. (2011). Effects of bacterial secondary symbionts on host plant use in pea aphids. *Proceedings of the Royal Society B-Biological Sciences, 278*, 760–766.
Medina, R. F., Nachappa, P., & Tamborindeguy, C. (2011). Differences in bacterial diversity of host-associated populations of *Phylloxera notabilis* Pergande (Hemiptera: Phylloxeridae) in pecan and water hickory. *Journal of Evolutionary Biology, 24*, 761–771.
Merville, A., Venner, S., Henri, H., Vallier, A., Menu, F., Vavre, F., … Bel-Venner, M.-C. (2013). Endosymbiont diversity among sibling weevil species competing for the same resource. *BMC Evolutionary Biology, 13*, 28. http://dx.doi.org/10.1186/1471-2148-13-28.
Miller, W. J., Ehrman, L., & Schneider, D. (2010). Infectious speciation revisited: Impact of symbiont-depletion on female fitness and mating behavior of *Drosophila paulistorum*. *PLoS Pathogens, 6*, e1001214.
Montllor, C. B., Maxmen, A., & Purcell, A. H. (2002). Facultative bacterial endosymbionts benefit pea aphids *Acyrthosiphon pisum* under heat stress. *Ecological Entomology, 27*, 189–195.
Moran, N. A., Degnan, P. H., Santos, S. R., Dunbar, H. E., & Ochman, H. (2005). The players in a mutualistic symbiosis: Insects, bacteria, viruses, and virulence genes. *Proceedings of the National Academy of Sciences of the United States of America, 102*, 16919–16926.
Moran, N. A., & Jarvik, T. (2010). Lateral transfer of genes from fungi underlies carotenoid production in aphids. *Science, 328*, 624–627.

Moran, N. A., McCutcheon, J. P., & Nakabachi, A. (2008). Genomics and evolution of heritable bacterial symbionts. *Annual Review of Genetics, 42*, 165–190.

Morin, S., Ghanim, M., Zeidan, M., Czosnek, H., Verbeek, M., & van den Heuvel, J. F. (1999). A GroEL homologue from endosymbiotic bacteria of the whitefly *Bemisia tabaci* is implicated in the circulative transmission of tomato yellow leaf curl virus. *Virology, 256*, 75–84.

Mueller, U. G., Gerardo, N. M., Aanen, D. K., Six, D. L., & Schultz, T. R. (2005). The evolution of agriculture in insects. *Annual Review of Ecology, Evolution and Systematics, 36*, 563–595. Retrieved from http://www.jstor.org/stable/30033817.

Naessens, E., Dubreuil, G., Giordanengo, P., Baron, O. L., Minet-Kebdani, N., Keller, H., & Coustau, C. A. (2015). Secreted MIF cytokine enables aphid feeding and represses plant immune responses. *Current Biology, 25*, 1898–1903.

Nikoh, N., McCutcheon, J. P., Kudo, T., Miyagishima, S. Y., Moran, N. A., & Nakabachi, A. (2010). Bacterial genes in the aphid genome: Absence of functional gene transfer from *Buchnera* to its host. *PLoS Genetics, 6*(2), e1000827.

Novotny, V., Basset, Y., Miller, S. E., Weiblen, G. D., Bremer, B., Cizek, L., & Drozd, P. (2002). Low host specificity of herbivorous insects in a tropical forest. *Nature, 416*, 841–844.

Oliver, K. M., Degnan, P. H., Burke, G. R., & Moran, N. A. (2010). Facultative symbionts in aphids and the horizontal transfer of ecologically important traits. *Annual Review of Entomology, 55*, 247–266. http://dx.doi.org/10.1146/annurev-ento-112408-085305.

Oliver, K. M., Russell, J. A., Moran, N. A., & Hunter, M. S. (2003). Facultative bacterial symbionts in aphids confer resistance to parasitic wasps. *Proceedings of the National Academy of Sciences of the United States of America, 100*, 1803–1807.

Oliver, K. M., Smith, A. H., & Russell, J. A. (2014). Defensive symbiosis in the real world -advancing ecological studies of heritable, protective bacteria in aphids and beyond. *Functional Ecology, 28*, 341–355.

Patot, S., Allemand, R., Fleury, F., & Varaldi, J. (2012). An inherited virus influences the coexistence of parasitoid species through behaviour manipulation. *Ecology Letters, 15*, 603–610.

Peccoud, J., Ollivier, A., Plantegenest, M., & Simon, J.-C. (2009). A continuum of genetic divergence from sympatric host races to species in the pea aphid complex. *Proceedings of the National Academy of Sciences of the United States of America, 106*, 7495–7500.

Perotti, M. A., Allen, J. M., Reed, D. L., & Braig, H. R. (2007). Host-symbiont interactions of the primary endosymbiont of human head and body lice. *The FASEB Journal, 21*, 1058–1066.

Pinheiro, P. V., Kliot, A., Ghanim, M., & Cilia, M. (2015). Is there a role for symbiotic bacteria in plant virus transmission by insects? *Current Opinion in Insect Science, 8*, 69–78.

Polin, S., Le Gallic, J. F., Simon, J.-C., Tsuchida, T., & Outreman, Y. (2015). Conditional reduction of predation risk associated with a facultative symbiont in an insect. *PLoS One, 10*(11), e0143728.

Poulsen, M., Hu, H. F., Li, C., Chen, Z. S., Xu, L. H., Otani, S., … Zhang, G. (2014). Complementary symbiont contributions to plant decomposition in a fungus-farming termite. *Proceedings of the National Academy of Sciences of the United States of America, 111*, 14500–14505.

Rao, Q., Rollat-Farnier, P.-A., Zhu, D. T., Santos-Garcia, D., Silva, F. J., Moya, A., … Liu, S. S. (2015). Genome reduction and potential metabolic complementation of the dual endosymbionts in the whitefly *Bemisia tabaci*. *BMC Genomics, 16*, 226. http://dx.doi.org/10.1186/s12864-015-1379-6.

Raubenheimer, D., Simpson, S. J., & Mayntz, D. (2009). Nutrition, ecology and nutritional ecology: Toward an integrated framework. *Functional Ecology, 23*, 4–16.

Robert, C. A. M., Frank, D. L., Leach, K. A., Turlings, T. C., Hibbard, B. E., & Erb, M. (2013). Direct and indirect plant defenses are not suppressed by endosymbionts of a specialist root herbivore. *Journal of Chemical Ecology, 39*, 507–515.

Rohfritsch, O. (2008). Plants, gall midges, and fungi: A three-component system. *Entomological Experimentalis et Applicata, 128*, 208–216.

Rollat-Farnier, P.-A., Santos-Garcia, D., Rao, Q., Sagot, M. F., Silva, F. J., Henri, H., … Liu, S. S. (2015). Two host clades, two bacterial arsenals: Evolution through gene losses in facultative endosymbionts. *Genome Biology and Evolution, 7*, 839–855.

Salem, H., Florez, L., Gerardo, N., & Kaltenpoth, M. (2015). An out-of-body experience: The extracellular dimension for the transmission of mutualistic bacteria in insects. *Proceedings of the Royal Society B-Biological Sciences, 282*, 20142957.

Sanders, D., Kehoe, R., van Veen, F. F., McLean, A., Godfray, H. C. J., Dicke, M., … Frago, E. (2016). Defensive insect symbiont leads to cascading extinctions and community collapse. *Ecology Letters, 19*, 789–799.

Schmid, M., Sieber, R., Zimmermann, Y. S., & Vorburger, C. (2012). Development, specificity and sublethal effects of symbiont-conferred resistance to parasitoids in aphids. *Functional Ecology, 26*, 207–215.

Schmitz, A., Anselme, C., Ravallec, M., Rebuf, C., Simon, J.-C., Gatti, J.-L., & Poirié, M. (2012). The cellular immune response of the pea aphid to foreign intrusion and symbiotic challenge. *PLoS One, 7*(7), e42114.

Schoonhoven, L. M., van Loon, J. J. A., & Dicke, M. (2005). *Insect–plant biology* (2nd ed.). Oxford: Oxford University Press.

Schulenburg, H., Boehnisch, C., & Michiels, N. K. (2007). How do invertebrates generate a highly specific innate immune response? *Molecular Immunology, 44*, 3338–3344.

Servedio, M. R., Doorn, G. S., van Kopp, M., Frame, A. M., & Nosil, P. (2011). Magic traits in speciation: 'magic' but not rare? *Trends in Ecology & Evolution, 26*, 389–397.

Sharon, G., Segal, D., Ringo, J. M., Hefetz, A., Zilber-Rosenberg, I., & Rosenberg, E. (2010). Commensal bacteria play a role in mating preference of *Drosophila melanogaster*. *Proceedings of the National Academy of Sciences of the United States of America, 107*, 20051–20056.

Shoemaker, D. D., Katju, V., & Jaenike, J. (1999). *Wolbachia* and the evolution of reproductive isolation between *Drosophilla recens* and *Drosophila subquinaria*. *Evolution; International Journal of Organic Evolution, 53*, 1157–1164.

Simon, J.-C., Boutin, S., Tsuchida, T., Koga, R., Gallic, J. F., Frantz, A., … Fukatsu, T. (2011). Facultative symbiont infections affect aphid reproduction. *PLoS One, 6*, e21831.

Slansky, F., Jr., & Rodriguez, J. G. (1987). *Nutritional ecology of insects, mites, spiders, and related invertebrates*. New-York: John Wiley.

Soucy, S. M., Huang, J., & Gogarten, J. P. (2015). Horizontal gene transfer: Building the web of life. *Nature Reviews Genetics, 16*, 472–482.

Stafford, C. A., Walker, G. P., & Ullman, D. E. (2011). Infection with a plant virus modifies vector feeding behavior. *Proceedings of the National Academy of Sciences of the United States of America, 108*, 9350–9355.

Su, Q., Oliver, K. M., Pan, H., Jiao, X., Liu, B., Xie, W., … Zhou, X. (2013). Facultative symbiont *Hamiltonella* confers benefits to *Bemisia tabaci* (Hemiptera: Aleyrodidae), an invasive agricultural pest worldwide. *Environmental Entomology, 42*, 1265–1271.

Su, Q., Oliver, K. M., Xie, W., Wu, Q., Wang, S., & Zhang, Y. (2015). The whitefly-associated facultative symbiont *Hamiltonella defensa* suppresses induced plant defences in tomato. *Functional Ecology, 29*, 1007–1018.

Sugio, A., Dubreuil, G., Giron, D., & Simon, J.-C. (2015). Plant-insect interactions under bacterial influence: Ecological implications and underlying mechanisms. *Journal of Experimental Botany, 66*, 467–478.

Sun, B. F., Xiao, J. H., He, S. M., Liu, L., Murphy, R. W., & Huang, D. W. (2013). Multiple ancient horizontal gene transfers and duplications in lepidopteran species. *Insect Molecular Biology, 22*, 72–87.

Toju, H., & Fukatsu, T. (2011). Diversity and infection prevalence of endosymbionts in natural populations of the chestnut weevil: Relevance of local climate and host plants. *Molecular Ecology, 20*, 853–868.

Toju, H., Tanabe, A. S., Notsu, Y., Sota, T., & Fukatsu, T. (2013). Diversification of endosymbiosis: Replacements, co-speciation and promiscuity of bacteriocyte symbionts in weevils. *ISME Journal, 7*, 1378–1390.

Tsuchida, T., Koga, R., Fujiwara, A., & Fukatsu, T. (2014). Phenotypic effect of 'Candidatus *Rickettsiella viridis*', a facultative symbiont of the pea aphid (*Acyrthosiphon pisum*), and its interaction with a coexisting symbiont. *Applied and Environmental Microbiology, 80*, 525–533.

Tsuchida, T., Koga, R., & Fukatsu, T. (2004). Host plant specialization governed by facultative symbiont. *Science, 303*, 1989.

Vavre, F., Fleury, F., Lepetit, D., Fouillet, P., & Bouletreau, M. (1999). Phylogenetic evidence for horizontal transmission of *Wolbachia* in host-parasitoid associations. *Molecular Biology and Evolution, 16*, 1711–1723.

Vavre, F., & Kremer, N. (2014). Microbial impacts on insect evolutionary diversification: From patterns to mechanisms. *Current Opinion in Insect Science, 4*, 29–34.

Visotto, L. E., Oliveira, M. G. A., Guedes, R. N. C., Ribon, A. O. B., & Good-God, P. I. V. (2009). Contribution of gut bacteria to digestion and development of the velvetbean caterpillar, *Anticarsia gemmatalis*. *Journal of Insect Physiology, 55*, 185–191.

Vorburger, C., Gehrer, L., & Rodriguez, P. (2010). A strain of the bacterial symbiont *Regiella insecticola* protects aphids against parasitoids. *Biology Letters, 6*, 109–111.

Wagner, S. M., Martinez, A. J., Ruan, Y. M., Kim, K. L., Lenhart, P. A., Dehnel, A. C., … White, J. A. (2015). Facultative endosymbionts mediate dietary breadth in a polyphagous herbivore. *Functional Ecology, 29*, 1402–1410.

Werren, J. H., Baldo, L., & Clark, M. E. (2008). Wolbachia: Master manipulators of invertebrate biology. *Nature Reviews Microbiology, 6*, 741–751.

Wybouw, N., Dermauw, W., Tirry, L., Stevens, C., Grbić, M., Feyereisen, R., & van Leeuwen, T. (2014). A gene horizontally transferred from bacteria protects arthropods from host plant cyanide poisoning. *eLife, 3*, e02365.

Xie, J., Vilchez, I., & Mateos, M. (2010). *Spiroplasma* bacteria enhance survival of *Drosophila hydei* attacked by the parasitic wasp *Leptopilina heterotoma*. *PLoS One, 5*, e12149.

Zabalou, S., Riegler, M., Theodorakopoulou, M., Stauffer, C., Savakis, C., & Bourtzis, K. (2004). *Wolbachia*-induced cytoplasmic incompatibility as a means for insect pest population control. *Proceedings of the National Academy of Sciences of the United States of America, 101*, 15042–15045.

Zarate, S. I., Kempema, L. A., & Walling, L. L. (2007). Silverleaf whitefly induces salicylic acid defenses and suppresses effectual jasmonic acid defenses. *Plant Physiology, 143*, 866–875.

Zhang, H., de Bernonville, T. D., Body, M., Glevarec, G., Reichelt, M., Unsicker, S., … Giron, D. (2016). Leaf-mining by *Phyllonorycter blancardella* reprograms the host-leaf transcriptome to modulate phytohormones associated with nutrient mobilization and plant defense. *Journal of Insect Physiology, 84*, 114–127.

Zhao, C., Escalante, L. N., Chen, H., Benatti, T. R., Qu, J., Chellapilla, S., … Batterton, M. (2015). A massive expansion of effector genes underlies gall-formation in the wheat pest *Mayetiola destructor*. *Current Biology, 25*, 613–620.

Zhu, F., Poelman, E. H., & Dicke, M. (2014). Insect herbivore-associated organisms affect plant responses to herbivory. *New Phytologist, 204*, 315–321.

CHAPTER EIGHT

How Host Plant and Fluctuating Environments Affect Insect Reproductive Strategies?

J. Moreau*[,1], E. Desouhant[§], P. Louâpre*, M. Goubault[¶], E. Rajon[§], A. Jarrige[¶], F. Menu[§], D. Thiéry[||]

*University of Bourgogne Franche-Comté, Evolutionary Ecology Team, UMR 6282 Biogéosciences, 6 Bd Gabriel, Dijon, France
[§]University of Lyon, Claude Bernard University, CNRS, Laboratory of Biometry and Evolutionary Biology UMR 5558, Villeurbanne, France
[¶]IRBI (Insect Biology Research Institute) – UMR 7261 CNRS/Université François-Rabelais de Tours, Tours, France
[||]INRA, UMR SAVE 1065, Bordeaux Sciences Agro, Centre INRA de recherches de Bordeaux- Aquitaine, Institut des Sciences de la Vigne et du Vin, Villenave d'Ornon Cedex, France
[1]Corresponding author: E-mail: jerome.moreau@u-bourgogne.fr

Contents

Advances in Botanical Research, Volume 81
ISSN 0065-2296
http://dx.doi.org/10.1016/bs.abr.2016.09.008

Abstract

Host plants possibly represent the strongest selection pressure for the evolution of reproductive traits in phytophagous insects. In a first part of this chapter, we review how plant quality affects both female and male life history traits and their respective reproductive success, and how the production and transfer to females of male sperm and associated nongametic substances (spermatophores as nuptial gifts) also depend on the host plant choice. At first glance, it seems that reproductive traits in phytophagous insects should be selected to maximize the success of this short-term interaction between host plant and phytophagous insects. This, however, ignores the fact that variation in reproductive success is detrimental to long-term fitness, which may explain that reproductive traits depart from their short-term expectation in unpredictable environments. Bet-hedging strategies — as exemplified by spatial or temporal dispersal (e.g., prolonged diapause) — can therefore evolve in such environments, as described in the second part of this chapter. The knowledge reviewed in this chapter is also integrated in the broader applied perspective of insect pest population management.

1. INTRODUCTION

Many aspects of the phytophagous insect's life history depend on the host plant on which they develop, emerge or reproduce. Temporal seasonal and intraday variations in quality and quantity of the plants may result from many external factors (e.g., climate or soil conditions such as nitrogen contents and water resource) and internal factors (i.e., plant health or injuries). In response, phytophagous insects have evolved life-history strategies to deal with spatial and temporal variability of the host plant quality. For example, the regulation of the offspring size by insects in response to the host plant quality, or the synchrony between egg hatching and host plant phenology are keypoints to understand insect pest outbreaks, or the dynamics of trophic networks in which phytophagous insects are involved [e.g., see chapter: Food Webs and Multiple Biotic Interactions in Plant-Herbivore Models by Corcket, Giffard, and Sforza (2016)]. Periodic environmental variation such as seasons (i.e., predictable component of environmental variability) plays a major role in the evolution of phytophagous insect life history traits. However, environment stochasticity (i.e., unpredictable component) must not be neglected since it can explain curious insect strategies as bet-hedging.

Even if several literature reviews on reproductive strategies in response to host plant quality exist (see Awmack & Leather, 2002 for an example), we attempt, in this chapter, to review the different effects of the host plants on insects' immediate fecundity, through the modulation of both female and male gamete production. We then focus on the impact of variation of host quality and availability in time and space (i.e., fluctuating environment) on their reproductive strategies, in particular through the evolution of bet-hedging strategies. We also discuss the literature in the light of crop protection against insect pests, especially in the context of climate change.

2. EFFECT OF HOST PLANT QUALITY ON MALE AND FEMALE REPRODUCTION

Reproduction is one of the most energy consuming activity for an individual during its lifetime. In most insect species, reproduction is a nutrient-limited process for both sexes, and largely relates to the individuals' energy reserves (e.g., Boggs & Freeman, 2005). The resources needed for reproduction can be acquired during the juvenile instar or adult stage. Species that can only acquire resources at the juvenile stage are referred to as *capital breeders*. For such species, the quality of the larval food plant [i.e., set of characteristics including levels of chemicals (nitrogen, carbon, defensive compounds...), reviewed by Awmack & Leather, 2002] is decisive as insect fitness is ultimately dependent on larval plant quality (Telang & Wells, 2004). Other species that can acquire additional resources during adulthood are called *income breeders*. Contrary to capital breeders, income breeders can compensate for the acquisition of poor-quality food at the larval stage; thus plant quality is generally considered more critical for capital breeders than for income breeders. However, both income and capital breeding females may complete the nutritive resource devoted to reproduction by nutrients contained in nuptial gifts or transferred by the partner during copulation (Lewis & South, 2012).

In this context, the nutrients obtained by phytophagous insects can be allocated to their somatic tissues and affect the general body condition of larvae and both adult males and females in many ways. For instance, these resources can be used to decrease their developmental duration (Cahenzli & Erhardt, 2012; Tigreros, 2013) or increase adult longevity (Cahenzli & Erhardt, 2012, 2013; Fritzsche & Arnqvist, 2015), body size (Cahenzli & Erhardt, 2012, 2013; Fritzsche & Arnqvist, 2015; Tigreros, 2013), and initial lipid, protein and glycogen contents (Vande Velde, Schtickzelle, &

Van Dyck, 2013). The reproductive output of most phytophagous insects is improved by an increase in the host plant's protein and carbon-based nutrient concentrations, and by a decrease in soluble carbohydrate concentration. The mineral content of the host plant also influences the reproduction of herbivorous insects, but this effect is complex, sometimes unclear and will not be addressed in this chapter. Among the different chemical substances contained in host plants, nitrogen has been identified as the key nutrient required by herbivorous insects (White, 1993). For instance, reproductive performance of aphid species is higher on high-nitrogen (Khan & Port, 2008). In addition to nutrients, secondary plant compounds can impact the performance of phytophagous insects (as reviewed by Hilker & Meiners, 2011). Overall, nutrient stress conditions occurring early in life (e.g., food deprivation or low plant quality) can greatly impede individual fitness (e.g., survival and fecundity) as well as less direct effects, such as a decrease in the ability to attract and locate mates (Muller, Arenas, Thiéry, & Moreau, 2016).

The host plant is thus an extremely important ecological factor for phytophagous insects of both sexes, but its gender-specific effect has been the object of scarce attention. Here we review the effect of host plant quality on female and male reproductive strategies.

2.1 Effects of the Host Plant Quality on Larval Development and Consequences on Female Reproductive Output

Host plant quality affects several female life history traits such as larval growth, diapause induction and larval defence against natural enemies (Coley, Bateman, & Kursar, 2006; Hunter & Mcneil, 1997). It is also known that the quality and quantity of plant tissues consumed during larval stages affects the reproduction of females, thus highlighting the importance of larval diet for future reproductive events (Awmack & Leather, 2002; Blanckenhorn, 2000; Honěk, 1993; Leather, 1994; Thiéry & Moreau, 2005).

For example, many essential compounds in butterfly eggs are exclusively provided by the larval diet (O'Brien, Boggs, & Fogel, 2013). When host plant consumed at the larval stage is of poor quality, females (especially income breeder females) can, to some extent, override this potential handicap by the use of high-quality plants as adults. Map butterflies [*Araschnia levana* (Linnaeus, 1758)] females, for instance, obtain amino acids required for egg production from nectar (Mevi-Schutz & Erhardt, 2005). In general, both larval residual energy derived from the host plant and the food ingested as adult are used for egg production, such that both income and capital breeder species rely on the quality of the host plant on which larvae develop.

The effects of host plant quality on the reproductive output of herbivorous insects are generally investigated through three main proxies: the number of eggs produced, their probability of hatching and their size. For example, egg size and clutch size depend on host plant quality: females feeding on plants of poor quality generally lay smaller and fewer eggs than those feeding on plants of high quality (Fox & Czesak, 2000). These proxies of the female fitness are, however, not synonymous, and they should be interpreted differently (see below).

2.1.1 Host Plant and Clutch Size

Fecundity is the number of eggs produced by a female during her lifetime. Potential and realized fecundities are usually considered, being defined as, respectively, the number of eggs in the reproductive tract and the number of eggs laid. The first one is a good estimate of the female potential fitness and is certainly the most studied reproductive life history trait due to the convenience of its estimation (Awmack & Leather, 2002; Leather, 1994). However, its interpretation is limited as egg maturation or fertilization can fail, thus resulting in sterile or aborting eggs. Moreover, when eggs are produced during the adult stage, the number of eggs in the reproductive tract may not be representative of the total number of eggs laid during a lifetime, which may vary depending on internal and external factors such as the quality and quantity of host plant consumed. For all these reasons, realized fecundity is a better fitness proxy than potential fecundity.

Because laid eggs may fail to develop, it is preferable to estimate fitness through fertility, defined as the number of hatching eggs. The relevance of this trait as a fitness proxy was demonstrated by Moreau, Benrey, and Thiery (2006) who studied the reproduction of one of the main vine pests, *Lobesia botrana* (Denis & Schiffermüller, 1775) (Lepidoptera: Tortricidae). In this species, the female fecundity is influenced by the variety of grapes they develop on, whereby females grown on Merlot lay more eggs than those grown on Riesling (Fig. 1A). Using fecundity as a fitness proxy, one would expect Merlot to be a better host for female development than Riesling. However, eggs of females grown on Riesling have higher hatching rates than those of females grown on Merlot (Fig. 1B). These two contradicting conclusions demonstrate that both fecundity and fertility should be considered carefully. In fact, the measure of the reproductive rate per female (number of larvae produced per female) shows no relationship with the host plant

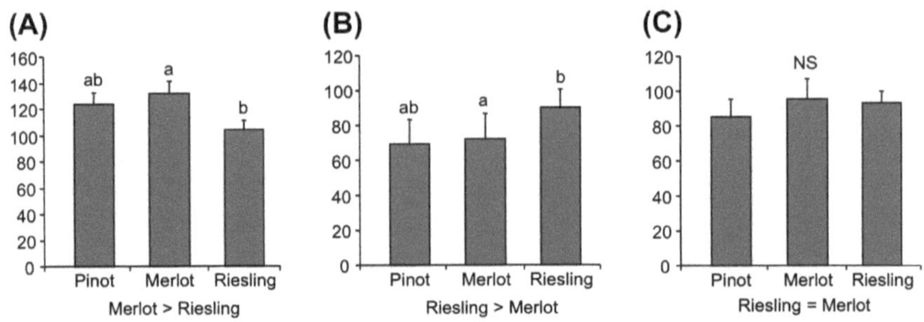

Figure 1 (A) Fecundity (number of eggs laid), (B) fertility (percentage of eggs hatched) and (C) reproductive rate (number of larvae produced per female) according to the cultivar where the larvae fed on. The text below each panel indicates the better cultivar for *L. botrana* females. *Modified from Moreau, J., Benrey, B., & Thiery, D. (2006). Assessing larval food quality for phytophagous insects: Are the facts as simple as they appear?* Functional Ecology, *20, 592–600.*

on which females developed (Fig. 1C). In this species, a high fecundity is counterbalanced by a low fertility. Unfortunately, fertility is difficult to measure under field conditions, thus is often neglected. Most studies investigating the quality of the host plant on herbivore reproduction only considered female fecundity, implicitly assuming that high-quality host plants will also result in a higher or similar hatching success as low-quality plants (see Tammaru, Esperk, & Castellanos, 2002 for an exception). This missing life history trait is, however, puzzling in our general understanding on how host plant quality affects the fitness of phytophagous insects.

2.1.2 Host Plant and Egg Size

As detailed above, the effect of host plant quality on female fitness cannot be limited to egg numbers but one should also take into account the hatching probability, which is directly linked to egg size. It is commonly accepted that females face a choice between the production of a large number of small eggs and the production of a small batch of large eggs (Fox & Czesak, 2000). Indeed, although the scientific community knew about these evolutionary trade-offs for a long time, most of the studies devoted to identify the effect of host plant quality on the fitness of phytophagous insects analyzed only one or independently a few life history traits. However, it is essential to study all fitness related life history traits together to fully understand the effects of larval food quality on herbivore fitness.

Egg size is recognized as a crucial reproductive life history trait for females. Egg size is related with host plant quality by at least two distinct mechanisms. On the one hand, females that developed on plants of high quality may accumulate enough resources to produce numerous large eggs. In this case, a direct relationship exists between the quality of host plant consumed by the female and offspring life history traits (Moreau, Arruego, Benrey, & Thiéry, 2006; Moreau, Benrey, et al., 2006). As stated above, the main sources of energy for egg production come from the resources accumulated at the larval stages, especially for capital breeder's species (Awmack & Leather, 2002; Kaspi, Mossinson, Drezner, Kamensky, & Yuval, 2002). Larger eggs are thus often associated with more nutritional provisions allocated by females (Berrigan, 1991; Fox & Czesak, 2000). On the other hand, females may express egg size plasticity in response to the host plant quality at the oviposition site. For example, females of the seed beetle, *Stator limbatus* (Horn, 1873) (Coleoptera: Chrysomelidae: Bruchinae), tend to deposit larger eggs on low rather than on high-quality host plants, thus increasing survival of larvae on low-quality host plants (Fox, 1997). In this case, the link between egg size and the host plant quality is indirect and does not involve a direct physiological relationship between the mother's condition and her reproductive output. Egg size is thus a fitness proxy linked with host plant quality by a complex relationship and depends on the quality of the plant on which the mother developed and the eggs are laid (Ekbom & Popov, 2004).

It is generally assumed that producing larger eggs gives them an advantage for they result in larger larvae. Indeed, under stressful conditions (lack of food, adverse environmental conditions…), a larger egg with more reserves is better equipped to resist desiccation or withstand stressful or variable conditions. Moreover, such large larvae should have bigger mandibles with obvious advantage in the perforation of plant tissues and thus in food acquisition (Awmack & Leather, 2002; Fox & Czesak, 2000). Ekbom and Popov (2004) thus suggested that large larvae emerging from large eggs are advantaged on poor-quality host plants in comparison with small larvae emerging from small eggs. Consequently, larger larvae may have a greater chance to successfully establish on a plant than smaller ones.

As part of an ecosystem, eggs and larvae also face many trophic threats, such as other phytophagous competitors or predators and parasitoids [e.g., chapters: Food Webs and Multiple Biotic Interactions in Plant-Herbivore Models by Corcket et al. (2016) and The Plant as a Habitat for Entomophagous Insects by Kaiser et al. (2016)]. The egg characteristics and the embryo development are influenced by secondary plant compounds, such as toxins incorporated into the eggs by females, which protect the embryo from parasitism and predation (Blum & Hilker, 2008). Considering offspring size, strong trophic pressure gives an advantage for intermediate larval size, resulting in the selection for nonmaximized larval growth. The classic explanation of such phenomenon is that larger eggs and larvae are known to be more apparent and susceptible to predation/parasitism than smaller ones but this effect is offset by the fact that they have more energy to escape and defence against their threats (Berger, Walters, & Gotthard, 2006).

2.2 Effect of Host Plant on Male Reproductive Output

There is increasing evidence that host plants also influence the reproductive output of insect males. In phytophagous species, male reproduction is generally affected by the quality or availability of nutrients acquired by feeding on plants as larvae (Muller, Thiéry, Moret, & Moreau, 2015; Takakura, 2004; Tigreros, 2013) or adults (Cahenzli & Erhardt, 2012, 2013; Fritzsche & Arnqvist, 2015). In some cases, plants can also affect male reproduction through the chemical compounds they emit (Ali, 2012; Shelly & Epsky, 2015). Altogether, the quality of host plants on which males feed can affect their reproductive output through variations in (1) the traits involved in mate acquisition and (2) the production of sperm and associated nongametic resources (spermatophore) that impacts female reproductive output, in terms of the number and quality of offspring males sire. Due to the costs associated with the production of sexual traits, sperm and spermatophores, males may further adjust their investment in such traits according to female quality and exhibit mate choice strategies.

2.2.1 Male Attractiveness and Competitive Ability

Males' ability to attract females and secure sexual partners and mating territories against rival males can be modulated by the resources obtained during host plant feeding. In the European grapevine moth (*L. botrana*), for instance, grape cultivars on which individuals develop influence their probability of acquiring a mate (Moreau, Thiery, Troussard, & Benrey, 2007). More precisely, females discriminate males of different qualities during the precopulatory phase and they preferentially call (i.e., via pheromones) those that fed on higher quality cultivars as larvae (Muller et al., 2015). In contrast to many species where females tend to prefer larger males, the preference observed in *L. botrana* females is unrelated to male size (Muller et al., 2015). While the cues or signals females use to discriminate males' host plant origin remain to be identified in this species, it has been demonstrated in others that host plant quality can affect male sexual signals. In particular, the production of sex pheromones often relies on bioactive chemicals or chemical precursors, as illustrated by two examples: pyrrolizidine alkaloids in arctiid moths, *Utetheisa ornatrix* (Linnaeus, 1758) (Lepidoptera: Noctuoidea) (Conner, Eisner, Vandermeer, Guerrero, & Meinwald, 1981), and methyl eugenol in the oriental fruit fly, *Bactrocera dorsalis* (Hendel, 1912) (Diptera: Tephritidae) (Shelly, 2000), that males acquire when feeding on plant as adults or sequester as larvae (see Landolt & Phillips, 1997 for a review). It then confers a mating advantage to males emitting pheromones with high levels of these compounds (Shelly, 2000).

Host plant feeding can also alter other male sexual traits associated with their mating success such as colouration and vibratory signals. In the small white butterfly, *Pieris rapae* (Linnaeus, 1758) (Lepidoptera: Pieridae), females prefer more colourful males (Morehouse & Rutowski, 2010) whereby wing colouration is based on pterins, pigments rich in nitrogen which is mainly acquired during larval feeding (Tigreros, 2013). In the *Enchenopa binotata* species complex of treehoppers (Hemiptera: Membracidae), males communicate with plant-borne vibrational signals which vary according to the clone plant on which they fed as larvae (Rebar & Rodríguez, 2014). This likely affects their reproductive success, with females exhibiting strong mate preference on the basis of the features of those signals (Rodriguez, Sullivan, & Cocroft, 2004). Thus the host plant quality can condition phytophagous male reproductive output by affecting the characteristics of their sexual traits, and recent studies show that it can additionally affect male sexual behaviour. In species where males display territorial behaviours, those in better physiological conditions generally have greater chances of accessing and holding territories (Briffa & Hardy, 2013). As said above, host plant can largely affect individuals' morphology (e.g., body mass/size, wing size, flight muscle ratio) and energy reserves (i.e., lipid, protein, glycogen contents). It results in variation in male territorial success and potentially in various male sexual strategies. Experimental work conducted in the speckled wood, *Pararge aegeria* (Linnaeus, 1758) (Lepidoptera: Nymphalydae), showed that larval food plant quality is a key determinant of male adult body mass, lipid content and flight muscle ratio (Vande Velde et al., 2013). These parameters further condition the performance of perching males when defending sunlit spots on the forest floor where they wait for females. As a result, low-quality males that developed on drought-stressed plants adopt an alternative nonterritorial searching behaviour (i.e., patrolling tactic), which is less energetically demanding (Vande Velde et al., 2013).

2.2.2 Sperm and Associated Substances Production

While numerous studies have examined the effect of variation in both larval and adult nutrition of the plant on female gamete production (see above), much less is known about the effect of plant production on sperm production. The few studies gave contrasting results and mainly concerned Lepidoptera, in which males produce two types of sperm: the eupyrene (i.e., fertile) sperm, and the apyrene (i.e., nonfertile) sperm, whose function remains unclear but could potentially play a role in sperm competition (Silberglied, Shepherd, & Dickinson, 1984). In *L. botrana*, host plant (grape cultivars) on which larvae feed affects the number of eupyrene sperm produced and males transferring more sperm have a greater reproductive output by fertilizing a greater number of eggs (Muller et al., 2015). Nevertheless, to our knowledge this result is unique since in two other cases, no link between larval nutrition on host plant and sperm production was found [in *P. aegeria* in Vande Velde et al., 2013; and *Bicyclus anynana* (Butler, 1879) (Lepidoptera: Nymphalidae) in Lewis & Wedell, 2007]. Complementary adult feeding on nectar or rotten fruit has also no apparent direct effect on eupyrene sperm production in *B. anynana* (Lewis & Wedell, 2007). This latter result is, however, not surprising because, in Lepidoptera, spermatogenesis ends before adult emergence (Friedländer, 1997). Only the production of apyrene sperm continues during the early adult life and might be affected by adult feeding. Further studies are needed to draw general conclusions on this topic.

In many insect species, not only phytophagous ones, males also provide females with nongametic resources prior to or during mating (Lewis & South, 2012; Vahed, 1998) (Fig 2). Such 'nuptial gifts' can take many forms such as preys captured by the male, oral secretion or spermatophores produced by the male, or body parts or the whole body of the male (see Vahed,

Figure 2 (A) *Ephippiger diurnus* male transferring his spermatophore to a female (on top) during mating. (B) *E. diurnus* female with a spermatophore [in orange (grey in print versions): the ampulla containing the sperm, in white: the spermatophylax containing various nutrients and water]. *Sonia Dourlot.*

1998 for a complete review). They contain a variety of compounds such as proteins, fats, carbohydrates, minerals, uric acid (Vahed, 1998) which are passed to the females and potentially affect both female and male reproductive outputs. Elegant radiolabelling experiments showed that spermatophore-derived substances (such as amino acids, zinc, phosphorus and sodium) passed to the female somatic tissue and eggs (Gwynne, 2001; Lewis & South, 2012; Vahed, 1998). In addition, other chemicals present in the spermatophore, such as plant-derived defensive compounds, such as pyrrolizidine alkaloids or cyanogenic glycosides, can also favour the protection of females, their eggs or both against predation (Eisner & Meinwald, 1995). Little is known about the impact of male diet on their spermatophore quality. One study demonstrated that, in the small white butterfly (*P. rapae*) the quality of the larval food [on two different host plants: garlic mustard, *Alliaria petiolata* (M.Bieb.) Cavara & Grande, and nasturtium, *Tropaeolum majus* L.] affected the size of the spermatophores (Cook & Wedell, 1996). Other experiments showed that supplementing male adult diet with electrolytes and amino acids enhance their spermatophore mass, size and contents (e.g., Lederhouse, Ayres, & Scriber, 1990). The subsequent consequences on the male reproductive output, in terms of the quality and quantity of eggs fertilized by the gift-giving male, have received even less attention (Delisle & Bouchard, 1995; Delisle & Hardy, 1997; Royer & McNeil, 1993; South & Lewis, 2011). In the rosaceous leaf roller, *Choristoneura rosaceana* (Harris, 1841) (Lepidoptera: Tortricidae), for instance, males fed as larvae with beaked hazel (*Corylus cornuta* Marshall), a host plant of low nutritional quality, produced smaller spermatophores than males fed with striped maple (*Acer pensylvanicum* L.), a plant of higher nutritional value (Delisle & Bouchard, 1995). In return, females mated once with hazel-fed males lay fewer eggs and produce fewer offspring than those mated once with maple-fed males. Therefore host plant quality affects male reproductive success in phytophagous species through its effect on the size and nutrient contents of the spermatophore that males transfer to females at mating (Delisle & Bouchard, 1995; Muller et al., 2015; South & Lewis, 2011). Nevertheless, the precise relationship between the quality of host plant on which males feed and the quality of their spermatophore and the associated fitness return, deserves to be extended to other species to draw general conclusions on these relationships in plant-eating insects.

Furthermore, spermatophores not always act as a paternal investment, by increasing the number or quality of males' offspring, but also as a mating effort by protecting the donors' sperm (Gwynne, 2001; Vahed,

1998). In such a case, nuptial-gift quality can modulate the female refractory period and consequently the risk of sperm competition that males face. Females receiving a large spermatophore usually show a long remating latency (Muller et al., 2016), ultimately increasing the male's fertilization success.

2.2.3 Consequences on Male Mate Choice

Spermatophores are highly costly to produce and can thus strongly limit males' reproductive rate (Gwynne, 2001; Lewis & South, 2012). Hence, it would be advantageous for males to adjust their reproductive investment to female quality. In line with this hypothesis, males of the bush cricket, *Ephippiger diurnus* (Dufour, 1841) (Orthoptera: Tettigoniidae), modify the size and composition of their spermatophore according to female body mass and age (Jarrige, Greenfield, & Goubault, 2013, 2015) (Fig 2). In this species presenting a last-male sperm precedence effect (i.e., the sperm of the last male to mate a female fertilizes a larger proportion of the female's eggs; Hockham, Graves, & Ritchie, 2004), gifts transferred to older females which are less likely to remate contained large amounts of nutritious protein-bound amino acids (Jarrige, Body, Giron, Greenfield, & Goubault, 2015). In contrast, younger large females, in which sperm competition risk is higher, received more diluted spermatophores, containing increased amounts of free glycine, a substance which increases female handling time of the spermatophore. This gift allocation has been suggested to represent a form of cryptic mate choice, allowing males to maximize their chances of paternity in relation to the risk of sperm competition associated with mate quality. Nevertheless, to our knowledge, the impact of host plant feeding on spermatophore quality and its subsequent consequences on male differential investment in female by manipulating its composition remains to be investigated.

Males can also show more direct mate preference, which can vary according to the plant they developed from. For instance, larval host plant experience modulates male attraction to the female sex pheromone in the cotton worm, *Spodoptera littoralis* (Boisduval, 1833) (Lepidoptera: Noctuidae). In this species, males were more attracted to the female sex pheromone when combined with the odour of the host plant species they fed on as larvae than to female sex pheromone combined with odours of host plant species they did not experience (Anderson, Sadek, Larsson, Hansson, & Thöming, 2013).

2.2.4 Effect of Exposure to Host Plant Volatiles on Male Reproductive Output

Host plant can also affect the male reproductive output without requiring males to feed on them. Plant volatiles can indeed favour male mate finding especially when females concentrate on host plants to feed and oviposit. In such a situation, plants act as 'sexual rendezvous' points (Bernays & Chapman, 1994). Additionally, the exposure to plant odours can modulate male emission of sex pheromone (e.g., Bachmann et al., 2015), male responsiveness to female sex pheromone (Binyameen, Hussain, Yousefi, Birgersson, & Schlyter, 2013), male competitiveness (Morató, Shelly, Rull, & Aluja, 2015) and, as a result, male mating success (Vera et al., 2013). Male responses to host plant volatiles are extensively studied in fruit flies, because a better understanding of such effects may permit the development of new control techniques of these pests of agronomic importance. *Ceratitis capitata* (Wiedemann, 1824) (Diptera: Tephritidae) males exposed to various *Citrus* species or common guava (*Psidium guajava* L.) show a mating advantage over nonexposed males (Shelly & Epsky, 2015). This advantage can be conferred by the increased calling behaviour and pheromone release displayed by exposed males, as observed in the South American fruit fly, *Anastrepha fraterculus* (Wiedemann, 1830) (Bachmann et al., 2015). Likewise, the reproductive success of males exposed to the essential oils extracted from different host plants is strongly enhanced (Morató et al., 2015; Shelly & Epsky, 2015). This effect could be more specifically triggered by specific compounds, such as terpenes (e.g., α-copaene; Shelly, 2001) present in the plant odour. These effects can be used to increase sterile males' competitiveness over wild males in pest management programs whereby sterile males exposed to plant oils during mass-rearing outcompete wild males for copulation with wild females up to 3 days after exposure (Shelly & Epsky, 2015).

3. INSECT REPRODUCTIVE STRATEGIES IN RISKY ENVIRONMENTS

Phytophagous insects live in varying environments and have evolved life-history strategies to deal with this variability. Only a part of this variability comes from host plant–insect interactions. Thus it would be misleading to consider this part alone in studies of the evolution of these strategies. In this section, we highlight cases where such strategies have been studied extensively or suggested in phytophagous insects. We also review bet-hedging strategies well known in other organisms but that require more empirical and theoretical attention in phytophagous insects.

3.1 Evolutionary Strategies in Risky Environments

Plants differ in their suitability as food for a given insect species. In a stable environment where plants are abundant, one may predict that natural selection favours the ability of a female to lay its eggs in a host plant that is most suitable for her offspring's development (Bernays, 1991; Mayhew, 1997). The fitness consequences of a wrong choice are particularly severe because developing phytophagous insects are often unable to switch from one host plant to another. This leads to a positive correlation between the preference of a female for a host plant and the performance of her offspring – often referred to as the preference–performance hypothesis (PPH; Gripenberg, Mayhew, Parnell, & Roslin, 2010). This relationship has been demonstrated in several examples, but dozens of studies report a weak or no relationship between preference and performance (for a review see Gripenberg et al., 2010). This suggests that females sometimes choose less suitable or unsuitable host plants, which raises the following question: why, in some contexts, females exhibit apparent suboptimal choices? One possible explanation may be that the abundance or quality of resources varies in time, such that highly specialized females may sometimes face bad conditions that result in low reproductive success.

Genotypes with reproductive successes that vary in time are generally counterselected, even if their (arithmetic) mean reproductive success is unaffected. For instance, a genotype whose females at a given generation lay no egg under certain environmental conditions would immediately go extinct, even if this loss could be compensated by laying many eggs in the future. The negative impact of varying growth rates on the selective value of a genotype are captured by calculating fitness as the geometric instead of the arithmetic mean of growth rates. The geometric mean decreases as the temporal variance in growth rates increases, such that a genotype that reduces this variance may have a selective advantage. This is generally known as a *bet-hedging strategy*, which maximizes the geometric mean even at the cost of a reduced arithmetic mean (Cohen, 1966; Philippi & Seger, 1989; Ripa et al., 2010; Seger & Brockmann, 1987; Slatkin, 1974). Bet-hedging theory permits to explain why generalist phytophagous may evolve in an environment where resource abundance or quality varies in time. Thereby a specialist may better exploit a given resource, but at a cost of maximum variance in its reproductive success, whereas a generalist would reduce this variance (Starrfelt & Kokko, 2012). In this case, one phenotype is expressed by the genotype, which is generally known as conservative bet-hedging.

Selection in fluctuating environments can yield more spectacular strategies where one genotype yields several phenotypes randomly — also known as *coin-flipping plasticity* (Cooper & Kaplan, 1982) — each of them being specialized to a given environmental condition that may occur. Likewise, this strategy — known as *diversifying bet-hedging* — reduces the temporal variance of the genotype's growth rates.

Another distinction between bet-hedging strategies differentiates those that buffer environmental variations 'here and now' versus 'elsewhere or later' (Hopper, 1999; Hopper, Rosenheim, Prout, & Oppenheim, 2003; Solbreck, 1978). Traits such as generalism, clutch and egg size or mating strategies can be interpreted as examples of the former, as we will show in next Section 3.2, whereas traits such as facultative dispersal or dormancy are potential examples of the latter (Section 3.3).

3.2 Dealing With Unpredictable Variation Here and Now

3.2.1 Generalism

In the context of phytophagous insects, generalism is the ability to exploit several host plant species. Such a strategy is generally considered costly, first because it requires being able to overcome the defence mechanisms (e.g., toxic compounds, physical barriers) of several host plant species and second because generalists exploit a large range of host species, such that sometimes they exploit less suitable hosts. Nonetheless, as we have seen, generalism may provide a fitness benefit when the quality and/or quantity of host plant vary temporally in an unpredictable fashion (Futuyma, 1979; Starrfelt & Kokko, 2012). Wiklund and Friberg (2009) surveyed the annual survival of the generalist orange-tip butterfly, *Anthocharis cardamines* (Linnaeus, 1758) (Lepidoptera: Pieridae) and observed that survival on its host plant species is indeed temporally uncorrelated. Generalism is nevertheless rather uncommon in phytophagous insects, with about 10% species exploiting several resources (Bernays & Graham, 1988). This may be due to the high cost of generalism, and to the existence of alternative — and possibly less costly — bet-hedging strategies that can allow specialists to buffer environmental variation (see below).

3.2.2 Egg Size and Number

These traits are often thought as being negatively correlated, or equivalently that they are part of a trade-off (see Part 2 of this chapter). In a stable environment, the evolutionarily expected combination of these two traits linked by a trade-off is the one that maximizes the average number of viable offspring produced. Einum and Fleming (2004) predict that by producing higher quality (larger) – but fewer – offspring, regardless of variation in environmental conditions, may be considered as conservative bet-hedging. Indeed, these offspring will be more able to survive and reproduce when resources are scarce. This strategy comes at the cost of a lower number of offspring under favourable conditions, which can be compensated by the advantage of a reduced variance in growth rates above some threshold of environmental variations. This has been studied in birds (e.g., Boyce & Perrins, 1987), but not in phytophagous insects to our knowledge.

The reasoning above on the evolution of egg size and number neglects the variation of these traits within a clutch, although egg sizes have been found to vary within clutches (Fox & Czesak, 2000). While this may appear as nonadaptive phenotypic variation, Olofsson, Ripa, Jonzén, and Jonze (2009) have suggested that this might be a case of adaptive diversifying bet-hedging. Indeed, a genotype producing both large and small eggs produces specialists of good and bad conditions, respectively (Kaplan & Cooper, 1984; McGinley, Temme, & Geber, 1987). Phytophagous insects – especially those who lay their eggs in extra-host batches (such as butterflies) – seem like a relevant biological model to test the hypothesis of adaptive bet-hedging (Box 1).

3.2.3 Traits That Are Often Described as Bet-Hedging but may not Be

Making many small clutches – instead of one large clutch – is often described as a bet-hedging strategy (Freese & Zwölfer, 1996). Intuitively, a female laying all her eggs in a single place risks losing all her offspring, for instance, if a predator feeds on this host plant. This is controversial because in a large well-mixed population the risk of predation would be spread among all the carriers of a genotype. In this context, making many small clutches does not buffer variations in the genotype's growth rate (Starrfelt & Kokko, 2012). The quality of a reproductive partner, possibly in interaction with host plant, may have strong impact on lifetime reproductive success. Multiple mating is therefore sometimes thought of as bet-hedging (e.g., Fox & Rauter, 2003), because it can reduce the risk that an individual mates with a low-quality partner. But similarly to the reasoning above, the risk is spread among all the carriers of a given genotype in a large population (Starrfelt & Kokko, 2012). However, Yasui and Garcia-Gonzalez (2016) have shown that a bet-hedging strategy consisting in mating

Box 1 Testing bet-hedging in phytophagous insects

One may think that observing such a curious strategy as a high level of iteroparity (i.e., reproducing several times) may be enough to identify a bet-hedging strategy. This is not the case for at least two reasons. First, a bet-hedging strategy is by definition a response to unpredictable environmental variation but similar strategies can evolve in response to other selection pressures; iteroparity, for instance, can provide a selective advantage if adults can acquire lots of resources between distant reproductive events, and thereby increase their net reproductive rate. Second, strategies that look like bet-hedging can evolve neutrally in the complete absence of any selection pressure (Verin, Menu, & Rajon, 2015).

The demonstration that an observed heritable strategy corresponds to an adaptive bet-hedging strategy thus requires formal testing. One way of performing such a test consists in comparing this observation with the prediction of a realistic model (Simons, 2011). Such a prediction can only be obtained through a modelling approach that incorporates observed distributions of time-varying parameters (e.g., survival rates, fecundities). Many theoretical studies also suggest that density dependency has a strong impact on the evolutionary outcome (Rajon, Venner, & Menu, 2009); thus such dependency should be quantified. The adaptive dynamics framework (Geritz, Kisdi, Meszéna, & Metz, 1998; Metz, Nisbet, & Geritz, 1992; Rajon et al., 2009) and individual-based modelling (Grimm, 1999) are appropriate in this context. The conclusion of the test would be even more robust if performed in several populations.

Experimental evolution may also be an efficient design to test if bet-hedging can evolve as an adaptation to a variable environment (Kawecki et al., 2012). The experimental setting should start with a non–bet-hedging genotype facing random sequences of some parameters – mimicking unpredictable environmental variation – and compare its evolutionary dynamics with that of a control placed in a stable environment. The de novo emergence of bet-hedging has been observed in similar experiments in microorganisms (Beaumont, Gallie, Kost, Ferguson, & Rainey, 2009). The rather short generation time of most phytophagous insects and their rearing convenience make them good models to perform such experimental evolution experiment in multicellular organisms. Combined with sequencing methods, this could provide information about the genetic and physiological mechanisms underlying bet-hedging traits (Box 2).

with different partners (i.e., being polyandrous or polygynous) may still evolve in small populations, or in structured populations with small effective populations sizes (N_e). Holman (2015) tested this prediction in 49 datasets in various taxa (including many phytophagous insects) and found that bet-hedging is highly unlikely to explain the evolution of these strategies.

3.3 Dealing With Unpredictable Variation Elsewhere or Later

3.3.1 *Spatial Dispersal*

In their model, Levin, Cohen, and Hastings (1984) consider that a part of the carriers of a genotype that adopt a bet-hedging dispersal strategy systematically disperses, while the others stay and reproduce in the locality they were born in. One can also consider these strategies in terms of distance: some carriers of a gene disperse far, while the others travel a shorter distance (Snyder, 2006). All these formulations for bet-hedging dispersal strategies have in commons that a single genotype systematically expresses a range of phenotypes differing in their ability to disperse; this is a form of diversifying bet-hedging.

Such a strategy may have a selective advantage if (1) the conditions for survival or reproduction may be bad locally without being predictable and (2) the probability of finding different conditions elsewhere is not zero (i.e., in a lack of spatial autocorrelation; Bulmer, 1984; Hopper, 1999). In such a context, the carriers of a bet-hedging genotype encounter many environmental conditions, independently of their natal locality, which reduces the impact of locally variable conditions on the genotype's growth rate. A metapopulation structure is necessary for a spatial dispersal strategy to evolve (Olivieri, Michalakis, & Gouyon, 1995), and the selective advantage of dispersal bet-hedging generally increases with the number of localities (Bulmer, 1984). This type of structure is often found in insect populations (Hopper, 1999). However, to our knowledge no empirical study has yet shown that dispersal has evolved as a bet-hedging strategy in insects.

3.3.2 *Prolonged Diapause*

A bet-hedging dormancy genotype produces phenotypes that differ in the length of their development cycle – hence a case of diversifying bet-hedging – such that its carriers reproduce at different reproductive seasons. Therefore the carriers of the genotype experience various environmental conditions in an unpredictable environment, which reduces the temporal variance in their growth rates (Gourbière & Menu, 2009; Hopper, 1999; Menu, Roebuck, & Viala, 2000; Seger & Brockmann, 1987). Many insect species display variance in the length of their life cycle, involving a facultative additional or prolonged diapause (e.g., Danks, 1987; Soula & Menu, 2005; Tauber, Tauber, & Masaki, 1986). Hopper (1999) found scarce evidence for diversifying bet-hedging in insects. Surprisingly, he neglected prolonged diapause strategies in his analysis. Simons (2011) considered these strategies, however, and found several prospective examples in phytophagous insects. Although many of these studies suggest the presence of bet-hedging, none at the time had compared observations to a realistic model's prediction. To our knowledge, only one study since has performed such a test, namely with the chestnut weevil, *Curculio elephas* (Gyllenhaal, 1836) (Coleoptera: Curculionidae) (Rajon, Desouhant, Chevalier, Débias, & Menu, 2014). They have clearly demonstrated that the observed diapause strategies in two populations of *C. elephas* can be predicted by a realistic bet-hedging evolutionary model. This model was parameterized using a long-term estimate of several survival rates, density-dependent fecundities, resource abundances and predation pressure. Research studies have also focused on the physiological mechanisms underlying bet-hedging in this species (Menu & Desouhant, 2002; Soula & Menu, 2005; see also Box 2).

3.3.3 Hatching Asynchrony

Hatching asynchrony (or 'hatching spread') has been observed in phytophagous insects where parents provide food for their offspring (Nalepa, 1988; Smiseth, Ward, & Moore, 2006). It may provide a selective advantage if, as a consequence of this strategy, various offspring encounter different, randomly occurring environmental conditions and if these conditions determine their survival or the success of their development as suggested in birds (Laaksonen, 2004).

3.3.4 Temporal Clutching

Iteroparity (i.e., reproducing several times in a lifetime) is typically considered as a conservative bet-hedging strategy, where long-lived adults, instead of offspring, disperse in time (Bulmer, 1985; Wilbur & Rudolf, 2006). In insects, however, mating is often uncoupled from egg laying. As we have seen, multiple mating as a bet-hedging strategy is controversial. Nonetheless, laying one's eggs in multiple clutches spread in time could provide a similar evolutionary advantage as iteroparity, regardless of the mating strategy.

3.3.5 Pausing in Social Insects

Bet-hedging dispersal in time might also explain the presence of inactive workers in social phytophagous insects (Charbonneau, Hillis, & Dornhaus, 2014), who can carry a part of the workload when catastrophic unpredictable event occurs. This is predicted to increase the long-term sustainability of the colony at the expense of decreasing short-term productivity (Hasegawa, Ishii, Tada, Kobayashi, & Yoshimura, 2016). However, it remains to be

Box 2 The genetic and physiological architecture of diversifying bet-hedging strategies

The precise mechanisms that may randomly produce several phenotypes from a single genotype are poorly known in insects, despite its crucial importance for our understanding of the evolution of these strategies. Two main candidate mechanisms have been proposed (Cooper & Kaplan, 1982; Simons & Johnston, 1997; Walker, 1986): (1) developmental instability, which may amplify small differences in gene expression into different developmental routes and eventually, different phenotypes and (2) the dependency of the offspring phenotype on the value of a physiological or environmental variable it experiences, this variable being uncorrelated with — and thus impossible to foresee — future environmental conditions. In both cases, the phenotype is determined by a comparison between an underlying variable and a genetically (or possibly epigenetically) determined threshold. The frequency of each phenotype among the offspring can be changed by changing the threshold (Rajon et al., 2014).

The former mechanism (1), based on gene networks exploiting gene expression noise, has been studied extensively theoretically (Kussell & Leibler, 2005). Levy, Ziv, and Siegal (2012) have shown experimentally in yeast that the expression of a slow-growing resistant phenotype (the equivalent of a facultative dormant phenotype in insects) can be predicted by the level of expression of a single gene, such that phenotype switching may occur as this level increases or decreases randomly. To our knowledge, no study has studied this mechanism in phytophagous insects.

In phytophagous insects, however, a few empirical studies suggest that the second mechanism may be at the origin of bet-hedging dormancy strategies. For instance, a hypothetical mechanism has been proposed for *C. elephas*, in which the choice of entering into facultative dormancy is only made by larvae with high lipid content (Soula & Menu, 2005). This mechanism might ensure that those larvae with high fat content choose dormancy, as they probably are the most likely to survive an additional dormancy and still have the energy required to finish developing and reproduce. Danforth (1999) found a similar relationship between body weight and dormancy frequency in the desert bee *Perdita portalis*. Interestingly, such a mechanism makes the dormancy frequency, at a given year, dependent on the environmental conditions that year — for instance, a warm winter will make larvae consume more of their resources — which are uncorrelated with future environmental conditions. Rajon et al. (2014) modelled the evolution of bet-hedging dormancy strategies in two populations of the chestnut weevil, and they could only explain the observed dormancy frequencies with a model that included such noise. Nonetheless, we are still far from the identification of a causal relationship, and further work on the precise physiological and genetic determinants of diversifying bet-hedging is needed.

demonstrated that a genotype adopting such a strategy indeed increases the geometric mean of its growth rates.

3.4 Future Challenges: Life-History Syndromes

In this review, we have focused on unpredictable environments, which selects for bet-hedging. However, evidence shows that a trait involved in a bet-hedging strategy can also respond to predictable environmental cues, thus combining predictive plasticity and bet-hedging (Clauss & Venable, 2000). Reality is further complicated by the fact that several bet-hedging strategies may evolve jointly, thus forming life-history syndromes.

We have described many strategies that have the same aim, namely buffering the negative impact of environmental variance. One may expect that in this situation the traits correlate negatively. This is a typical expectation for temporal (dormancy) and spatial dispersal strategies, described as the dispersal syndrome (Buoro & Carlson, 2014). The evolution of a combination of these traits should depend on ecological parameters; for instance, the patterns of spatial and temporal autocorrelation sets the probability that better conditions may be found elsewhere or later in case the natal locality offers bad conditions, thus setting the success of one or the other strategy. Syndromes go beyond dispersal in time or space, and any of the bet-hedging traits described above are redundant as buffers of environmental variation. Venable and Brown (1988) modelled the joint evolution of the size of offspring and of their ability to disperse spatially and temporally, and generally found negative correlations between them. Pelisson, Bernstein, Débias, Menu, and Venner (2013) have shown that four species of the *Curculio* genus in a guild exhibit various combinations of three bet-hedging strategies, which supposedly have favoured their coexistence.

Nonetheless, the general prediction that seemingly emerges from these studies — that bet-hedging strategies should be negatively correlated — ignores the specificity of the interaction between these strategies (Rubio de Casas, Donohue, Venable, & Cheptou, 2015). As we have seen, large larvae of *C. elephas* are also the most likely to disperse temporally. This is a counterintuitive observation as producing offspring that are bigger on average and more likely to enter a facultative diapause combines two bet-hedging strategies, conservative and diversifying respectively. This positively correlated occurrence of two bet-hedging strategies actually has an evolutionary explanation (see Box 2).

Likewise, generalism and other bet-hedging strategies respond to a common selection pressure, such that one might expect diversifying bet-hedgers to be more prone to specialism. However, Kisdi (2002) has made the exact opposite prediction, showing that local adaptation — a form of specialization in a spatial context — occurs conjointly with a decrease in the probability of dispersal — such that both bet-hedging strategies are lost when the advantage of being locally adapted overcomes that of buffering environmental variance. From these examples, we see that further theoretical developments are needed to reach a general theory of bet-hedging combining several traits.

4. CONCLUSIONS

By several aspects, the host plant affects the reproductive strategies of almost all phytophagous insects for which it has been studied. The corpus of literature attempting to characterize the insect's female life history traits affected by the plant and the mechanisms beyond thus regularly increases. However, conclusions are still limited in males by largely using artificial diets, laboratory-controlled conditions and often stock culture insects with a homogeneous genetic background. In addition, less is known concerning such relationships with plants as a natural food source varying in time and space on male reproductive outputs. Furthermore, insects adapt their fecundity, their spatial dispersive strategy or synchrony with their resource. However, we identified gaps in knowledge concerning intimate mechanisms to better understand and predict variations in phytophagous insect population dynamics. For example, understanding how plant quality tunes the plasticity in diapause duration or the variation in immune systems which regulates the resistance to natural enemies or infections has to be improved in future years. This is even more critical in insect crop pests whose damages are estimated to represent over 30% of the crop losses worldwide.

Progressing in this field of research has important for plant protection against crop insect pests and especially in the perspective of reducing pesticide use. Several alternative strategies to insecticide control exist but they all rely on a better prediction of pest populations' size and their variation. Biotechnical methods such as mating disruption or the modification of pests behavioural strategies based on the semiochemicals use, but also biological control by natural enemies, suppose to anticipate pest population variations and request a good knowledge of the their reproductive strategies. Developing agroecological practices in many crop productions also requires in depth knowledge of the interactions between the host plant and the insect reproduction. Our current knowledge will probably be modified rather soon by the climatic environment which will profoundly affect those interactions and the insect reproductive strategies. Thus studying pest insect life history traits in interaction with the host plant in the age of unpredictable climatic changes represents for scientists and people in charge of plant protection and biodiversity an exciting and challenging period.

ACKNOWLEDGEMENTS

We thank Sonia Durlot for the two pictures of *Ephippiger diurnus* and the referee for his helpful comments.

REFERENCES

Ali, M. Z. (2012). *Influence of herbivore-induced changes in host plants on reproductive behaviours in* Spodoptera littoralis (Ph.D. thesis). Swedish University of Agricultural Sciences.

Anderson, P., Sadek, M. M., Larsson, M., Hansson, B. S., & Thöming, G. (2013). Larval host plant experience modulates both mate finding and oviposition choice in a moth. *Animal Behaviour, 85*, 1169–1175.

Awmack, C. S., & Leather, S. R. (2002). Host plant quality and fecundity in herbivorous insects. *Annual Review of Entomology, 47*, 817–844.

Bachmann, G. E., Segura, D. F., Devescovi, F., Juárez, M. L., Ruiz, M. J., Vera, M. T., & Fernández, P. C. (2015). Male sexual behavior and pheromone emission is enhanced by exposure to guava fruit volatiles in *Anastrepha fraterculus*. *PLoS One, 10*, 1–17.

Beaumont, H. J. E., Gallie, J., Kost, C., Ferguson, G. C., & Rainey, P. B. (2009). Experimental evolution of bet hedging. *Nature, 462*, 90–93.

Berger, D., Walters, R., & Gotthard, K. (2006). What keeps insects small? – Size dependent predation on two species of butterfly larvae. *Evolutionary Ecology, 20*, 575–589.

Bernays, E. A. (1991). Evolution of insect morphology in relation to plants. *Philosophical Transactions of the Royal Society of London Series B Biological Science, 333*, 257–264.

Bernays, E. A., & Chapman, R. F. (1994). *Host-plant selection by phytophagous insects*. New York: Chapman & Hall.

Bernays, E. A., & Graham, M. (1988). On the evolution of host specificity in phytophagous arthropods. *Ecology, 69*, 886–892.

Berrigan, D. (1991). The allometry of egg size and number in insects. *Oikos, 60*, 313.

Binyameen, M., Hussain, A., Yousefi, F., Birgersson, G., & Schlyter, F. (2013). Modulation of reproductive behaviors by non-host volatiles in the polyphagous egyptian cotton leafworm, *Spodoptera littoralis*. *Journal of Chemical Ecology, 39*(10), 1273–1283.

Blanckenhorn, W. U. (2000). The evolution of body size: What keeps organisms small? *Quarterly Review of Biology, 75*, 385–407.

Blum, M. S., & Hilker, M. (2008). Chemical protection of insect eggs. In M. Hilker, & T. Meiners (Vol. Eds.), *Chemoecology of insect eggs and egg deposition* (pp. 61–90). Oxford: Blackwell Publishing Ltd.

Boggs, C. L., & Freeman, K. D. (2005). Larval food limitation in butterflies: Effects on adult resource allocation and fitness. *Oecologia, 144*, 353–361.

Boyce, M. S., & Perrins, C. M. (1987). Optimizing great tits clutch size in a fluctuating environment. *Ecology, 68*, 142–153.

Briffa, M., & Hardy, I. C. W. (2013). *Animal contests*. Cambridge: Cambridge University Press.

Bulmer, M. G. (1984). Delayed germination of seeds: Cohen's model revisited. *Theoretical Population Biology, 26*, 367–377.
Bulmer, M. G. (1985). Selection for iteroparity in a variable environment. *American Naturalist, 126*, 63–71.
Buoro, M., & Carlson, S. M. (2014). Life-history syndromes: Integrating dispersal through space and time. *Ecology Letters, 17*, 756–767.
Cahenzli, F., & Erhardt, A. (2012). Nectar sugars enhance fitness in male *Coenonympha pamphilus* butterflies by increasing longevity or realized reproduction. *Oikos, 121*, 1417–1423.
Cahenzli, F., & Erhardt, A. (2013). Nectar amino acids enhance reproduction in male butterflies. *Oecologia, 171*, 197–205.
Charbonneau, D., Hillis, N., & Dornhaus, A. (2014). 'Lazy' in nature: Ant colony time budgets show high. *Insectes Sociaux, 62*, 31–35.
Clauss, M. J., & Venable, D. L. (2000). Seed germination in desert annuals: An empirical test of adaptive bet hedging. *American Naturalist, 155*, 168–186.
Cohen, D. (1966). Optimizing reproduction in a randomly varying environment. *Journal of Theoretical Biology, 12*, 119–129.
Coley, P., Bateman, M. L., & Kursar, T. A. (2006). The effects of plant quality on caterpillar growth and defense against natural enemies. *Oikos, 115*, 219–228.
Conner, W. F., Eisner, T., Vandermeer, R. K., Guerrero, A., & Meinwald, J. (1981). Precopulatory sexual interaction in an arctiid moth (*Utetheisa ornatrix*): Role of a pheromone derived from dietary alkaloids. *Behavioral Ecology and Sociobiology, 9*, 227–235.
Cook, P. A., & Wedell, N. (1996). Ejaculate dynamics in butterflies: A strategy for maximizing fertilization success? *Proceedings of the Royal Society B: Biological Sciences, 263*, 1047–1051.
Cooper, W. S., & Kaplan, R. H. (1982). Adaptive 'coin-flipping': A decision-theoretic examination of natural selection for random individual variation. *Journal of Theoretical Biology, 94*, 135–151.
Corcket, E., Giffard, B., & Sforza, R. F. H. (2016). Food webs and multiple biotic interactions in plant-herbivore models. In N. Sauvion, D. Thiery, & P.-A. Calatayud (Eds.), *Insect-Plant Interactions in a Crop Protection Perspective* (Vol. 81, pp. 111–137).
Danforth, B. N. (1999). Emergence dynamics and bet hedging in a desert bee, *Perdita portalis*. *Proceedings of the Royal Society B: Biological Sciences, 266*, 1985–1994.
Danks, H. V. (1987). *Insect dormancy: An ecological perspective*. Ottawa: Biological Survey of Canada.
Delisle, J., & Bouchard, A. (1995). Male larval nutrition in *Choristoneura rosaceana* (Lepidoptera: Tortricidae): An important factor in reproductive success. *Oecologia, 104*, 508–517.
Delisle, J., & Hardy, M. (1997). Male larval nutrition influences the reproductive success of both sexes of the spruce budworm, *Choristoneura fumiferana* (Lepidoptera: Tortricidae). *Functional Ecology, 4*, 451.
Einum, S., & Fleming, I. A. (2004). Environmental unpredictability and offspring size: Conservative and diversified bet-hedging. *Evolutionary Ecology Research, 6*, 443–455.
Eisner, T., & Meinwald, J. (1995). The chemistry of sexual selection. *Proceedings of the National Academy of Sciences of the United States of America, 92*, 50–55.
Ekbom, B., & Popov, S. Y. A. (2004). Host plant affects pollen beetle (*Meligethes aeneus*) egg size. *Physiological Entomology, 29*, 118–122.
Fox, C. W. (1997). Egg-size manipulations in the seed beetle *Stator limbatus*: Consequences for progeny growth. *Canadian Journal of Zoology, 75*, 1465–1473.
Fox, C. W., & Czesak, M. E. (2000). Evolutionary ecology of progeny size in arthropods. *Annual Review of Entomology, 45*, 341–369.
Fox, C. W., & Rauter, C. M. (2003). Bet-hedging and the evolution of multiple mating. *Evolutionary Ecology Research, 5*, 273–286.

Freese, G., & Zwölfer, H. (1996). The problem of optimal clutch size in a tritrophic system: The oviposition strategy of the thistle gallfly *Urophora cardui* (Diptera, Tephritidae). *Oecologia, 108*, 293–302.

Friedländer, M. (1997). Control of the eupyrene-apyrene sperm dimorphism in Lepidoptera. *Journal of Insect Physiology, 43*, 1085–1092.

Fritzsche, K., & Arnqvist, G. (2015). The effects of male phenotypic condition on reproductive output in a sex role-reversed beetle. *Animal Behaviour, 102*, 209–215.

Futuyma, D. J. (1979). *Evolutionary biology*. Sunderland: Sinauer.

Geritz, S., Kisdi, E., Meszéna, G., & Metz, J. (1998). Evolutionarily singular strategies and the adaptive growth and branching of the evolutionary tree. *Evolutionary Ecology, 12*(1), 35–57.

Gourbière, S., & Menu, F. (2009). Adaptive dynamics of dormancy duration variability: Evolutionary trade-off and priority effect lead to suboptimal adaptation. *Evolution, 63*, 1879–1892.

Grimm, V. (1999). Ten years of individual-based modelling in ecology: What have we learned and what could we learn in the future? *Ecological Modelling, 115*, 129–148.

Gripenberg, S., Mayhew, P. J., Parnell, M., & Roslin, T. (2010). A meta-analysis of preference-performance relationships in phytophagous insects. *Ecology Letters, 13*, 383–393.

Gwynne, D. T. (2001). *Katydids and bush-crickets: Reproductive behavior and evolution of the Tettigoniidae*. Cornell: Cornell University Press.

Hasegawa, E., Ishii, Y., Tada, K., Kobayashi, K., & Yoshimura, J. (2016). Lazy workers are necessary for long-term sustainability in insect societies. *Scientific Reports, 6*, 20846.

Hilker, M., & Meiners, T. (2011). Plants and insect eggs: How do they affect each other? *Phytochemistry, 72*, 1612–1623.

Hockham, L., Graves, J., & Ritchie, M. (2004). Sperm competition and the level of polyandry in a bushcricket with large nuptial gifts. *Behavioral Ecology and Sociobiology, 57*, 149–154.

Holman, L. (2015). Bet hedging via multiple mating: A meta-analysis. *Evolution, 70*, 62–71.

Honěk, A. (1993). Intraspecific variation in body size and fecundity in insects: A general relationship. *Oikos, 66*, 483–492.

Hopper, K. R. (1999). Risk-spreading and bet-hedging in insect population biology. *Annual Review of Entomology, 44*, 535–560.

Hopper, K. R., Rosenheim, J. A., Prout, T., & Oppenheim, S. J. (2003). Within-generation bet hedging: A seductive explanation? *Oikos, 101*, 219–222.

Hunter, M. D., & Mcneil, J. N. (1997). Host-plant quality influences diapause and voltinism in a polyphagous insect herbivore. *Ecology, 78*, 977–986.

Jarrige, A., Body, M., Giron, D., Greenfield, M. D., & Goubault, M. (2015). Amino-acid composition of nuptial gift: A differential allocation toward female quality suggests a dynamic role of nuptial gift in a bushcricket. *Physiology & Behavior, 151*, 463–468.

Jarrige, A., Greenfield, M. D., & Goubault, M. (2013). Male song as a predictor of the nuptial gift in bushcrickets: On the confounding influence of male choice. *Animal Behaviour, 85*, 1427–1434.

Kaiser, L., Ode, P., van Nouhuys, S., Calatayud, P.-A., Colazza, S., Cortesero, A.-M., & van Baaren, J. (2016). The plant as a habitat for entomophagous insects. In N. Sauvion, D. Thiery, & P.-A. Calatayud (Eds.), *Insect-Plant Interactions in a Crop Protection Perspective* (Vol. 81, pp. 179–223).

Kaplan, R. H., & Cooper, W. S. (1984). The evolution of developmental plasticity in reproductive characteristics: An application of the adaptive coin-flipping principle. *American Naturalist, 123*, 393–410.

Kaspi, R., Mossinson, S., Drezner, T., Kamensky, B., & Yuval, B. (2002). Effects of larval diet on developmental rates and reproductive maturation of male and female Mediterranean fruit flies. *Physiological Entomology, 27*, 29–38.

Kawecki, T. J., Lenski, R. E., Ebert, D., Hollis, B., Olivieri, I., & Whitlock, M. C. (2012). Experimental evolution. *Trends in Ecology & Evolution, 27*, 547–560.
Khan, M., & Port, G. (2008). Performance of clones and morphs of two cereal aphids on wheat plants with high and low nitrogen content. *Entomological Science, 11*, 159–165.
Kisdi, E. (2002). Dispersal: Risk spreading versus local adaptation. *American Naturalist, 159*, 579–596.
Kussell, E., & Leibler, S. (2005). Phenotypic diversity, population growth, and information in fluctuating environments. *Science, 309*, 2075–2078.
Laaksonen, T. (2004). Hatching asynchrony as a bet-hedging strategy – an offspring diversity hypothesis. *Oikos, 104*, 616–620.
Landolt, P. J., & Phillips, T. W. (1997). Host plant influences on sex pheromone behavior of phytophagous insects. *Annual Review of Entomology, 42*, 371–391.
Leather, S. R. (1994). The effect of temperature on oviposition, fecundity and egg hatch in the pine beauty moth, *Panolis flammea* (Lepidoptera: Noctuidae). *Bulletin of Entomological Research, 84*, 515–520.
Lederhouse, R. C., Ayres, M. P., & Scriber, J. M. (1990). Adult nutrition affects male virility in *Papilio glaucus* L. *Functional Ecology, 4*, 743–751.
Levin, S. A., Cohen, D. H., & Hastings, A. (1984). Dispersal strategies in patchy environments. *Theoretical Population Biology, 26*, 165–191.
Levy, S. F., Ziv, N., & Siegal, M. L. (2012). Bet hedging in yeast by heterogeneous, age-correlated expression of a stress protectant. *PLoS Biology, 10*(5), e1001325.
Lewis, S., & South, A. (2012). The evolution of animal nuptial gifts. *Advances in the Study of Behavior, 44*, 53–97.
Lewis, Z., & Wedell, N. (2007). Effect of adult feeding on male mating behaviour in the butterfly, *Bicyclus anynana* (Lepidoptera: Nymphalidae). *Journal of Insect Behavior, 20*(2), 201–213.
Mayhew, P. J. (1997). Adaptive patterns of host plant selection by phytophagous insects. *Oikos, 79*, 417–428.
McGinley, M. A., Temme, D. H., & Geber, M. A. (1987). Parental investment in offspring in variable environments: Theoretical and empirical considerations. *American Naturalist, 130*, 370–398.
Menu, F., & Desouhant, E. (2002). Bet-hedging for variability in life cycle duration: Bigger and later-emerging chestnut weevils have increased probability of a prolonged diapause. *Oecologia, 132*, 167–174.
Menu, F., Roebuck, J., & Viala, M. (2000). Bet-hedging diapause strategies in stochastic environments. *American Naturalist, 155*, 724–734.
Metz, J., Nisbet, R., & Geritz, S. (1992). How should we define 'fitness' for general ecological scenarios? *Trends in Ecology and Evolution, 7*, 198–202.
Mevi-Schutz, J., & Erhardt, A. (2005). Amino acids in nectar enhance butterfly fecundity: A long-awaited link. *American Naturalist, 165*, 411–419.
Morató, S., Shelly, T., Rull, J., & Aluja, M. (2015). Sexual competitiveness of *Anastrepha ludens* (Diptera: Tephritidae): Males exposed to *Citrus aurantium* and *Citrus paradisi* essential oils. *Journal of Economic Entomology, 108*, 621–628.
Moreau, J., Arruego, X., Benrey, B., & Thiéry, D. (2006). Differences in nutritional quality of parts of *Vitis vinifera* berries affect fitness of the European grapevine moth. *Entomologia Experimentalis et Applicatta, 119*, 93–99.
Moreau, J., Benrey, B., & Thiery, D. (2006). Assessing larval food quality for phytophagous insects: Are the facts as simple as they appear? *Functional Ecology, 20*, 592–600.
Moreau, J., Thiery, D., Troussard, J. P., & Benrey, B. (2007). Grape variety affects female but also male reproductive success in wild European grapevine moths. *Ecological Entomology, 32*, 747–753.

Morehouse, N. I., & Rutowski, R. L. (2010). In the eyes of the beholders: Female choice and avian predation risk associate with an exaggerated male butterfly color. *American Naturalist, 176*, 768–784.

Muller, K., Arenas, L., Thiéry, D., & Moreau, J. (2016). Direct benefits from mating with a virgin male in the usually monandrous European grapevine moth (*Lobesia botrana*). *Animal Behaviour, 114*, 165–172.

Muller, K., Thiéry, D., Moret, Y., & Moreau, J. (2015). Male larval nutrition affects adult reproductive success in wild European grapevine moth (*Lobesia botrana*). *Behavioral Ecology and Sociobiology, 69*, 39–47.

Nalepa, C. A. (1988). Cost of parental care in the woodroach *Cryptocercus punctulatus* Scudder (Dictyoptera: Cryptocercidae). *Behavioral Ecology and Sociobiology, 23*, 135–140.

O'Brien, D. M., Boggs, C. L., & Fogel, M. L. (2013). The amino acids used in reproduction by butterflies: A comparative study of dietary sources using compound-specific stable isotope analysis. *Physiological and Biochemical Zoology, 78*, 819–827.

Olivieri, I., Michalakis, Y., & Gouyon, P. H. (1995). Metapopulation genetics and the evolution of dispersal. *American Naturalist, 146*, 202–228.

Olofsson, H., Ripa, J., Jonzén, N., & Jonze, N. (2009). Bet-hedging as an evolutionary game: The trade-off between egg size and number. *Proceedings of the Royal Society B: Biological Sciences, 276*, 2963–2969.

Pelisson, P. F., Bernstein, C., Débias, F., Menu, F., & Venner, S. (2013). Dispersal and dormancy strategies among insect species competing for a pulsed resource. *Ecological Entomology, 38*, 470–477.

Philippi, T., & Seger, J. (1989). Hedging one's evolutionary bets, revisited. *Trends in Ecology & Evolution, 4*, 41–44.

Rajon, E., Desouhant, E., Chevalier, M., Débias, F., & Menu, F. (2014). The evolution of bet hedging in response to local ecological conditions. *American Naturalist, 184*, E1–E15.

Rajon, E., Venner, S., & Menu, F. (2009). Spatially heterogeneous stochasticity and the adaptive diversification of dormancy. *Journal of Evolutionary Biology, 22*, 2094–2103.

Rebar, D., & Rodríguez, R. L. (2014). Trees to treehoppers: Genetic variation in host plants contributes to variation in the mating signals of a plant-feeding insect. *Ecology Letters, 17*(2), 203–210.

Ripa, J., Olofsson, H., & Jonzen, N. (2010). What is bet-hedging, really? Invited reply. *Proceedings of the Royal Society B: Biological Sciences, 277*, 1153–1154. http://dx.doi.org/10.1098/rspb.2009.2023.

Rodrıguez, R. L., Sullivan, L. M., & Cocroft, R. B. (2004). Vibrational communication and reproductive isolation in the *Enchenopa binotata* species complex of treehoppers (Hemiptera: Membracidae). *Evolution, 58*, 571–578.

Royer, L., & McNeil, J. N. (1993). Male investment in the European corn borer, *Ostrinia nubilalis* (Lepidoptera: Pyralidae): Impact on female longevity and reproductive performance. *Functional Ecology, 7*, 209–215.

Rubio de Casas, R., Donohue, K., Venable, D. L., & Cheptou, P. O. (2015). Gene-flow through space and time: Dispersal, dormancy and adaptation to changing environments. *Evolutionary Ecology, 29*, 813–831.

Seger, J., & Brockmann, H. J. (1987). What is bet-hedging? *Oxford Surveys in Evolutionary Biology, 4*, 182–211.

Shelly, T. E. (2000). Flower-feeding affects mating performance in male oriental fruit flies, *Bactrocera dorsalis*. *Ecological Entomology, 25*, 109–114.

Shelly, T. E. (2001). Exposure to α-copaene and α-copaene–containing oils enhances mating success of male Mediterranean fruit flies (Diptera: Tephritidae). *Annals of the Entomological Society of America, 94*, 497–502.

Shelly, T. E., & Epsky, N. D. (2015). Exposure to tea tree oil enhances the mating success of male Mediterranean fruit flies (Diptera: Tephritidae). *Florida Entomologist, 98*(4), 1127–1133.
Silberglied, R. E., Shepherd, J. G., & Dickinson, J. L. (1984). Eunuchs: The role of apyrene sperm in Lepidoptera? *American Naturalist, 123*, 255–265.
Simons, A. M. (2011). Modes of response to environmental change and the elusive empirical evidence for bet hedging. *Proceedings of the Royal Society B: Biological Sciences, 278*, 1601–1609.
Simons, A. M., & Johnston, M. O. (1997). Developmental instability as a bet-hedging strategy. *Oikos, 80*, 401–406.
Slatkin, M. (1974). Hedging ones evolutionary bets. *Nature, 250*, 704–705.
Smiseth, P. T., Ward, R. J. S., & Moore, A. J. (2006). Asynchronous hatching in *Nicrophorus vespilloides*, an insect in which parents provide food for their offspring. *Functional Ecology, 20*, 151–156.
Snyder, R. E. (2006). Multiple risk reduction mechanisms: Can dormancy substitute for dispersal? *Ecology Letters, 9*, 1106–1114.
Solbreck, C. (1978). Migration, diapause, and direct development as alternative life histories in a seed bug, *Neacoryphus bicrucis*. In H. Dingle (Ed.), *Evolution of insect migration and diapause* (pp. 195–217). New York: Springer.
Soula, B., & Menu, F. (2005). Extended life cycle in the chestnut weevil: Prolonged or repeated diapause? *Entomologia Experimentalis et Applicatta, 115*, 333–340.
South, A., & Lewis, S. M. (2011). The influence of male ejaculate quantity on female fitness: A meta-analysis. *Biological Reviews of the Cambridge Philosophical Society, 86*, 299–309.
Starrfelt, J., & Kokko, H. (2012). Bet-hedging-a triple trade-off between means, variances and correlations. *Biological Reviews of the Cambridge Philosophical Society, 87*, 742–755.
Takakura, K. I. (2004). Variation in egg size within and among generations of the bean weevil, *Bruchidius dorsalis* (Coleoptera, Bruchidae): Effects of host plant quality and paternal nutritional investment. *Annals of the Entomological Society of America, 97*, 346–352.
Tammaru, T., Esperk, T., & Castellanos, I. (2002). No evidence for costs of being large in females of *Orgyia* spp. (Lepidoptera, Lymantriidae): Larger is always better. *Oecologia, 133*, 430–438.
Tauber, M. J., Tauber, C. A., & Masaki, S. (1986). *Seasonal adaptations of insects*. Oxford: Oxford University Press.
Telang, A., & Wells, M. A. (2004). The effect of larval and adult nutrition on successful autogenous egg production by a mosquito. *Journal of Insect Physiology, 50*, 677–685.
Thiéry, D., & Moreau, J. (2005). Relative performance of European grapevine moth (*Lobesia botrana*) on grapes and other hosts. *Oecologia, 143*, 548–557.
Tigreros, N. (2013). Linking nutrition and sexual selection across life stages in a model butterfly system. *Functional Ecology, 27*, 145–154.
Vahed, K. (1998). The function of nuptial feeding in insects: A review of empirical studies. *Biological Reviews of the Cambridge Philosophical Society, 73*, 43–78.
Vande Velde, L., Schtickzelle, N., & Van Dyck, H. (2013). Effect of larval food stress on male adult behaviour, morphology and reproductive investment in the butterfly *Pararge aegeria*. *Evolutionary Ecology, 27*(1), 221–234.
Venable, D. L., & Brown, J. S. (1988). The selective interactions of dispersal, dormancy, and seed size as adaptations for reducing risk in variable environments. *The American Naturalist, 131*, 360–384.
Vera, M. T., Ruiz, M. J., Oviedo, A., Abraham, S., Mendoza, M., Segura, D. F., & Willink, E. (2013). Fruit compounds affect male sexual success in the South American fruit fly, *Anastrepha fraterculus* (Diptera: Tephritidae). *Journal of Applied Entomology, 137*, 2–10.

Verin, M., Menu, F., & Rajon, E. (2015). *The biased evolution of generation time*. arXiv, 1502.05508.

Walker, T. J. (1986). Stochastic polyphenism: Coping with uncertainty. *The Florida Entomologist, 69*, 46–62.

White, T. C. R. (1993). *The inadequate environment: Nitrogen and the abundance of animals*. Springer.

Wiklund, C., & Friberg, M. (2009). The evolutionary ecology of generalization: Among-year variation in host plant use and offspring survival in a butterfly. *Ecology, 90*, 3406–3417.

Wilbur, H. M., & Rudolf, V. H. W. (2006). Life history evolution in uncertain environments: Bet-hedging in time. *American Naturalist, 168*, 398–441.

Yasui, Y., & Garcia-Gonzalez, F. (2016). Bet-hedging as a mechanism for the evolution of polyandry, revisited. *Evolution, 70*, 385–397.

CHAPTER NINE

Plant–Insect Interactions in a Changing World

S. Pincebourde*[,1], J. van Baaren[§], S. Rasmann[¶], P. Rasmont[||], G. Rodet[#], B. Martinet[||], P.-A. Calatayud**[,§§]
*IRBI (Insect Biology Research Institute) – UMR 7261 CNRS/Université François-Rabelaisde Tours, Tours, France
[§]UMR-CNRS 6553 Ecobio, Université de Rennes 1, Université Bretagne-Loire, Rennes, France
[¶]University of Neuchâtel, Neuchâtel, Switzerland
[||]INRA, UR0406 Abeilles et Environnement, Domaine St Paul, Avignon, France
[#]AE, INRA, Avignon, France
**IRD UMR EGCE (Evolution, Génome, Comportement, Ecologie), CNRS-IRD-Univ. Paris-Sud, IDEEV, Université Paris-Saclay, Gif-sur-Yvette Cedex, France
[§§]IRD c/o ICIPE, Nairobi, Kenya
[1]Corresponding author: E-mail: sylvain.pincebourde@univ-tours.fr

Contents

Advances in Botanical Research, Volume 81
ISSN 0065-2296
http://dx.doi.org/10.1016/bs.abr.2016.09.009

Abstract

Global change is resetting the spatial and ecological equilibrium of complex co-evolutionary relationships between plants and their insect herbivores. We review the mechanisms at play in the responses of plant–insect interactions to global changes, including increased temperature and atmospheric CO_2 concentrations, modification of land use and pollution. We distinguish between the direct effects of global changes on each partner from the indirect impacts on insects via the responses of plants. The indirect effects include a change in the nutritional quality of the plant tissues for herbivore insects, as well as a change in the microclimatic conditions at the leaf surface. Pollinators are involved in a close symbiotic relationship with their favourite plants, and any depression caused by climate stress can lead to pollination deficit. Pollinators are, indeed, quite sensitive to global changes. Furthermore, although species are connected by trophic links, all species respond differently to global changes. We highlight that more research is needed to elucidate the plant-mediated indirect effects of climate change on insects. Then, other human activities, such as land transformations and release of pollutants, are likely to modulate these links between climate and plant–insect relationships. We argue that predicting the net effect of global change on plant–insect relationships requires a comprehensive understanding of the mechanisms that modulate the interaction strength between the plants and the insects, rather than on focusing on each partner individually.

1. INTRODUCTION

Earth is experiencing rapid changes in both climatic conditions and landscape structure. CO_2 concentration in the atmosphere has risen from about 280 ppm during pre-industrial times to the current 397 ppm. The atmospheric CO_2 concentration is predicted to reach up to 600 ppm towards the end of the 21st century if the rate of fossil fuel combustion does not change (IPCC, 2014). As a consequence, global average temperature would increase from 1.7°C to 6°C by the year 2100 (IPCC, 2014). In addition, an increase in the frequency of extreme events is expected (Hance, van Baaren, Vernon, & Boivin, 2007). Meanwhile, European landscapes have undergone a simplification through agricultural intensification leading to the removal of hedges, groves, woods and natural grasslands, and tending to reduce the effectiveness of ecosystem services such as biological pests control (Thies et al., 2011). Thus, even if their combined influences on the ecosystems are still poorly valued, there is increasing evidence that working on both landscape properties and climate aspects can greatly advance our understanding of their impacts on plant–insect interactions (Tougeron, van Baaren, Burel, & Alford, 2016). These changes in climate and landscapes are disrupting the energetic budget of plants, ultimately affecting their eco-physiological responses against insect herbivores. Therefore, global change is resetting the spatial and ecological equilibrium of complex co-evolutionary relationships between plants and their insect herbivores (Coley, 1998).

Humans disturb the ecosystems in which species live and evolve, leading to ecological and evolutionary consequences for plant–insect interactions (Mitter, Farrell, & Futuyma, 1991; Rundle & Nosil, 2005). Intentionally, humans create totally new environments, such as agriculture fields and cities, in which fauna and flora are subjected to new selection pressures. These new environments are maintained only through human actions. On the other hand, accidentally, humans disturb natural ecosystems due to proximity, needs, pollution and unexpected consequences of their actions. These effects can occur over large areas, at great distances from their anthropogenic origin. Humans are ingenious but they seem to be impotent when faced with the fact that all their innovations are accompanied with the emergence of new accidents (Virilio, 2005). Global climate change is probably the greatest accident. Overall, humans influence ecological processes in three ways: (1) by altering environmental conditions through modifications and transformations of the landscape, (2) by creating new niches in anthropized environments and (3) by rearranging ecological communities through species introduction or removal.

Since more than 500 million years ago, plants and insect herbivores have been engaged in an evolutionary arms race that has been the source of major diversification events (e.g., Currano, Labandeira, & Wilf, 2010; for a review see chapters: Plant-Insect Interactions: A Paleontological and an Evolutionary Perspective by Schatz, Sauvion, Kjellberg, & Nel, 2017 and Evolution of Plant–Insect Interactions: Insights From Macroevolutionary Approaches in Plants and Herbivorous Insects by Kergoat, Meseguer, Jousselin, 2016), which has resulted in plants covering more than half of the biodiversity described today (Strong, Lawton, & Southwood, 1984). Due to their sessile nature, plants evolved sophisticated strategies to protect themselves from herbivore attacks, including structural and chemical means (Schoonhoven, van Loon, & Dicke, 2005). In turn, insects responded to this incredible chemical diversity by evolving detoxification mechanisms (Feyereisen, 1999; Heckel, 2014; for a review see chapter: From Plant Exploitation to Mutualism by Lieutier et al., 2016), behavioural avoidance mechanisms (Dussourd & Denno, 1991) and temporal and spatial specialization for different plant organs (Jaenike, 1990). Because the equilibrium of the complex interactions between plants and insects depends on environmental conditions, current global change is likely to induce profound changes in the strength of the plant–insect interactions (Tylianakis, Didham, Bascompte, & Wardle, 2008).

Species are linked by trophic links (see chapters: Food Webs and Multiple Biotic Interactions in Plant–Herbivore Models by Corcket, Giffardx, & Sforza, 2017 and The Plant as a Habitat for Entomophagous Insects by Kaiser et al., 2016). In a given community, plants are attacked by herbivores which are themselves attacked by predators and parasitoids. Insect herbivores such as aphids contain different obligatory and facultative endosymbiotic bacteria (see chapter: Influence of Microbial Symbionts on Plant–Insect Interactions by Giron et al., 2016). In ecosystems, these multitrophic interactions are the result of a long co-evolutionary process that unfolds within a particular environment. Any modification occurring at one trophic level will have repercussions on the others. Indeed, each species of the multitrophic network has its own tolerance to temperature, CO_2 concentration or humidity. The preferendum may differ between species and trophic levels, causing various responses to climate change. For example, *Trichogramma chilonis* Ishii 1941 (Hymenoptera: Trichogrammatidae) has an optimal response at temperatures between 25°C and 35°C, whereas it is between 30°C and 35°C for its competitor *Cotesia vestalis* (Haliday, 1834) (Hymenoptera: Braconidae) (Reddy et al., 2002). During the quaternary period, global warming induced distribution shifts with different intensities and speed according to the species: the individual responses of each population or species produced new communities (Graham & Grimm, 1990). To predict the impact of climate change at the community level, it is therefore necessary to understand what will happen to every single species, at each trophic level (guild level), at the level of the synchronism between trophic levels and finally at the community level (Tylianakis et al., 2008).

A well-discussed case of how climate change might impact plant–insect interactions is the effect of environmental temperature. Global warming causes variation in the composition of the plants and their herbivore communities, particularly through differential colonization abilities from more sessile plants to more mobile animal herbivores (Burrows et al., 2011; Parmesan, 1996; Pateman, Hill, Roy, Fox, & Thomas, 2012; Pauli et al., 2012; Stange & Ayres, 2001), and also through shifts in phenological events such as the flowering date or emergence date of overwintering insects

(Parmesan & Yohe, 2003). Such phenological mismatch should have profound effects on the co-evolved plant–herbivore interactions (Harrington, Woiwod, & Sparks, 1999). These effects will be largely mediated by plant defence traits (Rasmann & Pellissier, 2015). Nevertheless, the magnitude and direction of change still remains to be fully elucidated, since the mechanisms at play are highly context dependent and variable (Zavala, Nabity, & DeLucia, 2013; Zvereva & Kozlov, 2006).

Here, we review the mechanisms by which global climate change and human activities impact the way plants and insects interact with each other. We do not pretend to exhaustively cover the various effects of global change on plant–insect interactions (for a review see Björkman & Niemelä, 2015).

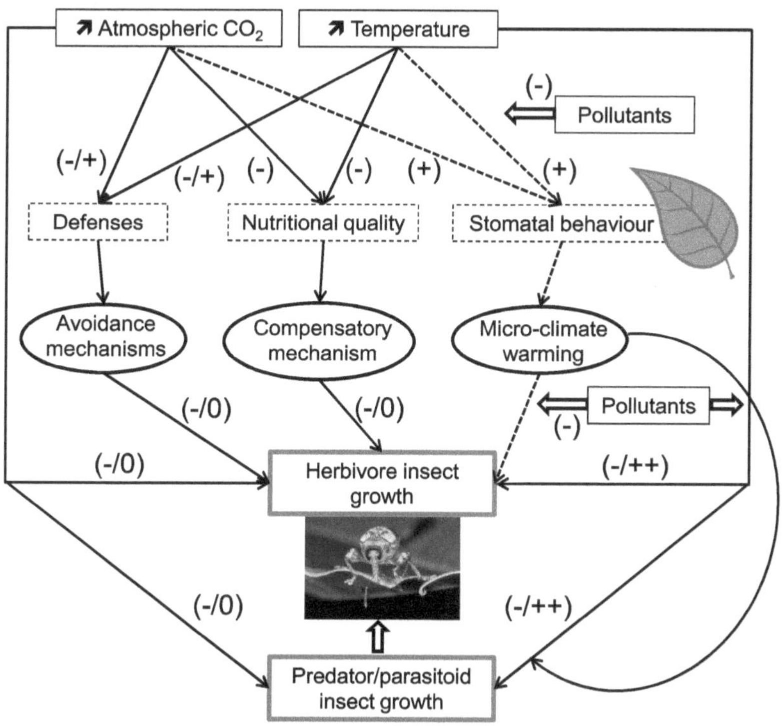

Figure 1 Diagram illustrating the complexity of the multiple interactions between some of the abiotic variables associated with global change (CO_2, temperature, pollution), some plant key processes (structural and chemical defences against herbivores, nutritional quality of plant tissues, stomatal ecophysiology), some herbivore insect responses (avoidance behaviour, compensatory responses) and the third trophic level, the predators and parasitoids. Signs (−/+/0) indicate the possible effect, negative, positive or no effect, respectively. This diagram is not exhaustive, and other mechanisms may play a significant role.

Our aim is to emphasize the mechanisms behind the responses of plants and insects, and how they interact (Fig. 1). We specifically focus on the role of rising atmospheric CO_2 concentration and increasing temperature. The direct effects on both the plants and the insects are reviewed separately from the indirect effects of these factors on the insects via their influence on the plants. In particular, we highlight that more research is needed to elucidate the plant-mediated indirect effects of climate change on insects. Finally, we detail the effects of human activities on plant–insect relationships by focusing on land use and on pollution. We argue that predicting the net effect of global change on plant–insect relationships requires a comprehensive understanding of the mechanisms that modulate the interaction strength between the plants and the insects, rather than focusing on each partner individually.

2. DIRECT EFFECTS OF CLIMATE CHANGE ON PLANT–INSECT INTERACTIONS

2.1 The Plant Side: Impact of Climate Change on Plant Defence Mechanisms

The arms race between plants and insects has led to the evolution of plant mechanisms to minimize damage by insects (Futuyma & Agrawal, 2009; Mitter et al., 1991), including the deployment of chemical and physical defence traits (Schoonhoven et al., 2005). More specifically, plant resistance against herbivores is mediated by structural attributes, such as thick and tough epidermal layers, spines, trichomes, as well as toxic phytochemicals (also called secondary metabolites) that can inhibit, intoxicate and/or kill the herbivore (Schoonhoven et al., 2005). The main classes of secondary metabolites include the terpenoids, phenolics and alkaloids, and also other nitrogen- or sulphur-containing compounds such as the glucosinolates in the Brassicaceae or sugar-bound molecules such as the iridoids in the Plantagineae (Rosenthal & Berenbaum, 1991).

In addition to directly deploying physical and chemical defences, it has become a general agreement that plants can benefit from attracting predators near the site of herbivore attack (Dicke & Baldwin, 2010). More specifically, indirect defences involve the production of features that provide shelter (e.g., domatia), reward (e.g., extrafloral nectar) or information on herbivore presence, such as the release of volatile organic compounds (VOCs), for the natural enemies of herbivores (Kessler & Heil, 2011). These defence traits mediate herbivore–enemy interactions by increasing herbivore suppression, and ultimately potentially increasing plant fitness (Romero & Koricheva, 2011; Schmitz, Hamback, & Beckerman, 2000).

While most of the variations in plant's defence arsenal are attributed to a variety of genetic, ontogenetic and phenologic components, including evolutionary history (Futuyma & Agrawal, 2009), plants also have to constantly cope with variations in environmental conditions, generally imposing a reshuffling of the carbon/energy balance, and thus the allocation between growth and defence (Coley, Bryant, & Chapin, 1985; Herms & Mattson, 1992). In the next section, we will discuss the general patterns emerging from several years of research on the two major climate change drivers – elevated CO_2 and temperature – on plant direct and indirect defences and how they might affect plant's resistance to herbivore attack.

2.1.1 Effect of Elevated CO_2 on Plant Defence and Resistance to Herbivore

The effect of elevated CO_2 on insect herbivores is mainly mediated by changes in plant chemistry (Cornelissen, 2011; Lincoln, Fajer, & Johnson, 1993). An overall conclusion drawn from summarizing the effect of CO_2 addition on plants suggests that changes in primary metabolism are in fact relatively predictable, whereas changes in plant's secondary metabolite production, and their corresponding effects on plant direct and indirect defences, are highly variable (Bidart-Bouzat & Imeh-Nathaniel, 2008), with subsequent strong variation in herbivore performance (Coviella & Trumble, 1999; Robinson, Ryan, & Newman, 2012). Elevated CO_2 typically increases the carbon:nitrogen ratio (Robinson et al., 2012). In addition, it has dramatic but variable impact on the general leaf secondary chemistry (Bidart-Bouzat & Imeh-Nathaniel, 2008; Cornelissen, 2011; Lindroth, 2010; Stiling & Cornelissen, 2007). Perhaps the only exception is the production of VOCs, which was consistently shown to increase in concentration in the headspace surrounding the plant under elevated CO_2 conditions (Peñuelas & Staudt, 2010). Nevertheless, such increase does not automatically translate into higher herbivore suppression by predators in the field, and again, context dependency applies when analyzing the community-wide impacts of elevated CO_2 (Facey, Ellsworth, Staley, Wright, & Johnson, 2014). This variation poses a clear challenge for developing credible predictions of how further CO_2 addition into the atmosphere will impact the plant resistance against herbivores. Therefore, one has to move away from the highly variable secondary metabolism and look at other, more predictable traits. For instance, a relatively small literature which is rapidly gaining attention shows a direct effect of elevated CO_2 on plant hormones (phytohormones), such as jasmonic acid (JA) and salicylic acid (SA) (see below).

Plants attacked by herbivores generate specific hormonal cascades which ultimately elicit downstream changes in plant biochemistry and secondary metabolism through complex signalling networks (Browse & Howe, 2008; Wu & Baldwin, 2010). Both JA and SA are involved in the orchestration of plant defence after herbivore attack (Farmer, Alméras, & Krishnamurthy, 2003; Thaler, Agrawal, & Halitschke, 2010). Exposure to elevated CO_2 can modulate hormone production, which might explain some of the variations in the observed responses of allelochemicals (Ode, Johnson, & Moore, 2014; Zavala et al., 2013). To summarize, the emerging picture suggests that (1) elevated CO_2 down-regulates the constitutive herbivore-induced expression of several key genes associated with the JA pathway, resulting in an increased herbivore damage in field and laboratory conditions (Sun et al., 2011; Zavala, Casteel, DeLucia, & Berenbaum, 2008); (2) because of the inherent cross-talks between JA- and SA-signalling pathways (Thaler, Humphrey, & Whiteman, 2012), elevated CO_2 enhances induced defences derived from the SA-signalling pathway (Ghasemzadeh, Jaafar, & Rahmat, 2010), such as the pathogenesis-related protein (Sun et al., 2011) and (3) the differential responses among phytohormonal pathways provide new insights into how elevated CO_2 modulates plant defences against herbivory.

2.1.2 Effect of Elevated Temperature on Plant Defence and Resistance to Herbivore

Temperature affects the plant physiological processes, growth, reproduction and survival (Jamieson, Trowbridge, Raffa, & Lindroth, 2012; Julkunen-Tiitto, Nybakken, Randriamanana, & Virjamo, 2015). In addition, warming modifies the biosynthesis of plant secondary metabolites (Bidart-Bouzat & Imeh-Nathaniel, 2008; Loreto & Schnitzler, 2010). A general trend was reported for the responses of phytochemicals to enhanced temperature (Zvereva & Kozlov, 2006). Phenolics tended to decrease, whereas terpenes tended to increase at higher temperatures. However, any generalization remains difficult since several examples showed no, or even the opposite effects of elevated temperature on phenolic-based compounds, alkaloids, terpenoids, or glucosinolates (Bidart-Bouzat & Imeh-Nathaniel, 2008; Julkunen-Tiitto et al., 2015). When looking at the effect of increased temperature on the production of VOCs, the picture seems to be clearer. VOC production generally increases with temperature up to the physiological maximum of the plant, beyond which enzymatic degradation inhibits emission (Guenther, Zimmerman, Harley, Monson, & Fall, 1993). The mechanisms of increased emission reside particularly on enhanced enzymatic activities, and increased VOC vapour pressure (Peñuelas & Llusià, 2003). Interestingly, Peñuelas and Llusià, (2003) suggested that increased VOC production should not only be related to enhanced biotic interactions (pollinator and/or predator attraction), which might be also less predictable (de Sassi, Staniczenko, & Tylianakis, 2012; Tylianakis et al., 2008), but also to an increased thermal tolerance for plants. In other words, increased temperatures favour VOC emissions, which in turn enable the plant to better withstand the heat stress.

Another approach for studying the effect of temperature changes on plant defences and plant–insect interactions is to use elevation gradients as natural source of variation in biotic and abiotic factors (Körner, 2007; Rasmann, Alvarez, & Pellissier, 2014; Rasmann, Pellissier, Defossez, Jactel, & Kunstler, 2014). Recent examples show that high-altitude adapted *Plantago lanceolata* L. or *Vicia sepium* L. genotypes produce lower levels of secondary metabolites (iridoid glycosides or VOCs, respectively). While such decline could be attributed to a reduction in herbivore pressure at high elevation, colder temperatures also inhibit iridoid glycoside production in *P. lanceolata* (Pellissier, Roger, Bilat, & Rasmann, 2014). A general decline in plant resistance at high altitude seems to be the rule (Pellissier et al., 2012; for exceptions see Rasmann, Alvarez, et al., 2014; Rasmann, Pellissier, et al., 2014). Specifically, we can hypothesize a general decline of plant defences at high elevation for the predominant plants in the community, while the rarest plants benefit from having their slow growing leaves well protected (Fine, Mesones, & Coley, 2004; Pellissier et al., 2016).

According to predictions, climate warming will reshuffle plant and herbivore communities along altitudinal gradients through high elevation colonization mismatch (Pauli et al., 2012). Therefore, we suggest that high elevation plants will be more exposed to herbivore pressure in the upcoming years (Rasmann, Alvarez, et al., 2014; Rasmann, Pellissier, et al., 2014). Plants' tolerance to such predicted increase in herbivory will be the outcome of several factors, including genetic (and epigenetic) variations, the ability to change their phenotype rapidly in the presence of herbivore attack (i.e. phenotypic plasticity), and ultimately the rate to which plants can adapt (Rasmann & Pellissier, 2015). In other words, climate warming will indirectly impose a strong habitat-specific selection pressure on plants, in turn favouring evolution for novel defence syndromes at high elevation.

2.2 The Insect Side: Impacts of Climate Change on Herbivore Performance

2.2.1 *Effect of CO_2 Addition on Insect Performance*

The direct effect of changing CO_2 concentration in the atmosphere on the eco-physiology of insects has been largely overlooked till date. Recently, Kerr, Phelan, and Woods (2013) showed that the developmental rate of *Manduca sexta* Linnaeus 1763 caterpillars was only slightly affected when reared under high (above 1200 ppm) CO_2 concentrations. More importantly, rising atmospheric CO_2 concentration may be linked to changes in the acidification of insect body fluids. This may be especially true for the eggs that cannot ventilate like the larvae and adult stages, but can only exchange gases via passive diffusion (Woods, Bonnecaze, & Zrubek, 2005). Nevertheless, exposing *M. sexta* eggs to high CO_2 concentrations for several hours did not influence the pH in the egg yolk (Kerr et al., 2013). Undoubtedly, more experimental evidence is needed to generalize these findings.

Rising atmospheric CO_2 concentration may be more likely detrimental to insect species that already live in hypercarbic environments such as cambium miners, stem borers and insects that induce large galls (Pincebourde & Casas, 2016). These endophagous insects live deep into the plant tissues and gases diffuse slowly between outside the plant and the insect location. Therefore, an increase in atmospheric CO_2 concentration may slow down the diffusion of CO_2 outward, enhancing the risk of reaching dangerous asphyxiating levels (Pincebourde & Casas, 2016). Indeed, the hypercarbic conditions in the galleries of these insects coincide with hypoxia, i.e. low oxygen levels. But, again, a lack of knowledge on the direct physiological effects of CO_2 on insect herbivores precludes any firm conclusion.

Finally, changes in CO_2 could also affect the behaviour of insects. Particularly, several soil-dwelling herbivores utilize root-emitted CO_2 as a cue for finding their hosts (Nicolas & Sillans, 1989). Several species can be quite sensitive to variations in CO_2 levels, as they can detect CO_2 concentration increase as small as 0.003% and locate their host plant from a distance of 20 cm (Doane, Lee, Klingler, & Westcott, 1975). This peculiarity makes these insects especially sensitive to rising atmospheric CO_2 concentration.

2.2.2 Effect of Temperature on Insect Performance

Most insects are ectotherms and temperature variations were shown to directly influence their eco-physiology including: metabolic rate, growth, feeding rates and life-history traits, such as fecundity and longevity (Angilletta, 2009; Bale et al., 2002). Generally, the thermal response of insects follows the classic thermal performance curve. An insect can increase the performance above a threshold (minimal critical temperature, CT_{min}) up to the optimal temperature, after which performance decreases sharply and reaches the upper threshold for performance (maximal critical temperature, CT_{max}) (Angilletta, 2009). Therefore, a warming climate should generally improve the eco-physiological performance of insects, as long as the temperature does not surpass the temperature optimum, and does not reach the CT_{max}. The ecological consequences of temperature-mediated improved performance include the expansion of species distribution ranges, as new thermal niches appear northwards (Parmesan, 2006; Parmesan & Yohe, 2003), and the increase of insect population outbreaks (Jepsen, Hagen, Ims, & Yoccoz, 2008). Understanding these processes, however,

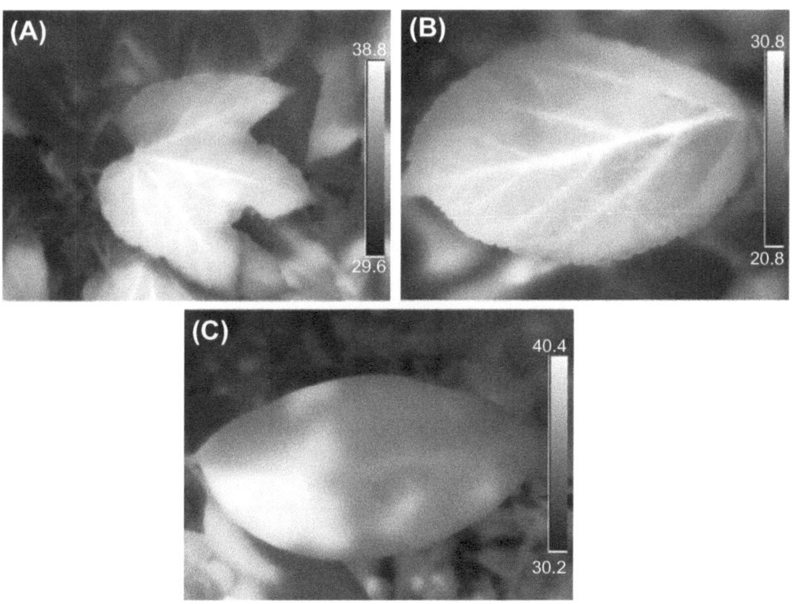

Figure 2 Thermographic images of leaf surfaces exposed to solar radiation for (A) grape ivy (France), (B) apple (France) and (C) *Clusia* sp. (French Guiana). During these measurements with an infrared camera, air temperature was 28.6°C, 24°C and 29.2°C, respectively.

necessitates clarifying the predictions of when and where the temperature falls above the optimal temperature and reaches the CT_{max} of species (Sunday et al., 2014). Specifically, this requires the determination of the exact temperature experienced by insects at the leaf surface or within the leaf tissues (Pincebourde & Woods, 2012).

The temperature at the surface of plants can deviate from ambient air temperature (Fig. 2). For instance, the leaf temperature excess (i.e. temperature deviation between leaf surface and ambient air) can be up to 10°C in temperate plants (Cook, Dixon, & Leopold, 1964; Pincebourde, Sinoquet, Combes, & Casas, 2007) and even up to 20°C in alpine plants in full sunlight (Linacre, 1967). By contrast, in the arid ecosystem of Arizona (USA), *Datura* leaves are colder than the surroundings during the day (Potter, Davidowitz, & Woods, 2009). The developmental and tolerance thresholds of *Manduca sexta* eggs are adapted to the *Datura* leaf temperature patterns, as they would die if they experience the high desert ambient air temperature (Potter et al., 2009). Generally, the excess temperature of leaves decreases when ambient air increases (Linacre, 1967; Michaletz et al., 2015; Pincebourde & Woods, 2012). The shape of the relationships between leaf temperature and ambient air temperature, however, may depend on the plant system and on the spatial scale at which the process is investigated (Pincebourde, Murdock, Vickers, & Sears, 2016).

The degree to which leaf temperature influences insects depends on their body size. The main reason is purely physical: any organism small enough to remain totally embedded into the leaf boundary layer is subjected to leaf temperature directly (Kaspari, Clay, Lucas, Yanoviak, & Kay, 2015). By contrast, organisms larger than the leaf boundary layer mainly depend on the ambient conditions (Woods, 2013). Across their ontogeny, most herbivore insects experience leaf temperature fluctuation during the first stages of their development until they reach a body size, beyond which their thermal budget shifts as they are influenced by ambient air conditions. For example, early instars of the *Manduca sexta* caterpillar experience temperature about 5°C below ambient air at the *Datura* leaf surface, while the last larval stage reaches body temperatures that are several degrees above ambient air (Woods, 2013).

Understanding how warmer climates affect insects thus requires quantifying the effect of global warming on leaf surface temperatures. From an evolutionary perspective, the convergence of mean leaf temperature at all latitudes towards a value of ~22°C (Helliker & Richter, 2008) indicates that climate change may have a relatively small effect on leaf surface temperature, as long as the whole plant can compensate. Nevertheless, it is not clear how global change will cause changes at the leaf surface, and several hypotheses were developed recently (Pincebourde & Woods, 2012). Depending on the level of adaptation to environmental fluctuations that are larger in the temperate zone compared to lower latitudes, the amplitude of warming may be buffered or by contrast be reported to the leaf surface temperature—more empirical research is needed. In addition, the ability of arthropods for behavioural thermoregulation could modify the leaf surface temperature pattern they experience (Pincebourde & Suppo, 2016). Spider mites, for example, make use of the within-leaf surface thermal heterogeneity to thermoregulate and keep track of the optimal temperature for their development (Caillon, Suppo, Casas, Woods, & Pincebourde, 2014). However, warming homogenizes leaf surface temperatures, thereby disrupting the behavioural thermo-compensation of mites, and ultimately leading them to overheat (Caillon et al., 2014).

2.3 Impact of Climate Change on Pollinators

Recently, climate change was pointed out as one of the major causes of regression of insect pollinator populations. While other groups may be involved, most of the pollination service is provided by the ~20,000 bee species (Michener, 2007; Proctor, Yeo, & Lack, 1996). Interestingly, these insects, and in particular bumblebees, generally show an elaborated endothermic behaviour, which allowed them to colonize a vast number of habitats (Heinrich, 1979; Owen, Bale, & Hayward, 2013). This specialized endothermy, however, exposes bumblebees to climate risks.

The modelling of eco-climatic envelope for bumblebee species indicated that many species will be significantly rarefied due to global warming (Rasmont et al., 2015). The climatic risk is more or less accentuated depending on the scenarios considered by IPCC (2007) (Fig. 3). In Europe, the distribution of 30, 47 and 53 species among the 69 bumblebee species will be reduced according to the scenario SEDG, BAMBU and GRAS, respectively (Settele et al., 2005). The population disruptions due to climate change can be severe. Specifically, for areas such as Madrid, Paris, London, and Helsinki, only one to three species can survive while several dozen species were present in the 20th century (Rasmont et al., 2015). By contrast, pollinators may find refuges in mountain areas which will conserve their current diversity. Therefore, population movements will likely determine the survival of these species. Comparable trends are expected for North America, except that the Northern margin of bumblebee distribution will

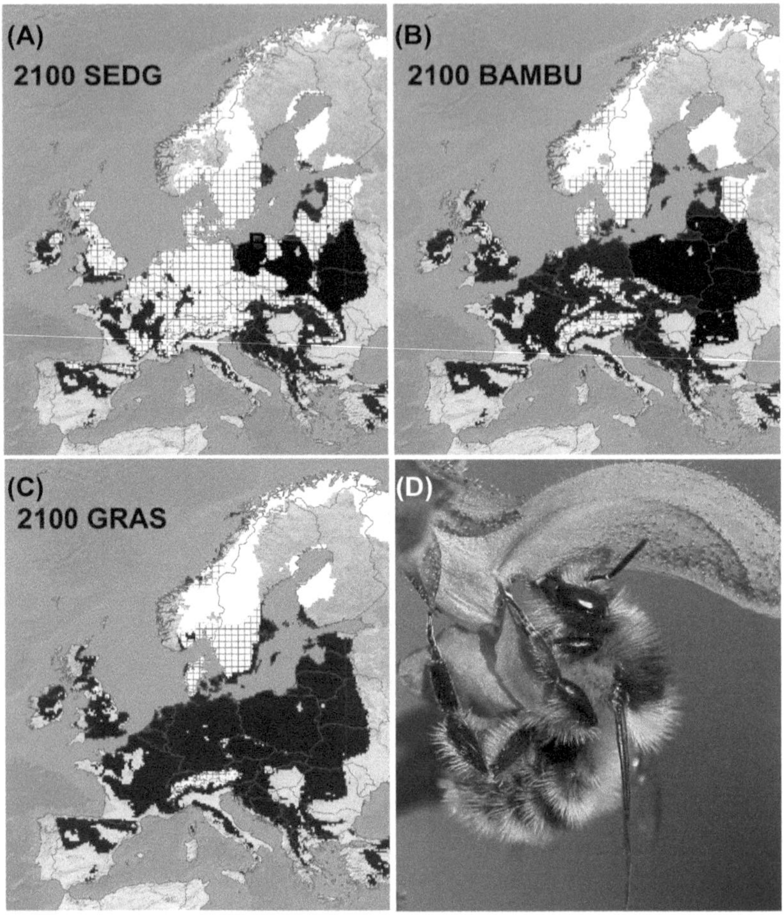

Figure 3 Projected suitable climatic area of *Bombus sylvarum* in 2100 in Europe. (A) SEDG scenario, (B) BAMBU scenario, (C) GRAS scenario and (D) *Bombus sylvarum* visiting a flower. Photo by J. Carteron. Areas in black will be lost; hatched areas will be conserved; areas in white will be gained. *After Rasmont, P., Franzen, M., Lecocq, T., Harpke, A., Roberts, S.P.M., Biesmeijer, K., …, Schweiger O. (2015). Climatic risk and distribution atlas of European bumblebees.* Biorisk, 10, *1–246. http://dx.doi.org/10.3897/biorisk.10.4749.*

not shift as much as the Southern limits (Kerr et al., 2015). North American taxa may have smaller dispersion ability than European species. Indeed, the postglacial history of species in Europe may have induced a strong selection of the most mobile taxa (Hewitt, 1999; Lecocq et al., 2013).

Some isolated mesoclimatic zones in southern European regions can play the role of 'Noah's Ark' by becoming a source-bumblebee population for this entire geographic region (Rasmont et al., 2015). Unfortunately, this perspective could be compromised by the genetic drift of the small surviving isolates. Indeed, the mechanism of sex determination in bumblebees may contribute to their decline in a warmer world. The male, normally haploid, is determined by the homozygosity of a small number of loci (Duchateau, Hoshiba, & Velthuis, 1994). During inbreeding, a large number of diploid males emerge with a drastically reduced fertility. Therefore, the monomorphism of these loci induced by genetic drift could depress the reproductive potential (Whitehorn, Tinsley, Brown, Darvill, & Goulson, 2009). Only large mountain areas, which will shelter large and genetically diverse residual populations, could ensure the survival of these pollinator species in the future.

The modelling studies of Kerr et al. (2015) and Rasmont et al. (2015) are mainly based on temperature averages. These models do not include extreme climatic events such as heat waves which will increase in frequency and duration in the next decades (Meehl & Tebaldi, 2004; Robinson, 2001) and which can cause local extinctions of some bumblebee species (Rasmont & Iserbyt, 2012). Martinet, Lecocq, Smet, and Rasmont (2015) showed that a typical heat wave temperature (40°C in temperate regions) rapidly kills the tested specimens (Fig. 4). Some bumblebee species, however, are more sensitive to thermal stress than others (e.g., arctic and top mountain species). Moreover, critical thermal limits vary according to altitudinal distribution of bumblebees (Oyen, Susma, & Dillon, 2016). By contrast, a species such as *Bombus terrestris* (Linnaeus, 1758) seems to tolerate well heat wave conditions. The particular thermal tolerance of this species may explain its recent Northward expansion in Europe despite the repeated heat waves (Martinet et al., 2016).

Some wild bee species are linked to hot and dry climates (Michener, 1979). For instance, the honeybee *Apis mellifera* Linnaeus 1758 or *Xylocopa varipuncta* Patton 1879 are able to fly in very hot conditions (Heinrich, 1979; Heinrich & Buchmann, 1986). Numerous wild bees are living in desert and other dry habitats, with their maximum diversity in these environments (Patiny & Michez, 2007). As of today, however, we have no evaluation of the thermal tolerance in these thermophilic taxa or on their resistance to climate change (Nieto et al., 2014, p. 84). For a small number of species, global warming is the best explanation for their northward expansion, e.g., *Xylocopa (Koptortosoma) pubescens* Spinola 1838 which is now present in Europe (Terzo & Rasmont, 2014).

Beyond the fate of pollinator insects during climate change, the question of pollination service is crucial. Both the plant and its pollinators may respond differently to climate change. For example, the large shift towards

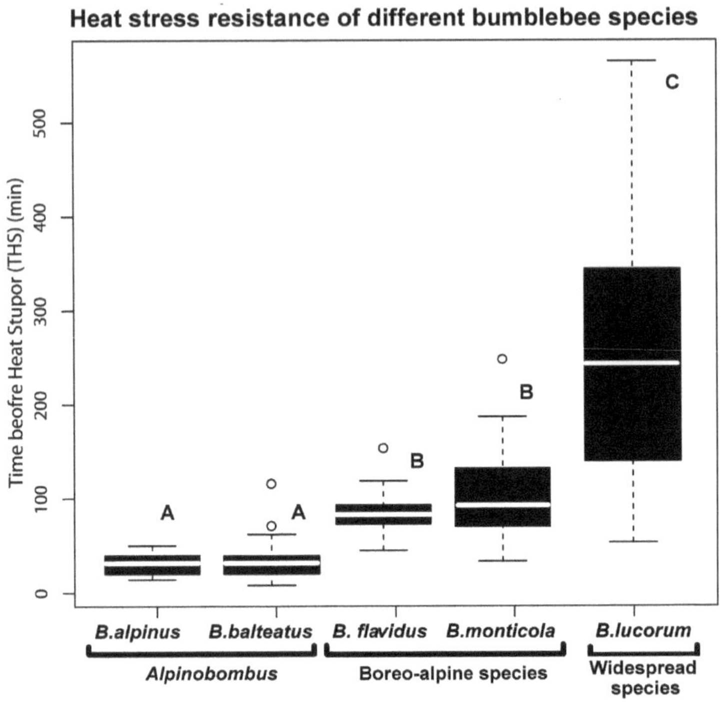

Figure 4 Boxplots of the time before heat stupor for six bumblebees species (*Bombus* sp.). (A) Arcto-alpine species: *Bombus alpinus* and *Bombus balteatus*, (B) boreo-alpine species: *Bombus flavidus* and *Bombus monticola* and (C) widespread species: *Bombus lucorum*. *Circles* are extreme values. *After Martinet, B., Lecocq, T., Smet, J., & Rasmont, P. (2015). A protocol to assess insect resistance to heat waves, applied to bumblebees (*Bombus Latreille, *1802).* PLoS One, 10, *e0118591.*

the North of orchards relative to the small shift of the associated pollinators may generate an extensive pollination deficit area (Polce et al., 2014). In general, pollinators are involved in a close symbiotic relationship with their favourite plants, and any depression caused by climate stress can lead to pollination deficit (Franzén & Ockinger, 2012). This deficit can significantly reduce the reproductive potential of both domesticated and wild plants. In turn, this reduction can lower biodiversity by generating food deficiency for the species associated with these plants. This interaction chain still escapes our modelling capabilities.

3. INDIRECT EFFECTS OF CLIMATE CHANGE ON PLANT–INSECT INTERACTIONS

3.1 Plants Responses Matter for Insects

3.1.1 Impacts of Climate Change on the Plant Nutritional Quality for Herbivores

Rising atmospheric CO_2 concentration influences the development of herbivores indirectly by inducing changes in the plant nutritional quality (Lincoln et al., 1993). In general, plants growing in a CO_2-enriched atmosphere have higher levels of non-structural sugars, which are beneficial to insect herbivores (Bezemer & Jones, 1998). Nevertheless, higher CO_2 levels also lead to a lower concentration of proteins (Ehleringer, Cerling, & Dearing, 2002; Whittaker, 2001) and some amino acids (Docherty, Wade, Hurst, Whittaker, & Lea, 1997) in leaves. In addition, the concentration of secondary compounds, which are toxic to herbivore insects, increases in a CO_2-enriched atmosphere (Bidart-Bouzat, Mithen, & Berenbaum, 2005; Cornelissen, Stiling, & Drake, 2003; Stiling & Cornelissen, 2007). Therefore, the nutritional quality of plants is lower when grown under high CO_2 levels compared to current concentrations. Negative impacts of increasing CO_2 levels were also found on the development rate and survival of phytophagous insects (Smith & Jones, 1998; Whittaker, 2001). Herbivore insects may develop compensatory mechanisms by increasing their food intake quantitatively or by enhancing assimilation efficiency (Barbehenn, Karowe, & Chen, 2004; Stiling & Cornelissen, 2007). For example, leaf miner caterpillars compensate for the lower quality of their plant food by increasing the amount of leaf tissues eaten (Cornelissen et al., 2003; Salt, Brooks, & Whittaker, 1995). Compensatory mechanisms were not found in all species (Smith & Jones, 1998), however, and even when compensation occurs it does not cover all the negative impacts of low quality food (Hunter, 2001).

Temperature changes can also influence the nutritional quality of host plants. The net effect is highly plant species specific. A long-term increase in temperature induces a decrease in the water content of leaves (Williams, Norby, & Lincoln, 2000). In herbaceous, warming induces an increase in nitrogen concentration of plant tissues (Bezemer & Jones, 1998). By contrast, in woody species, warming negatively impacts the nitrogen concentration and increases the concentration of tannins (Dury, Good, Perrins, Buse, & Kaye, 1998). The concentration of sugar can also be affected by the increase in temperature (Morison & Lawlor, 1999).

Interactive effects of temperature increase and CO_2 concentration change were reported, but the net effect varies according to the insect–plant system (Zvereva & Kozlov, 2006). In some systems, only one of the two factors has an impact on the development rate of the insect, either negative (Williams et al., 2000) or positive (Buse & Good, 1996). In other systems, however, temperature and CO_2 interact such that the negative effect of one is suppressed by the positive effect of the second (Fajer, Bowers, & Bazzaz, 1991). Finally, in most cases, compensatory effects happen resulting in the compensation of the negative effect from one factor by the positive effect of the second (Johns, Beaumont, & Hughes, 2003; Johns & Hughes, 2002). Overall, the predicted negative effects of CO_2 elevation on herbivores are likely to be mitigated by temperature increase (Zvereva & Kozlov, 2006), but more experimental data are needed on different biological systems to generalize these findings (Robinson et al., 2012).

3.1.2 Cascading Effects of Changing the Plant Eco-Physiology on the Insect Microclimate

When feeding on their host plant, insects modify the plant tissues physically and/or physiologically. In turn, these modifications can induce shifts in the insect microclimate — an effect called the physical feedback of herbivory (Pincebourde & Casas, 2006a). This indirect effect is especially prominent in endophagous insects such as leaf miners and gallers. These organisms alter the structure and the properties of the leaf (Pincebourde & Casas, 2016). In general, the outcome of these modifications is an increase in the microclimatic temperature for the insect herbivore. For example, the leaf miner *Phyllonorycter blancardella* (Fabricius, 1781) (Lepidoptera: Gracillariidae) induces stomatal closure in apple leaf tissues (Pincebourde, Frak, Sinoquet, Régnard, & Casas, 2006) and generates white patches at the leaf surface (Pincebourde & Casas, 2006a). These two modifications contribute to an elevation of the temperature within the mine by up to 12°C above ambient air (Pincebourde & Casas, 2006b, 2015). These elevated temperatures can speed up the insect development, but this strategy is at risk during extreme climatic events such as heat waves (Pincebourde & Casas, 2015). During heat waves, the mosaic of favourable and risky microclimates is reshuffled (Pincebourde et al., 2007).

External feeders also induce variations in leaf surface temperatures during their feeding activity. These variations are in the order of few degrees, and are generally quite local around the leaf portion attacked by the herbivore (Nabity, Hillstrom, Lindroth, & DeLucia, 2012). These temperature changes at the leaf surface are the consequence of alteration in stomatal conductance and evapotranspiration rates. Nevertheless, little is known on these interactive effects during warming. Indeed, insect herbivores tend to increase their feeding rate in a warmer environment (see above), suggesting that more local impacts on leaf transpiration rates are expected, thereby reinforcing the general warming effect. Overall, the plant responses to climate change should have consequences on the microclimatic conditions at the leaf surface. Many parameters impact the leaf heat budget, including its size and shape, stomatal conductance and its orientation towards the sun (Gates, 1980). Stomata have an important role in the plant responses: they regulate the overall energy and gas budget of the plant, and they react to several environmental (e.g., climatic variables) and endogenous factors (e.g., plant water status). Climate change can be expected to induce variations in these factors, with likely cascading effects on surface temperatures (Pincebourde & Woods, 2012). Finally, changing plant architecture may also partially buffer the amplitude of warming for insects at the leaf surface, but this effect seems to be quite limited given the magnitude of warming (Saudreau et al., 2013).

3.2 Biotic Interactions Matter for Insects

Species are linked by trophic links. Therefore, a series of indirect effects are expected to cascade through the entire food web, thereby increasing the difficulty to anticipate the effects, especially in complex multitrophic systems (Tylianakis et al., 2008).

3.2.1 Thermal Traits Diverge Across the Different Components of a Multitrophic System

3.2.1.1 Development Time and Growth Rate

The speed at which organisms respond to climate change is partially determined by their generation time. In general, temperature increases the development rate of plants and insects inside their tolerance range. For plants, the period of vegetation increases in length, allowing insects to feed on them for a longer period throughout the year. Some multivoltine species can realize an additional generation per year due to both the longest period of vegetation and their shorter development time. For example, in the European grapevine moth *Lobesia* (*Lobesia*) *botrana* (Denis & Schiffermüller, 1775) in South-West of France, Marchal and Feytaud (1911) observed three generations of this insect in 1910, but a century later Martin-Vertedor, Ferrero-Garcia, and Torres-Vila (2010) noted a fourth generation at the end of fall. The increased reproduction rate influences the population dynamics at the other trophic levels. The parasitism rate of the eastern spruce budworm, *Choristoneura fumiferana* Clemens 1865 (Lepidoptera: Tortricidae) by its parasitoids decreases when the temperature increases (Harrington, Fleming, & Woiwod, 2001). By contrast, the predation rate of the seven-spot ladybird, *Coccinella septempunctata* Linnaeus 1758 on the aphid *Acyrthosiphum pisum* (Harris, 1776) increases at high temperatures. The impact of temperature increase cannot be generalized without the understanding of the specific thermal responses.

3.2.1.2 Metabolic Rate, Longevity and Fecundity

These traits are driven by temperature in natural enemies such as parasitoids, thereby modifying the parasitism rate and then the impact of phytophagous insects on plants. An increase in temperature increases metabolism and activity (May, 1979). For example, the walking speed of the parasitoid *Aphelinus asychis* Walker 1839 (Hymenoptera: Aphelinidae) increases with temperature (Mason & Hopper, 1997). These traits are associated with an increase in the instantaneous rate of parasitism. However, the increased metabolic rate also results in a reduced longevity due to the accelerated use of energy (Huey & Stevenson, 1979; Trotta et al., 2006). Therefore, the impact at the life span level is difficult to determine. In addition, above a given temperature threshold, the negative effects of high temperature exceed the advantages. This complexity of the relationship between traits at the individual scale is found at each trophic level. An increase of temperature can increase the fecundity of the phytophagous insect and decrease that of the parasitoid, or vice versa.

3.2.1.3 Sex Allocation

Temperature can influence sex allocation in insects. This was observed for arrhenotokous hymenopteran parasitoids, in which unfertilised eggs develop into haploid males and fertilised eggs develop into diploid females. Females are able to allocate male or female eggs in one host according to internal or external factors, thus affecting the sex ratio. For example, the sex ratio of *Aphelinus varipes* (Förster, 1841) (Hymenoptera: Aphelinidae) varied from 92% female when reared at 25°C, to 70% at 20°C (Rohne, 2002). By contrast, in the parasitoid *Diaeretiella rapae* (M'Intosh, 1855) (Hymenoptera: Aphidiidae), the percentage of females was lowest at 26.7°C compared to 21°C (Bernal & Gonzalez, 1997). The effect of temperature on sex allocation is complicated by the fact that the sex ratio can be influenced by host size, with males more likely to emerge from smaller hosts. Host size is also linked to temperature. For example, the body size of the black bean aphid, *Aphis fabae* Scopoli 1763, is larger when reared at low temperatures (Li & Mills, 2004).

3.2.1.4 Diapause

Diapause is an environmentally pre-programmed and hormonally mediated state of low metabolic activity associated with arrested development and increased resistance to environment extremes (Tauber, Tauber, & Masaki, 1986). Warmer and shorter winters act on the diapause strategy of insects. For example, the parasitoid *Aphidius avenae* Haliday 1834 was detected for the first time in its non-diapausing form on cereal fields in Brittany in the winter of 2011–12, whereas during the previous 20 years, it was observed in the diapausing phase outside cereal crops during winters and it only colonized the crops in spring (van Baaren et al., 2004; Krespi, Dedryver, & Creach, 1997; Le Lann et al., 2011). These modifications in the host-parasitoid relationship were linked to an increase in winter temperatures (Andrade, Krespi, Bonnardot, van Baaren, & Outreman, 2016). This shift in diapause strategy modified the food web, with *A. avenae* becoming a dominant species in this system and with an increase in the parasitism rate in winter. The plasticity of diapause induction may allow an efficient exploitation of available resources and, consequently, non-diapausing parasitoids should only be associated with anholocyclic aphids (i.e. aphids with incomplete life cycle, or which do not alternate parthenogenesis and sexual reproduction) that remain reproductive throughout winter. More generally, the parasitoid should remain synchronized with the host cycle (Godfray, 1994). The parasitoid either follows the diapause strategy of its host, and in this case the parasitoid does not kill its host before entering diapause, or the diapause strategy is decoupled from the host. In the context of climate change, the challenge for the parasitoid is to remain synchronized with the host strategy and to avoid the end of its diapause when the host is still diapausing.

3.2.1.5 Endosymbionts

Both parasitoids and herbivore insects may host endosymbiotic bacteria that can influence various life-history traits (for a review see chapter: Influence of Microbial Symbionts on Plant–Insect Interactions by Giron et al., 2016). For example, aphids have a variety of bacterial symbionts including an obligate association with the mutualist *Buchnera aphidicola* (Shigenobu, Watanabe, Hattori, Sakaki, & Ishikawa, 2000). They may also possess secondary symbionts (e.g., *Wolbachia*) which are conditionally beneficial or deleterious (Degnanj & Moran, 2008). These bacteria have major effects on aphid biology, including enhanced tolerance to heat stress and changes in host plant range (Moran, Degnan, Santos, Dunbar, & Ochman, 2005; Tsuchida, Koga, & Fukatsu, 2004). Both major groups of endosymbionts, *Wolbachia* and *Buchnera*, may be negatively affected or eliminated by short exposures to high temperature (Thomas & Blanford, 2003). Moreover, the effect of secondary endosymbionts varies with temperature. Endosymbionts protect the aphid *Acyrthosiphum pisum* against heat stress at 25°C, whereas they decrease the fecundity at 20°C (Cheng, Montllor, & Purcell, 2000). The defensive immunity conferred to aphids by the endosymbiont *Candidatus Hamiltonella defensa* fails under heat stress (Bensadia, Boudreault, Guaya, Michaud, & Cloutier, 2006). In conclusion, climate change is expected to have broad-ranging indirect impacts on insect communities via direct effects on their associated endosymbionts.

3.2.1.6 Virus Transmission

Climate change is expected to modify the incidence of phytophagous-borne virus infections since temperature influences virus transmission (Thomas & Blanford, 2003). For example, barley yellow dwarf (BYD) is one of the world's most severe viral diseases of autumn-sown cereals. BYD viruses are transmitted by several aphid species. Environmental conditions govern the pattern and timing of primary infections by viruliferous alates and the speed of subsequent virus dissemination by apterae (secondary spread). Temperature influences both the population dynamics of the vector (e.g., 5°C is the developmental threshold for the aphid *Rhopalosiphum padi* (Linnaeus, 1758)) and the virus transmission process (e.g., 15°C is the take-off threshold for alatae). Warm January to August periods generate a high percentage of viruliferous aphids the following autumn, perhaps due to an increased rate of population growth and virus transmission in agroecosystems (Fabre et al., 2005).

3.2.2 Synchronization Between Trophic Levels

3.2.2.1 Phenological Synchronization

Numerous studies have reported climate change-induced phenological modifications at one particular trophic level. But few studies quantified the consequences of phenological shifts at several trophic levels. In UK, *Anthocharis cardamines* (Linneaus, 1758) (Lepidoptera: Pieridae) exactly follows the phenology of its host plant, for which the phenology is advanced by 2–3 weeks with warming, but this example is an exception (Parmesan, 2006). The phenological synchrony between plants and insects, and between hosts and parasitoids, can be decoupled if the species are affected by temperature differently. For example, Harrington et al. (2001) showed in the system *Picea sitchensis* Carrière (Pinaceae)–*Operophtera brumata* (Linnaeus, 1758) (Lepidoptera: Geometridae) that the onset of bud development is not modified by warming, whereas the emergence date of the insects that feed on them is advanced. Nevertheless, the development of the caterpillars collapses because buds are not yet available when they emerge.

Few studies focused on the second and third trophic levels, and they suggest that their synchrony could be affected too. Annual variation in phenological asynchrony may be an important destabilizing factor (Godfray, Hassell, & Holt, 1994). Similarly, the time of parasitoid arrival in an aphid population influences the growth rate of the parasitoid population and its impact on the host population (Hoover & Newman, 2004). If the parasitoid arrives too early (before the aphid population has entered the exponential growth phase), the parasitoid population disappears before the aphid population increases or the parasitoid eliminates the small aphid population before it reaches the exponential growth phase. By contrast, if the parasitoid arrives during the exponential growth phase of the host, large parasitoid populations will be produced. Even small climatic changes can affect the synchrony of parasitoid activity with host populations, with large effects on the population dynamics of both.

3.2.2.2 Spatial Synchronization

The geographical distribution of plants, herbivores and parasitoids may change during climate change as a result of various processes, including changes in the dispersal ability of insects and their host plants, and the destabilization of resident ecosystems (Parmesan, 2006). In the last century, the geographical distribution of species was modified by climate changes in 75% and 81% of the studied species in tropical and temperate areas, respectively (Parmesan, 1996). Phytophagous species often extend their geographic distribution northwards in temperate regions following an increase in temperature (Parmesan, 2006). For example, *Pararge aegeria* (Linnaeus, 1758) (Lepidoptera: Nymphalidae) extended its repartition towards the North since 1940 (Hill, Thomas, & Huntley, 1999). Extending the distribution becomes easier when the phytophagous insect succeeds to adapt to a new host plant. Also, a phytophagous species extending its repartition can reach zones which are free of any natural enemies, as it is the case for the processionary pine moth, *Thaumetopoea pityocampa* (Denis & Schiffermüller, 1775) (Battisti et al., 2005).

3.2.3 Impacts at the Guild or at the Community Levels

Communities are mixtures of specialist and generalist species. It is predicted that specialists will suffer from climate change more than generalists, because generalists exploit a large variety of host species which can be adapted to different climatic conditions (Stireman et al., 2005). The trophic rank hypothesis predicts that higher-trophic level organisms are more negatively affected by environmental changes and disturbance than species in the lower trophic levels (Holt, Lawton, Polis, & Martinez, 1999; Tscharntke & Brandl, 2004). Moreover, at each trophic level, species succeed each other along the season, or they can be active at the same time and compete with each other. Climate change can affect the phenology of species and then the strength of the competition between species.

For phytophagous insects exploiting different plants along the season, or for natural enemies exploiting different host/prey along the season, the success requires the temporal synchronization between the different species. Changes in temperature can modify the degree to which this synchronization occurs. For example, the ant *Iridomyrmex humilis* (Mayr, 1868) (Hymenoptera: Dolichoderinae) is parasitized by several parasitoids which appear successively during the season. The first species is active below 14°C, the followers are active between 14°C and 23°C and the latest species emerges above 23°C. Temperature changes will induce discontinuity or reinforce competition between these species (Folgareit, Bruzzone, & Gilbert, 2003). However, studies at this scale are rare, impairing general conclusions.

4. IMPACT OF HUMAN ACTIVITIES ON PLANT–INSECT INTERACTIONS

4.1 The Main Anthropogenic Drivers Affecting Plant–Insect Interactions

Here, we focus on the ecological and evolutionary effects of anthropogenic factors on plant–insect interactions. We suggest a distinction between two categories of human influences: acts on purpose and acts by accident.

4.1.1 *Effects of Human Eco-Engineering*

4.1.1.1 Influence of Agrosystems

Agriculture is mankind's great project from the Neolithic Revolution. Creating agro-ecosystems means reducing ecosystems to what is required for the primary production of biomass: an edible plant in a favourable environment. Agrosystems differ from ecosystems in their instability and the high density of plants in monospecific stands. In addition, agrosystems show rapid and contrasting changes of the milieu, at least alternating bare soil and synchronized plant growths. Agrosystems are also characterized by the weakness of physical and chemical defences of plants towards the herbivores (Chen, Gols, & Benrey, 2015).

Agriculture and forestry select pest insects that are adapted to these agricultural conditions, in particular to the abundance and short-time availability of food resources (Bianchi, Booij, & Tscharntke, 2006; Carnus et al., 2006). Insect pests usually combine high fertility and multivoltinism with great voracity and ability to access the crops. These adaptations contribute to a high population growth rate. Half of the million described species of insects are herbivorous. Among them, around 10,000 species are pests (Herrera & Pellmyr, 2002). Pests are rather rare in natural ecosystems (Pimentel et al., 1992) because herbivorous insect populations are generally regulated by the quantity and accessibility of the plants that they exploit. But 'natural pests' exist too. For example, they are represented by crickets (Orthoptera: Acrididae) in dry tropical ecosystems. Pest insects may devastate temperate and Nordic forests, for example the oak processionary, *T. processionea*. Existence of these 'natural pests' can be explained by the similarity between these ecosystems and some agrosystems: shortness of plant growth periods in the dry tropics, homogeneous forest stands over large surfaces in cold regions (Schvester, 1985).

Agriculture causes biogeographical disruptions. Thus, it is a source of sympatric or ecological speciation (Orr & Smith, 1998). Crop plants are established anywhere possible until they become cosmopolitan (wheat, rice, corn etc.). Therefore plant crops come into contact with new insects, and new trophic relationships can take place. Adaptations of insects to new plant hosts are considered as the key mechanism of sympatric speciation. The first case of speciation via host plant shift was described by Walsch around 1860 (Berlocher & Feder, 2002). The fruit fly *Rhagoletis pomonella* Walsh 1867 (Diptera: Tephritidae) shifted from the hawthorn, *Crataegus* sp. (Rosaceae), to the common apple tree, *Malus pumila* (Rosaceae). Agriculture and the transport of plants and insects amplify this type of diversification. This is probably why examples of sympatric diversification are less observed in natural ecosystems. For example, the European species *Ostrinia nubilalis* (Hübner, 1796) (Lepidoptera: Crambidae) recently split into two sympatric sibling species which eat different wild plants (Malausa, Delacky, et al., 2007; Malausa, Leniaud, et al., 2007).

4.1.1.2 Making Landscapes From Ecosystems

Humans destroy natural habitats to build new anthropized environments, the landscapes. The main ecosystem that was destroyed in Europe was forest (Klemm, 1996, pp. 17—34). In ecology, landscapes are often analyzed as systems deviating from the original ecosystem. They are characterized by their deviation 'distance', which integrates the difference between landscape areas (open spaces) and relics of forests (woodlots), or a proportion of wooded areas and the degree of their connections (wooded hedgerows) compared to open spaces. This metric has led to major developments in modelling (With & Crist, 1995), in particular by making the analogy between the fragmented ecosystem and island biogeography (Fahrig, 2003; Ricketts, 2001; Young, Boyle, & Brown, 1996), and by the duality in the landscape between cultivated and uncultivated areas (Carré et al., 2009). However, authors such as Turner (2005) criticized the reference to the original ecosystem, represented by the less anthropic environments, which prevents considering the landscape itself as a new ecosystem (Manel, Schwartz, Luikart, & Taberlet, 2003). Landscapes are more heterogeneous than the original systems. They abound in contact zones between anthropic and semi-natural environments. This is very conducive to the diversification of relations between insects and plants.

4.1.1.3 Urbanization Effects

The urban environment can be considered as an extreme form of landscape, often with no precise boundary between the city and rural landscapes (McKinney, 2002). Even the most densely urbanized areas offer opportunities for spontaneous life which represent, however, small biomass. Small and interstitial habitats (borders, wasteland), and mini-landscapes (parks, cemeteries) and particular niches may host adapted or synanthropic species (commensal or parasitic). Cities are places where cultivated exotic plants abound, although native plants are surprisingly well represented (Frankie & Ehler, 1978; Kühn, Brandl, & Klotz, 2004) despite the difficulty to access soil, water and light. Plant—insect relationships are subjected to the same evolutionary influences than in other anthropized environments, but the communities and ecological conditions differ. The herbivorous insects that are tolerated by their host plants in natural landscapes become multivoltine pests in urban areas, such as the aphid *Aphis nerii* Boyer de Fonscolombe 1841 on the shrub *Nerium oleander* L. (Apocynaceae) (Frankie & Ehler, 1978).

4.1.2 Evolutionary Responses to Accidental Anthropogenic Drivers

4.1.2.1 The Effects of Non-Intentional Anthropogenic Drivers

Human actions have secondary, not intended outcomes with effects without apparent spatial relationship with their cause. Unintentional disturbances are mainly changes in environmental conditions (pollution, dust, night lighting, urbanization etc.) and community (introductions and depopulation). This phenomenon is best illustrated by the historical example known under the generic term 'industrial melanism of the peppered moth, *Biston betularia*'. In the 1950s, the British biologist Kettlewell worked on the assumption of an ecological factor of selection specific to industrialized regions and which should favour the black form of the butterfly rather than the white form. He discovered that the mortality of the white form was greatly increased because they were more visible to birds on trees darkened by industrial dust (Grant, Owen, & Clarke, 1996; Howlett & Majerus, 1987).

4.1.2.2 Community Disturbances (Introductions, Population Declines and Species Extinctions)

Species introductions are a worldwide problem. Human activities multiply the transportation of species (Kenis, Rabitsch, Auger-Rozenberg, & Roques, 2007). Symmetrically, we observe populations declining or disappearing, and extinction of species. In Europe, insect introductions accelerated from 4 introductions per year between 1971 and 1990 to 11 between 1991 and 2007, while higher plant introductions decreased from 27 to 18 during the same period (Hulme, Pysek, Nentwig, & Vila, 2009). Part of these taxa settle down on arrival, either through their pre-adaptations or because they are human commensals (cultivated plants and their pests, honeybees etc.). Introduced species have to contest a specific niche to indigenous species, but they take advantage of losing their main competitors in the native ecosystem (Bossdorf et al., 2005). The introduction, however, corresponds to a genetic bottleneck (Estoup et al., 2016). Some herbivore insect populations have a genetic trait that makes them invasive, and sometimes this trait is linked to a single gene (Lee, 2002). Thus, the invasive aphid *Acyrthosiphum pisum* loose the sexual part of its life cycle, thereby gaining the ability to switch to new host plants (Via, 2001).

Species extinction is a major concern in biodiversity conservation. Declines and extinction can trigger concatenated disturbances because of the interdependence of species in ecosystems. It is certainly the case for species associated in the mutualism of the entomophilous pollination (Berenbaum et al., 2006). For example, the decline of wild bees (non-*Apis*) is linked to the decline of some entomogamous flowering plants (Biesmeijer et al., 2006). In Europe, an unknown factor, likely of anthropogenic origin, is widely unfavourable to the biodiversity of pollinators and flowering plants.

4.2 Impact of Pollution on Plant—Insect Interactions

The number of pollutants is high and the plant—insect responses are highly diverse. Here, we address the major common characteristics that govern the bottom-up (host plant quality) and the top-down (natural enemies) processes that influence the fitness and population dynamics of herbivore insects. Inorganic and organic pollutants include air pollutants such as ozone, sulphur oxides (SO_x), nitrogen oxides (NO_x), carbon oxides (CO_x), fluorides and acidic precipitations, as well as soil pollutants such as metalloids and heavy metals (Butler & Trumble, 2008). Pollutants are more or less toxic to the environment according to their chemical composition, and this toxicity depends on the dose (concentration $\times$ time) (see Calatayud et al., 2013 for review). Indeed, pollutants can have more impact on the environment when applied for a short period than when they are present for a long time, at equal dose. In plants, this peak effect is explained by plant response time relative to the pollutant residence time: plants do not have time to initiate their defence systems towards pollutants.

After penetration of the pollutant into plant tissues via the stomata or the roots, the pollutant generates a stress, such as oxidative stress for oxide pollutants and disturbance of calcium metabolism in the case of fluorides pollution (Calatayud, Garrec, & Nicole, 2013). Classically, the plant limits the absorption of pollutant and increases tolerance to it by implementing physical processes (e.g., stomatal closure, falling leaves) as well as chemical and biochemical processes (e.g., production of insoluble precipitates, enzymatic degradation by P450). The plant resistance to pollutants depends on the combination of the defence mechanisms already present in plant tissues and the defence processes that the plant can activate after the stress occurs. Resistance, however, also depends on other abiotic factors (e.g., temperature, humidity, light) and biotic factors (e.g., age, disease, genotype), which can have positive or negative impacts on plant response to air and soil pollution. In situ, there is generally an increase in insect populations on plants in polluted areas through both bottom-up and top-down processes (Hain, 1987).

4.2.1 Bottom–Up Effects of Pollution

Pollutants have a crucial impact on the key parameters of plants that govern plant–insect relationships by changing the physiology and the biochemistry of plant tissues (Nicole, 2002). In particular, pollutants alter the parameters related to recognition by insects, nutritional quality and plant defences.

4.2.1.1 Location and Recognition of Plants

Pollutants can cause changes in the colour of plants, thereby influencing the colour of associated insects. The most famous example was reported above by the melanism mutation in British peppered moths during the industrial revolution and recently shown to be due to a transposable element (van't Hof et al., 2016). In addition, pollutants disrupt the chemical communication in plant–insect relationships. In general, the alteration of VOCs emission occurs in polluted plants, impacting orientation behaviour in herbivore insects (Blande, Holopainen, & Ninemets, 2014). Pollutants act on the plant physiology (e.g., limiting volatile emissions by stomatal closure, modifying the plant physico-chemical characteristics), or by direct degradation of VOCs in the air.

4.2.1.2 Nutritional Quality of Plants

Accumulations of toxic pollutants in plant organs such as heavy metals, arsenic and fluorine are often the cause of poisoning of plant-eating insects (Führer, 1985). In addition, pollution leads to changes in primary and secondary metabolites in plants. In particular, there is often an increase in leaf concentration of amino acids (proline), soluble protein and sugars, thereby increasing the nutritional quality of plants for insects (Kainulainen, Holopainen, & Holopainen, 2000; Warrington, 1989). Indeed, pollutants such as SO_2 and NO_x increase the concentrations of S and N in plants, with positive effects on insects. This is especially true for plants on roadsides which are major sources of NO_x (Bolsinger & Flückiger, 1987; Braun & Flückiger, 1985). But conversely, CO_2 pollution associated with NO_x leads to lower nitrogen concentrations in leaves with negative consequences for herbivore insects (Bezemer & Jones, 1998; Hättenschwiler & Schhafellner, 1999). Heavy metals have in general a negative impact on the fitness of herbivores (Butler & Trumble, 2008).

4.2.1.3 Changes in Chemical and Physical Plant Defences

If secondary compounds (e.g., phenolic compounds) are part of the chemical defence systems of plants, the cuticle in turn becomes an effective barrier to insects. Pollutants such as ozone and CO_2 induce oxidative stress, and are the source of increased concentrations of phenolic compounds in plants with a negative impact on the nutritional quality of leaves for insects (Bolsinger, Lier, & Hughes, 1992; Bolsinger, Lier, Lansky, & Hughes, 1991). Meanwhile, ozone and elevated CO_2 promote the production of cuticular waxes, improving the characteristics of the physical barrier of the cuticle (Percy et al., 2002). In addition, the fitness of herbivores declines when they feed on plants contaminated by metalloids such as selenium and fluoride (Butler & Trumble, 2008).

4.2.2 Top—Down and Guild Effects of Pollution

Little is known about the impacts of pollutants on natural enemies, compared to herbivore insects. Butler and Trumble (2008) highlighted likely trends of pollution effects on insect parasitoids, predators and pathogens. The most common responses were either no effects or negative effects on natural enemies. This was reported in environments with elevated CO_2, ozone, heavy metals, metalloids and acidic precipitation. Butler and Trumble (2008) found that foliage feeders and miners often exhibit negative responses to pollutants through bottom-up process in CO_2 and heavy metal enriched environments. By contrast, phloem feeders exhibit greater fitness in environments polluted by SO_2 and NO_2. No conclusive pattern of pollution effects on bottom-up or top-down processes can be done on xylem feeders, mesophyll feeders, seed feeders and gallers; and no information is available on the fitness of borers, root feeders and, more surprisingly, pollinators (Butler & Trumble, 2008).

5. CONCLUSION AND PERSPECTIVES

The complex interactions between direct and indirect effects of global change makes particularly difficult to predict its net impact on plant—insect relationships (Fig. 1). The complexity of the mechanisms at play combines with the extreme level of diversification of life-history traits in both plants and insects. For example, insects demonstrate a high diversity of feeding modes which transposes into a high variability of plant eco-physiological responses to herbivory (Welter, 1989). Another layer of variability should be expected on the effects of global change on these specific plant responses, as reflected by the difficulty to generalize on the impacts of temperature and CO_2 on insect herbivores (Zvereva & Kozlov, 2006). Therefore, it seems illusive to extract a single general scheme depicting the net effect of global change on plant–insect relationships.

The intricate network of direct and indirect impacts of global change on plant–insect relationships is probably even more complex than it seems because several influential processes are still understudied. For example, the biochemical pathways of heat tolerance in plants are relatively well understood (Wahid, Gelani, Ashraf, & Foolad, 2007), but it remains to elucidate if these paths interconnect with the chemical machinery involved in plant defences against herbivore insects. A link can be expected because several molecules and chemicals are involved in the two processes, such as abscisic and salicylic acids, ethylene or phenolic compounds like flavonoids. In addition, pollutants may modulate these links between plant heat tolerance and plant defences. Pollutants decrease the ability of insects to tolerate heat (Slotsbo et al., 2009), but it is not clear if the same effect applies to plants.

Urban systems may provide good models to study these complex relationships because trophic webs are simplified, and also because cities may simulate the future abiotic conditions for natural habitats (Youngsteadt, Dale, Terando, Dunn, & Frank, 2014). Urban areas are often warmer and drier on average compared to natural ecosystems (Oke, 1982). The spatial heterogeneity in surface temperature in urban zones can be used as a 'laboratory' to study climate change impacts on the behaviour of insects (Pincebourde et al., 2016). Given the numerous stressors experienced by plants in cities, it seems appropriate to explore the complex links between plant tolerance to heat and drought, plant defences to herbivores and emission of VOC by plants in cities. However, we are not aware of such studies. Recently, it was observed that scale insects were more abundant on trees in cities as a direct effect of the urban heat island (Youngsteadt et al., 2014), suggesting that tree species may be less protected from pests in urban areas.

Nowadays, humans develop methods to increase the resilience of plant–insect communities to global changes, in particular in the agronomic context. Among them, the Conservation Biological Control (CBC, see chapter: Conservation Biological Control in Agricultural Landscapes by Rusch, Bommarco, & Ekbom, 2017 for a review) increases the efficiency of natural enemies and decreases the impacts of pests through modification of the environment (Eilenberg, Hajek, & Lomer, 2001). The success of the CBC, however, requires good knowledge of the plant–insect interactions. Several studies have shown that more diverse plant communities are functionally less susceptible to environmental stress (Steudel et al., 2012). This emphasizes the need to maintain biodiversity as an insurance against impacts of changing environmental conditions and sets the stage for exploring the mechanisms underlying biodiversity effects in stressed ecosystems. Indeed, biodiversity has unique effects such as complementarities and redundancies among species, dilution of plant species, or improving the stability of food webs (McCann, 2000; Naeem, 1998). Higher plant diversity, through association of plant species supplying alternative food and/or shelters at different seasons of the year, can increase the performance and fitness of natural enemies (Bompard, Jaworski, Bearez, & Desneux, 2013; Wratten, Gillespie, Decourtye, Mader, & Desneux, 2012) thus increasing their resistance to climate stresses. More diverse vegetation can promote associational resistance of host plants. This associational resistance may result from dilution of (plant) hosts for insect herbivores and from a complementarity of (insect) hosts for phytophagous-enemies, as well as a redundancy among these phytophagous-enemies. This hypothesis will be particular important for plants whose defences are weakened by climate stress. Alternatively, the associational resistance may emerge from a decrease in climate resistance of herbivores (e.g., delayed phenology of their host plants) and an increase in climate resistance of phytophagous-enemies (e.g., climate shelters, nectar). This hypothesis will be particularly important for phytophagous insects and phytophagous-enemies under climate stress.

ACKNOWLEDGEMENTS

We warmly thank the volume editors Nicolas Sauvion, Paul-André Calatayud and Denis Thiéry for giving us the opportunity to contribute. This work was partially supported by the grant ANR Blanc MicroCliMite (ANR-2010 BLAN-1706-02) to SP, and by a Swiss National Science Foundation grant 31003A_159869 to SR.

REFERENCES

Andrade, T. O., Krespi, L., Bonnardot, V., van Baaren, J., & Outreman, Y. (2016). Impact of climate conditions on the diversity and function of a guild of aphid parasitoids. *Oecologia, 180*, 877–888.

Angilletta, M. J. (2009). *Thermal adaptation—A theoretical and empirical synthesis*. Oxford, UK: Oxford University Press.

van Baaren, J., Héterier, V., Hance, T., Krespi, L., Cortesero, A.-M., Poinsot, D., … Outreman, Y. (2004). Playing the hare or the tortoise in parasitoids: Could different oviposition strategies have an influence in host partitioning in two *Aphidius* species? *Ethology, Ecology and Evolution, 16*, 231–242.

Bale, J. S., Masters, G. J., Hodkinson, I. D., Awmack, C., Bezemer, T. M., Brown, V. K., … Whittaker, J. B. (2002). Herbivory in global climate change research: Direct effects of rising temperature on insect herbivores. *Global Change Biology, 8*, 1–16.

Barbehenn, R. V., Karowe, D. N., & Chen, Z. (2004). Performance of a generalist grasshopper on a C3 and a C4 grass: Compensation for the effects of elevated CO_2 on plant nutritional quality. *Oecologia, 140*, 96–103.

Battisti, A., Stastny, M., Netherer, S., Robinet, C., Schopf, A., Roques, A., … Larsson, S. (2005). Expansion of geographic range in the pine processionary moth caused by increased winter temperatures. *Ecological Applications, 15*, 2084–2096.

Bensadia, F., Boudreault, S., Guaya, F., Michaud, D., & Cloutier, C. (2006). Aphid clonal resistance to a parasitoid fails under heat stress. *Journal of Insect Physiology, 52*, 146–157.

Berenbaum, M., Bernhardt, P., Buchmann, S., Calderone, N. W., Goldstein, P., Inouye, D. W., … Thompson, F. C. (2006). *Status of pollinators in North America*. Washington, DC, USA: National Academic Press.

Berlocher, S. H., & Feder, J. L. (2002). Sympatric speciation in phytophagous insects: Moving beyond controversy? *Annual Review of Entomology, 47*, 773–815.

Bernal, J., & Gonzalez, D. (1997). Reproduction of *Diaeretiella rapae* on Russian wheat aphid hosts at different temperatures. *Entomologia Experimentalis et Applicata, 82*, 159–166.

Bezemer, T. M., & Jones, T. H. (1998). Plant-insect herbivore interactions in elevated atmospheric CO_2: Quantitative analyses and guild effects. *Oikos, 82*, 212–222.

Bianchi, F. J. J. A., Booij, C. J. H., & Tscharntke, T. (2006). Sustainable pest regulation in agricultural landscapes: A review on landscape composition, biodiversity and natural pest control. *Proceedings of the Royal Society of London B, 273*, 1715–1727.

Bidart-Bouzat, M. G., & Imeh-Nathaniel, A. (2008). Global change effects on plant chemical defenses against insect herbivores. *Journal of Integrative Plant Biology, 50*, 1339–1354.

Bidart-Bouzat, M. G., Mithen, R., & Berenbaum, M. R. (2005). Elevated CO_2 influences herbivory-induced defense responses of *Arabidopsis thaliana*. *Oecologia, 145*, 415–424.

Biesmeijer, J. C., Roberts, S. P. M., Reemer, M., Ohlemüller, R., Edwards, M., Peeters, T., … Settele, J. (2006). Parallel declines in pollinators and insect-pollinated plants in Britain and The Netherlands. *Science, 313*, 351–354.

Björkman, C., & Niemelä, P. (2015). *Climate change and insect pests*. London, UK: CABI Climate Change Series, CAB International.

Blande, J. D., Holopainen, J. K., & Ninemets, U. (2014). Plant volatiles in a polluted atmosphere: Stress response and signal degradation. *Plant, Cell & Environment, 37*, 1892–1904.

Bolsinger, M., & Flückiger, W. (1987). Enhanced aphid infestation at motorways: The role of ambient air pollution. *Entomologia Experimentalis et Applicata, 45*, 237–243.

Bolsinger, M., Lier, M. E., & Hughes, P. R. (1992). Influence of ozone air pollution on plant-herbivore interactions. Part 2: Effects of ozone on feeding preference, growth and consumption rates of monarch butterflies (*Danaus plexippus*). *Environmental Pollution, 77*, 31–37.

Bolsinger, M., Lier, M. E., Lansky, D. M., & Hughes, P. R. (1991). Influence of ozone air pollution on plant- herbivore interactions. Part 1: Biochemical changes in ornamental milkweed (*Asclepia curassavica* L.; asclepiadaceae) induced by ozone. *Environmental Pollution, 72*, 69–83.

Bompard, A., Jaworski, C. C., Bearez, P., & Desneux, N. (2013). Sharing a predator: Can an invasive alien pest affect the predation on a local pest? *Population Ecology, 55*, 433–440.

Bossdorf, O., Auge, H., Lafuma, L., Rogers, W. E., Siemann, E., & Prati, D. (2005). Phenotypic and genetic differentiation between native and introduced plant populations. *Oecologia, 144*, 1–11.

Braun, S., & Flückiger, W. (1985). Increased population of the aphid *Aphis pomi* at a motorway. Part 3 – The effect of exhaust gases. *Environmental Pollution, 56*, 209–216.

Browse, J., & Howe, G. A. (2008). New weapons and a rapid response against insect attack. *Plant Physiology, 146*, 832–838.

Burrows, M. T., Schoeman, D. S., Buckley, L. B., Moore, P., Poloczanska, E. S., Brander, K. M., ... Holding, J. (2011). The pace of shifting climate in marine and terrestrial ecosystems. *Science, 334*, 652–655.
Buse, A., & Good, J. E. G. (1996). Synchronisation of larval emergence in winter moth (*Operophtera brumata* L.) and budburst in pedunculate oak (*Quercus robur* L.) under simulated climate change. *Ecological Entomology, 21*, 335–343.
Butler, C. D., & Trumble, J. T. (2008). Effects of pollutants on bottom-up and top-down processes in insect-plant interactions. *Environmental Pollution, 156*, 1–10.
Caillon, R., Suppo, C., Casas, J., Woods, H. A., & Pincebourde, S. (2014). Warming decreases thermal heterogeneity of leaf surfaces: Implications for behavioural thermoregulation by arthropods. *Functional Ecology, 28*, 1449–1458.
Calatayud, P.-A., Garrec, J.-P., & Nicole, M. (2013). Adaptation des plantes aux stress environnementaux. In N. Sauvion, P.-A. Calatayud, D. Thiéry, & F. Marion-Poll (Vol. Eds.), *Interactions insectes–plantes* (pp. 229–245). Paris: Co-edition IRD-Quae.
Carnus, J. M., Parrotta, J., Brockerhoff, E., Arbez, M., Jactel, H., Kremer, A., ... Walters, B. (2006). Planted forests and biodiversity. *Journal of Forestry, 104*, 65–77.
Carré, G., Roche, P., Chifflet, R., Morison, N., Bommarco, R., Harrison-Cripps, J., ... Settele, J. (2009). Landscape context and habitat type as drivers of bee biodiversity in European annual crops. *Agriculture, Ecosystems and Environment, 133*, 40–47.
Chen, Y. H., Gols, R., & Benrey, B. (2015). Crop domestication and its impact on naturally selected trophic interactions. *Annual Review of Entomology, 60*, 35–58.
Cheng, D. Q., Montllor, C. B., & Purcell, A. H. (2000). Fitness effects of two facultative endosymbiotic bacteria on the pea aphid, *Acyrthosiphon pisum*, and the blue alfalfa aphid, *A. kondoi*. *Entomologia Experimentalis et Applicata, 95*, 315–323.
Coley, P. (1998). Possible effects of climate change on plant/herbivore interactions in moist tropical forests. *Climatic Change, 39*, 455–472.
Coley, P. D., Bryant, J. P., & Chapin, F. S. (1985). Resource availability and plant antiherbivore defense. *Science, 230*, 895–899.
Cook, G. D., Dixon, J. R., & Leopold, A. C. (1964). Transpiration: Its effects on plant leaf temperature. *Science, 144*, 546–547.
Corcket, E., Giffardx, B., & Sforza, R. F. H. (2017). Food webs and multiple biotic interactions in plant–herbivore models. In N. Sauvion, D. Thiery, & P. A. Calatayud (Eds.), *Insect-Plant Interactions in a Crop Protection Perspective* (Vol. 81, pp. 111–137).
Cornelissen, T. (2011). Climate change and its effects on terrestrial insects and herbivory patterns. *Neotropical Entomology, 40*, 155–163.
Cornelissen, T., Stiling, P., & Drake, B. (2003). Elevated CO_2 decreases leaf fluctuating asymmetry and herbivory by leaf miners on two oak species. *Global Change Biology, 10*, 27–36.
Coviella, C. E., & Trumble, J. T. (1999). Effects of elevated atmospheric carbon dioxide on insect-plant interactions. *Conservation Biology, 13*, 700–712.
Currano, E. D., Labandeira, C. C., & Wilf, P. (2010). Fossil insect folivory tracks paleotemperature for six million years. *Ecological Monographs, 80*, 547–567.
Degnanj, P. H., & Moran, N. A. (2008). Evolutionary genetics of a defensive facultative symbiont of insects: Exchange of toxin-encoding bacteriophage. *Molecular Ecology, 17*, 916–929.
Dicke, M., & Baldwin, I. T. (2010). The evolutionary context for herbivore-induced plant volatiles: Beyond the 'cry for help'. *Trends in Plant Science, 15*, 167–175.
Doane, J. F., Lee, Y. W., Klingler, J., & Westcott, N. D. (1975). Orientation response of *Ctenicera destructor* and other wireworms (Coleoptera: Elateridae) to germinating grain and to carbon dioxide. *Canadian Entomologist, 107*, 1233–1252.
Docherty, M., Wade, F. A., Hurst, D. K., Whittaker, J. B., & Lea, P. J. (1997). Responses of tree sap-feeding herbivores to elevated CO_2. *Global Change Biology, 3*, 51–59.

Duchateau, M. J., Hoshiba, H., & Velthuis, H. H. W. (1994). Diploid males in the bumble bee *Bombus terrestris*, sex determination, sex alleles and viability. *Entomologia Experimentalis et Applicata, 71*, 263–269.

Dury, S. J., Good, J. E. G., Perrins, C. M., Buse, A., & Kaye, T. (1998). The effects of increasing CO_2 and temperature on oak leaf palatability and the implications for herbivorous insects. *Global Change Biology, 4*, 55–61.

Dussourd, D. E., & Denno, R. F. (1991). Deactivation of plant defense: Correspondence between insect behavior and secretory canal architecture. *Ecology, 72*, 1383–1396.

Ehleringer, J. R., Cerling, T. E., & Dearing, M. D. (2002). Atmospheric CO_2 as a global change driver influencing plant-animal interactions. *Integrative and Comparative Biology, 42*, 424–430.

Eilenberg, J., Hajek, A., & Lomer, C. (2001). Suggestions for unifying the terminology in biological control. *BioControl, 46*, 387–400.

Estoup, A., Ravigné, V., Hufbauer, R., Vitalis, R., Gautier, M., & Facon, B. (2016). Is there a genetic paradox of biological invasion? *Annual Review of Ecology, Evolution, and Systematics, 47*. http://dx.doi.org/10.1146/annurev-ecolsys-121415-032116.

Fabre, F., Plantegenest, M., Mieuzet, L., Dedryver, C. A., Leterrier, J. L., & Jacquot, E. (2005). Effects of climate and land use on the occurrence of viruliferous aphids and the epidemiology of barley yellow dwarf disease. *Agriculture, Ecosystems and Environment, 106*, 49–55.

Facey, S. L., Ellsworth, D. S., Staley, J. T., Wright, D. J., & Johnson, S. N. (2014). Upsetting the order: How climate and atmospheric change affects herbivore–enemy interactions. *Current Opinion in Insect Science, 5*, 66–74.

Fahrig, L. (2003). Effects of habitat fragmentation on biodiversity. *Annual Review of Ecology, Evolution, and Systematics, 34*, 487–515.

Fajer, E. D., Bowers, M. D., & Bazzaz, F. A. (1991). The effects of enriched CO_2 atmospheres on the buckeye butterfly, *Junonia coenia*. *Ecology, 72*, 751–754.

Farmer, E. E., Alméras, E., & Krishnamurthy, V. (2003). Jasmonates and related oxylipins in plant responses to pathogenesis and herbivory. *Current Opinion in Plant Biology, 6*, 372–378.

Feyereisen, R. (1999). Insect P450 enzymes. *Annual Review of Entomology, 44*, 507–533.

Fine, P. V. A., Mesones, I., & Coley, P. D. (2004). Herbivores promote habitat specialization by trees in amazonian forests. *Science, 305*, 663–665.

Folgareit, P. J., Bruzzone, O. A., & Gilbert, L. E. (2003). Seasonal patterns of activity among species of black fire ant parasitoid flies (*Pseudacteon*: Phoridae) in Argentina explained by analysis of climatic variables. *Biological Control, 28*, 368–378.

Frankie, G. W., & Ehler, L. E. (1978). Ecology of insects in urban environments. *Annual Review of Entomology, 23*, 367–387.

Franzén, M., & Ockinger, E. (2012). Climate-driven changes in pollinator assemblages during the last 60 years in an Arctic mountain region in Northern Scandinavia. *Journal of Insect Conservation, 16*, 227–238.

Führer, E. (1985). Air pollution and the incidence of forest insect problems. *Zeitschrift für Angewandte Entomologie, 99*, 371–377.

Futuyma, D. J., & Agrawal, A. A. (2009). Macroevolution and the biological diversity of plants and herbivores. *Proceedings of the National Academy of Sciences of the United States of America, 106*, 18054–18061.

Gates, D. M. (1980). *Biophysical ecology*. New York, USA: Springer-Verlag.

Ghasemzadeh, A., Jaafar, H. Z., & Rahmat, A. (2010). Elevated carbon dioxide increases contents of flavonoids and phenolic compounds, and antioxidant activities in Malaysian young ginger (*Zingiber officinale* Roscoe) varieties. *Molecules, 15*, 7907–7922.

Giron, D., Dedeine, F., Dubreuil, G., Huguet, E., Mouton, L., Outreman, Y., Vavre, F., & Simon, J.-C. (2016). Influence of microbial symbionts on plant–insect interactions. In

N. Sauvion, D. Thiery, & P. A. Calatayud (Eds.), *Insect-Plant Interactions in a Crop Protection Perspective* (Vol. 81, pp. 225–257).

Godfray, H. C. J. (1994). *Parasitoids: Behavioural and evolutionary ecology*. Princeton, New Jersey, USA: Princeton University Press.

Godfray, H. C. J., Hassell, M. P., & Holt, R. D. (1994). The population dynamic consequences of phenological asynchrony between parasitoids and their hosts. *Journal of Animal Ecology, 63*, 1–10.

Graham, R. W., & Grimm, E. C. (1990). Effects of global climate change on the patterns of terrestrial biological communities. *Trends in Ecology & Evolution, 5*, 289–292.

Grant, B. S., Owen, D. F., & Clarke, C. A. (1996). Parallel rise and fall of melanic peppered moths in America and Britain. *Journal of Heredity, 87*, 351–357.

Guenther, A. B., Zimmerman, P. R., Harley, P. C., Monson, R. K., & Fall, R. (1993). Isoprene and monoterpene emission rate variability: Model evaluations and sensitivity analyses. *Journal of Geophysical Research, 98*, 12609–12617.

Hain, F. P. (1987). Interactions of insects, trees and air pollutants. *Tree Physiology, 3*, 93–102.

Hance, T., van Baaren, J., Vernon, P., & Boivin, G. (2007). Impact of extreme temperatures on parasitoids in a climate change perspective. *Annual Review of Entomology, 52*, 107–126.

Harrington, R., Fleming, R. A., & Woiwod, I. P. (2001). Climate change impacts on insect management and conservation in temperate regions: Can they be predicted? *Agricultural and Forest Meteorology, 3*, 233–240.

Harrington, R., Woiwod, I., & Sparks, T. (1999). Climate change and trophic interactions. *Trends in Ecology & Evolution, 14*, 146–150.

Hättenschwiler, S., & Schhafellner, C. (1999). Opposing effects of elevated CO_2 and N deposition on *Lymantria monacha* larvae feeding on spruce trees. *Oecologia, 118*, 210–217.

Heckel, D.G. (2014). Insect detoxification and sequestration strategies. In C. Voelckel, & G. Jander (Vol. Eds.), *Annual plant reviews: Vol. 47. Insect-plant interactions* (pp. 77–114). Chichester, UK: John Wiley & Sons, Ltd. http://dx.doi.org/10.1002/9781118829783.ch3.

Heinrich, B. (1979). *Bumblebee economics*. Cambridge, London, England: Harvard University Press.

Heinrich, B., & Buchmann, S. L. (1986). Thermoregulatory physiology of the carpenter bee, *Xylocopa varipuncta*. *Journal of Comparative Physiology B, 156*, 557–562.

Helliker, B. R., & Richter, S. L. (2008). Subtropical to boreal convergence of tree-leaf temperatures. *Nature, 454*, 511–515.

Herms, D. A., & Mattson, W. J. (1992). The dilemma of plants — to grow or defend. *Quarterly Review of Biology, 67*, 283–335.

Herrera, C. M., & Pellmyr, O. (2002). *Plant — Animal interactions. An evolutionary approach*. Oxford, UK: Blackwell Science.

Hewitt, G. (1999). Post-glacial re-colonization of European biota. *Biological Journal of the Linnean Society, 68*, 87–112.

Hill, J. K., Thomas, C. D., & Huntley, B. (1999). Climate and habitat availability determine 20th century changes in a butterfly's range margin. *Proceedings of the Royal Society of London B, 266*, 1197–1206.

van't Hof, A. E., Campagne, P., Rigden, D. J., Yung, C. J., Lingley, J., Quail, M. A., … Saccheri, I. J. (2016). The industrial melanism mutation in British peppered moths is a transposable element. *Nature, 534*, 102–105.

Holt, R. D., Lawton, J. H., Polis, G. A., & Martinez, N. D. (1999). Trophic rank and the species-area relationship. *Ecology, 80*, 1495–1504.

Hoover, J. K., & Newman, J. A. (2004). Tritrophic interactions in the context of climate change: A model of grasses, cereal aphids and their parasitoids. *Global Change Biology, 10*, 1197–1208.

Howlett, R. J., & Majerus, M. E. N. (1987). The understanding of industrial melanism in the peppered moth (*Biston betularia*) (Lepidoptera: Geometridae). *Biological Journal of the Linnean Society, 30*, 31–44. http://dx.doi.org/10.1111/j.1095-8312.1987.tb00286.x.

Huey, R. B., & Stevenson, R. D. (1979). Integrating physiology and ecology of ectotherms: A discussion of approaches. *American Zoologist, 19*, 357–366.

Hulme, P. E., Pysek, P., Nentwig, W., & Vila, M. (2009). Will threat of biological invasions unite the European Union? *Science, 324*, 40–41.

Hunter, M. D. (2001). Effects of elevated atmospheric carbon dioxide on insect-plant interactions. *Agricultural and Forest Entomology, 3*, 153–159.

IPCC (2007). Contribution of Working Group I to the Fourth Assessment Report of the Intergovernmental Panel on Climate Change. In S. Solomon, D. Qin, M. Manning M. Marquis, K. Averyt, M.M.B. Tignor, ..., Z. Chen (Eds.), *Climate Change 2007: The Physical Science Basis* Cambridge, UK.

IPCC. (2014). *Climate change 2014: Impacts, adaptation, and vulnerability*. Cambridge, UK: Cambridge University Press.

Jaenike, J. (1990). Host specialization in phytophagous insects. *Annual Review of Ecology, Evolution, and Systematics, 21*, 243–273.

Jamieson, M. A., Trowbridge, A. M., Raffa, K. F., & Lindroth, R. L. (2012). Consequences of climate warming and altered precipitation patterns for plant-insect and multitrophic interactions. *Plant Physiology, 160*, 1719–1727.

Jepsen, J. U., Hagen, S. B., Ims, R. A., & Yoccoz, N. G. (2008). Climate change and outbreaks of the geometrids *Operophtera brumata* and *Epirrita autumnata* in subarctic birch forest: Evidence of a recent outbreak range expansion. *Journal of Animal Ecology, 77*, 257–264.

Johns, C. V., Beaumont, L. J., & Hughes, L. (2003). Effects of elevated CO_2 and temperature on development and consumption rates of *Octotoma championi* and *O. scabripennis* feeding on *Lantana camara*. *Entomologia Experimentalis et Applicata, 108*, 169–178.

Johns, C. V., & Hughes, L. (2002). Interactive effects of elevated CO_2 and temperature on the leaf-miner *Dialectica scalariella* Zeller (Lepidoptera: Gracillariidae) in Paterson's Curse, *Echium plantagineum* (Boraginaceae). *Global Change Biology, 8*, 142–152.

Julkunen-Tiitto, R., Nybakken, L., Randriamanana, T., & Virjamo, V. (2015). Boreal woody species resistance affected by climate change. In C. Björkman, & P. Niemelä (Vol. Eds.), *Climate change and insect pests* (pp. 54–73). Wallingford: CABI.

Kainulainen, P., Holopainen, J. K., & Holopainen, T. (2000). Combined effect of ozone and nitrogen on secondary compounds, amino acids, and aphid performance in Scots pine. *Journal of Environmental Quality, 29*, 334–342.

Kaiser, L., Ode, P., van Nouhuys, S., Calatayud, P.-A., Colazza, S., Cortesero, A.-M., Thiel, A., & van Baaren, J. (2016). The Plant as a habitat for entomophagous insects. In N. Sauvion, D. Thiery, & P. A. Calatayud (Eds.), *Insect-Plant Interactions in a Crop Protection Perspective* (Vol. 81, pp. 179–223).

Kaspari, M., Clay, N. A., Lucas, J., Yanoviak, S. P., & Kay, A. (2015). Thermal adaptation generates a diversity of thermal limits in a rainforest ant community. *Global Change Biology, 21*, 1092–1102.

Kenis, M., Rabitsch, W., Auger-Rozenberg, M.-A., & Roques, A. (2007). How can alien species inventories and interception data help us prevent in sect invasions? *Bulletin of Entomological Research, 97*, 489–502.

Kergoat, G. J., Meseguer, A. S., & Jousselin, E. (2016). Evolution of plant–insect interactions: insights from macroevolutionary approaches in plants and herbivorous insects. In N. Sauvion, D. Thiery, & P. A. Calatayud (Eds.), *Insect-Plant Interactions in a Crop Protection Perspective* (Vol. 81, pp. 25–54).

Kerr, E. D., Phelan, C., & Woods, H. A. (2013). Subtle direct effects of rising atmospheric CO_2 on insect eggs. *Physiological Entomology, 38*, 302–305.

Kerr, J. T., Pindar, A., Galpern, P., Packer, L., Potts, S. G., Roberts, S. M., ... Wagner, D. L. (2015). Climate change impacts on bumblebees converge across continents. *Science, 349*, 177–180.

Kessler, A., & Heil, M. (2011). The multiple faces of indirect defences and their agents of natural selection. *Functional Ecology, 25*, 348–357.

Klemm, M. (1996). Man-made bee habitats in the anthropogenous landscape of central Europe – Substitutes for threatened or destroyed riverine habitats?. In *Linnean society symposium series* (Vol. 18)Academic Press Limited.

Körner, C. (2007). The use of altitude in ecological research. *Trends in Ecology & Evolution, 22*, 569–574.

Krespi, L., Dedryver, C.-A., & Creach, V. (1997). Variability in the development of cereal aphid parasitoids and hyperparasitoids in oceanic regions as a response to climate and abundance of hosts. *Environmental Entomology, 26*, 545–551.

Kühn, L., BrandI, R., & Klotz, S. (2004). The flora of German cities is naturally species rich. *Evolutionary Ecology Research, 6*, 749–764.

Le Lann, C., Roux, O., Serain, N., van Alphen, J. J. M., Vernon, P., & van Baaren, J. (2011). Thermal tolerance of sympatric hymenopteran parasitoid species: Does it match seasonal activity? *Physiological Entomology, 36*, 21–28.

Lecocq, T., Dellicour, S., Michez, D., Lhomme, P., Vanderplanck, M., Valterova, I., ... Rasmont, P. (2013). Scent of a break-up: Phylogeography and reproductive trait divergences in the red-tailed bumblebee (*Bombus lapidarius*). *BMC Evolutionary Biology, 13*, 263.

Lee, C. E. (2002). Evolutionary genetics of invasive species. *Trends in Ecology & Evolution, 17*, 386–391.

Li, B. P., & Mills, N. (2004). The influence of temperature on size as an indicator of host quality for the development of a solitary koinobiont parasitoid. *Entomologia Experimentalis et Applicata, 110*, 249–256.

Lieutier, F., Bermudez-Torres, K., Cook, J., Harris, M. O., Legal, L., Sallé, A., Schatz, B., & Giron, D. (2016). From plant exploitation to mutualism. In N. Sauvion, D. Thiery, & P. A. Calatayud (Eds.), *Insect-Plant Interactions in a Crop Protection Perspective* (Vol. 81, pp. 55–109).

Linacre, E. T. (1967). Further notes on a feature of leaf and air temperatures. *Theoretical and Applied Climatology, 15*, 422–436.

Lincoln, D. E., Fajer, E. D., & Johnson, R. H. (1993). Plant-insect herbivore interactions in elevated CO_2 environments. *Trends in Ecology & Evolution, 8*, 64–68.

Lindroth, R. L. (2010). Impacts of elevated atmospheric CO_2 and O_3 on forests: Phytochemistry, trophic interactions, and ecosystem dynamics. *Journal of Chemical Ecology, 36*, 2–21.

Loreto, F., & Schnitzler, J.-P. (2010). Abiotic stresses and induced BVOCs. *Trends in Plant Science, 15*, 154–166.

Malausa, T., Dalecky, A., Ponsard, S., Audiot, P., Streiff, R., Chaval, Y., & Bourguet, D. (2007). Genetic structure and gene flow in French populations of two *Ostrinia* taxa: Host races or sibling species? *Molecular Ecology, 16*, 4210–4222.

Malausa, T., Leniaud, L., Martin, J.-F., Audiot, P., Bourguet, D., Ponsard, S., ... Dopman, E. (2007). Molecular differentiation at nuclear loci in French host races of the European corn borer *(Ostrinia nubilalis)*. *Genetics, 176*, 2343–2355.

Manel, S., Schwartz, M. K., Luikart, G., & Taberlet, P. (2003). Landscape genetics: Combining landscape ecology and population genetics. *Trends in Ecology & Evolution, 18*, 189–197.

Marchal, P., & Feytaud, J. (1911). On the parasite of the eggs of the grape berry moth and the vine moth. *Comptes Rendus Hebdomadaires des Séances de l'Académie des Sciences, 153*, 633–636.

Martin-Vertedor, D., Ferrero-Garcia, J. J., & Torres-Vila, L. M. (2010). Global warming affects phenology and voltinism of *Lobesia botrana* in Spain. *Agricultural and Forest Entomology, 12*, 169–176.

Martinet, B., Lecocq, T., Smet, J., & Rasmont, P. (2015). A protocol to assess insect resistance to heat waves, applied to bumblebees (*Bombus Latreille*, 1802). *PLoS One, 10*, e0118591.

Martinet, B., Rasmont, P., Cederberg, B., Evrard, D., Ødegaard, F., Paukkunen, J., & Lecocq, T. (2016). Forward to the North: Two Euro-Mediterranean bumblebee species now cross the Arctic Circle. *Annales de la Société entomologique de France (N.S.), 51*, 303–309.

Mason, P. G., & Hopper, K. R. (1997). Temperature dependence in locomotion of the parasitoid *Aphelinus asychis* (Hymenoptera: Aphelinidae) from geographical regions with different climates. *Environmental Entomology, 26*, 1416–1423.

May, M. L. (1979). Insect thermoregulation. *Annual Review of Entomology, 24*, 313–349.

McCann, K. S. (2000). The diversity–stability debate. *Nature, 405*, 228–233.

McKinney, M. L. (2002). Urbanization, biodiversity, and conservation. *BioScience, 52*, 883–890.

Meehl, G. A., & Tebaldi, C. (2004). More intense, more frequent and longer lasting heat waves in the 21st century. *Science, 305*, 994–997.

Michaletz, S. T., Weiser, M. D., Zhou, J., Kaspari, M., Helliker, B. R., & Enquist, B. J. (2015). Plant thermoregulation: Energetics, trait–environment interactions, and carbon economics. *Trends in Ecology & Evolution, 30*, 714–724.

Michener, C. D. (1979). Biogeography of the bees. *Annals of the Missouri Botanical Garden, 66*, 277–342.

Michener, C. D. (2007). *The bees of the world* (2nd ed.). Baltimore, USA: The Johns Hopkins University.

Mitter, C., Farrell, B., & Futuyma, D. J. (1991). Phylogenetic studies of insect-plant interactions: Insights into the genesis of diversity. *Trends in Ecology & Evolution, 6*, 290–293.

Moran, N. A., Degnan, P. H., Santos, S. R., Dunbar, H. E., & Ochman, H. (2005). The players in a mutualistic symbiosis: Insects, bacteria, viruses, and virulence genes. *Proceedings of the National Academy of Sciences of the United States of America, 102*, 16919–16926.

Morison, J. I. L., & Lawlor, D. W. (1999). Interactions between increasing CO_2 concentration and temperature on plant growth. *Plant, Cell & Environment, 22*, 659–682.

Nabity, P. D., Hillstrom, M. L., Lindroth, R. L., & DeLucia, E. H. (2012). Elevated CO_2 interacts with herbivory to alter chlorophyll fluorescence and leaf temperature in *Betula papyrifera* and *Populus tremuloides*. *Oecologia, 169*, 905–913.

Naeem, S. (1998). Species redundancy and ecosystem reliability. *Conservation Biology, 12*, 39–45.

Nicolas, G., & Sillans, D. (1989). Immediate and latent effects of carbon dioxide on insects. *Annual Review of Entomology, 34*, 97–116.

Nicole, M. C. (2002). Les relations des insectes phytophages avec leurs plantes hôtes. *Antennae, 9*, 6.

Nieto, A., Roberts, S. P. M., Kemp, J., Rasmont, P., Kuhlmann, M., García Criado, M., … De Meulemeester, T. (2014). *European red list of bees*. Brussels: International Union for Conservation of Nature, 03/2015; Publication Office of the European Union.

Ode, P. J., Johnson, S. N., & Moore, B. D. (2014). Atmospheric change and induced plant secondary metabolites — Are we reshaping the building blocks of multi-trophic interactions? *Current Opinion in Insect Science, 5*, 57–65.

Oke, T. R. (1982). The energetic basis of the urban heat island. *Quarterly Journal of the Royal Meteorological Society, 108*, 1–24.

Orr, M. R., & Smith, T. B. (1998). Ecology and speciation. *Trends in Ecology & Evolution, 16*, 502–506.

Owen, E. L., Bale, J. S., & Hayward, S. A. L. (2013). Can winter-active bumblebees survive the cold? Assessing the cold tolerance of *Bombus terrestris* audax and the effects of pollen feeding. *PLoS One, 8*(11), e80061.

Oyen, J. O., Susma, G., & Dillon, M. E. (2016). Altitudinal variation in bumblebee (*Bombus*) critical thermal limits. *Journal of Thermal Biology, 59*, 52–57.

Parmesan, C. (1996). Climate and species' range. *Nature, 382*, 765–766.

Parmesan, C. (2006). Ecological and evolutionary responses to recent climate change. *Annual Review of Ecology, Evolution, and Systematics, 37*, 637–669.

Parmesan, C., & Yohe, G. (2003). A globally coherent fingerprint of climate change impacts across natural systems. *Nature, 421*, 37–42.

Pateman, R. M., Hill, J. K., Roy, D. B., Fox, R., & Thomas, C. D. (2012). Temperature-dependent alterations in host use drive rapid range expansion in a butterfly. *Science, 336*, 1028–1030.

Patiny, S., & Michez, D. (2007). Biogeography of bees (Hymenoptera, Apoidea) in Sahara and the Arabian deserts. *Insect Systematics & Evolution, 38*, 19–34.

Pauli, H., Gottfried, M., Dullinger, S., Abdaladze, O., Akhalkatsi, M., Alonso, J. L. B., … Ghosn, D. (2012). Recent plant diversity changes on Europe's mountain summits. *Science, 336*, 353–355.

Pellissier, L., Fiedler, K., Ndribe, C., Dubuis, A., Pradervand, J.-N., Guisan, A., & Rasmann, S. (2012). Shifts in species richness, herbivore specialization, and plant resistance along elevation gradients. *Ecology and Evolution, 2*, 1818–1825.

Pellissier, L., Moreira, X., Danner, H., Serrano, M., Salamin, N., van Dam, N. M., & Rasmann, S. (2016). The simultaneous inducibility of phytochemicals related to plant direct and indirect defences against herbivores is stronger at low elevation. *Journal of Ecology, 104*, 1116–1125.

Pellissier, L., Roger, A., Bilat, J., & Rasmann, S. (2014). High elevation *Plantago lanceolata* plants are less resistant to herbivory than their low elevation conspecifics: Is it just temperature? *Ecography, 37*, 950–959.

Peñuelas, J., & Llusià, J. (2003). BVOCs: Plant defense against climate warming? *Trends in Plant Science, 8*, 105–109.

Peñuelas, J., & Staudt, M. (2010). BVOCs and global change. *Trends in Plant Science, 15*, 133–144.

Percy, K. E., Awmack, C. S., Lindroth, R. L., Kubiske, M. E., Kopper, B. J., Isebrands, J. G., … Oksanen, E. (2002). Altered performance of forest pests under atmospheres enriched by CO_2 and O_3. *Nature, 420*, 403–407.

Pimentel, D., Stachow, U., Takacs, D. A., Brubaker, H. W., Dumas, A. R., Meaney, J. J., … Corzilius, D. B. (1992). Conserving biological diversity in agricultural/forestry systems. *BioScience, 42*, 354–362.

Pincebourde, S., & Casas, J. (2006a). Leaf miner-induced changes in leaf transmittance cause variations in insect respiration rates. *Journal of Insect Physiology, 52*, 194–201.

Pincebourde, S., & Casas, J. (2006b). Multitrophic biophysical budgets: Thermal ecology of an intimate herbivore insect–plant interaction. *Ecological Monographs, 76*, 175–194.

Pincebourde, S., & Casas, J. (2015). Warming tolerance across insect ontogeny: Influence of joint shifts in microclimates and thermal limits. *Ecology, 96*, 986–997.

Pincebourde, S., & Casas, J. (2016). Hypoxia and hypercarbia in endophagous insects: Larval position in the plant gas exchange network is key. *Journal of Insect Physiology, 84*, 137–153.

Pincebourde, S., Frak, E., Sinoquet, H., Régnard, J. L., & Casas, J. (2006). Herbivory mitigation through increased water use efficiency in a leaf mining moth–apple tree relationship. *Plant, Cell & Environment, 29*, 2238–2247.

Pincebourde, S., Murdock, C. C., Vickers, M., & Sears, M. W. (2016). Fine-scale microclimatic variation can shape the responses of organisms to global change in both natural and urban environments. *Integrative and Comparative Biology*, icw016.
Pincebourde, S., Sinoquet, H., Combes, D., & Casas, J. (2007). Regional climate modulates the canopy mosaic of favourable and risky microclimates for insects. *Journal of Animal Ecology, 76*, 424—438.
Pincebourde, S., & Suppo, C. (2016). The vulnerability of tropical ectotherms to warming is modulated by the microclimatic heterogeneity. *Integrative and Comparative Biology, 56*, 85—97.
Pincebourde, S., & Woods, H. A. (2012). Climate uncertainty on leaf surfaces: The biophysics of leaf microclimates and their consequences for leaf-dwelling organisms. *Functional Ecology, 26*, 844—853.
Polce, C., Garratt, M. P., Termansen, M., Ramirez-Villegas, J., Challinor, A. J., Lappage, M. G., … Somerwill, K. E. (2014). Climate-driven spatial mismatches between British orchards and their pollinators: Increased risks of pollination deficits. *Global Change Biology, 20*, 2815—2828.
Potter, K. A., Davidowitz, G., & Woods, H. A. (2009). Insect eggs protected from high temperatures by limited homeothermy of plant leaves. *Journal of Experimental Biology, 212*, 3448—3454.
Proctor, M., Yeo, P., & Lack, A. (1996). *The natural history of pollination*. Portland, Oregon, USA: Timber Press Inc.
Rasmann, S., Alvarez, N., & Pellissier, L. (2014a). The altitudinal niche-breadth hypothesis in insect-plant interactions. *Annual Plant Reviews, 47*, 338—359.
Rasmann, S., & Pellissier, L. (2015). Adaptive responses of plants to insect herbivores under climate change. In C. Björkman, & P. Niemelä (Vol. Eds.), *Climate change and insect pests* (pp. 38—53). Wallingford, UK: CABI.
Rasmann, S., Pellissier, L., Defossez, E., Jactel, H., & Kunstler, G. (2014b). Climate-driven change in plant—insect interactions along elevation gradients. *Functional Ecology, 28*, 46—54.
Rasmont, P., Franzen, M., Lecocq, T., Harpke, A., Roberts, S. P. M., Biesmeijer, K., … Schweiger, O. (2015). Climatic risk and distribution atlas of European bumblebees. *Biorisk, 10*, 1—246. http://dx.doi.org/10.3897/biorisk.10.4749.
Rasmont, P., & Iserbyt, S. (2012). The bumblebees scarcity syndrome: Are heat waves leading to local extinctions of bumblebees (Hymenoptera: Apidae: *Bombus*)? *Annales de la Société Entomologique de France (N.S.), 48*, 275—280.
Reddy, G. V. P., Holopainen, J. K., & Guerrero, A. (2002). Olfactory responses of *Plutella xylostella* natural enemies to host pheromone, larval frass, and green leaf cabbage volatiles. *Journal of Chemical Ecology, 28*, 131—143.
Ricketts, T. H. (2001). The matrix matters: Effective isolation in fragmented landscapes. *American Naturalist, 158*, 87—99.
Robinson, E. A., Ryan, G. D., & Newman, J. A. (2012). A meta-analytical review of the effects of elevated CO_2 on plant—arthropod interactions highlights the importance of interacting environmental and biological variables. *New Phytologist, 194*, 321—336.
Robinson, P. J. (2001). On the definition of a heat wave. *Journal of Applied Meteorology, 40*, 762—775.
Rohne, O. (2002). Effect of temperature and host stage on performance of *Aphelinus varipes* Forster (Hym., Aphelinidae) parasitizing the cotton aphid, *Aphis gossypii* Glover (Hom., Aphididae). *Journal of Applied Entomology, 126*, 572—576.
Romero, G. Q., & Koricheva, J. (2011). Contrasting cascade effects of carnivores on plant fitness: A meta-analysis. *Journal of Animal Ecology, 80*, 696—704.

Rosenthal, G. A., & Berenbaum, M. R. (1991). *Herbivores: Their interactions with secondary plant metabolites*. San Diego, USA: Academic Press, Inc.
Rusch, A., Bommarco, R., & Ekbom, B. (2017). Conservation biological control in agricultural landscapes. In N. Sauvion, D. Thiery, & P. A. Calatayud (Eds.), *Insect-Plant Interactions in a Crop Protection Perspective* (Vol. 81, pp. 333–360).
Rundle, H. D., & Nosil, P. (2005). Ecological speciation. *Ecology Letters, 8*, 336–352.
Salt, D. T., Brooks, G. L., & Whittaker, J. B. (1995). Elevated carbon dioxide affects leafminer performance and plant growth in docks (*Rumex* spp.). *Global Change Biology, 1*, 153–156.
de Sassi, C., Staniczenko, P. P. A., & Tylianakis, J. M. (2012). Warming and nitrogen affect size structuring and density dependence in a host-parasitoid food web. *Philosophical Transactions of the Royal Society B, 367*, 3033–3041.
Saudreau, M., Pincebourde, S., Dassot, M., Adam, B., Loxdale, H. D., & Biron, D. G. (2013). On the canopy structure manipulation to buffer climate change effects on insect herbivore development. *Trees, Structure and Function, 27*, 239–248.
Schatz, B., Sauvion, N., Kjellberg, F., & Nel, A. (2017). Plant–insect interactions: a palaeontological and an evolutionary perspective. In N. Sauvion, D. Thiery, & P. A. Calatayud (Eds.), *Insect-Plant Interactions in a Crop Protection Perspective* (Vol. 81, pp. 1–53).
Schmitz, O. J., Hamback, P. A., & Beckerman, A. P. (2000). Trophic cascades in terrestrial systems: A review of the effects of carnivore removals on plants. *American Naturalist, 155*, 141–153.
Schoonhoven, L. M., van Loon, J. J. A., & Dicke, M. (2005). *Insect-plant biology*. Oxford, UK: Oxford University Press.
Schvester, D. (1985). Regard sur la santé de nos forêts. *Revue Forestière Française (Sp. Regard sur la santé de nos forêts)*, 45–64.
Settele, J., Hammen, V., Hulme, P., Karlson, U., Klotz, S., Kotarac, M., … Peterson, K. (2005). ALARM: Assessing large-scale environmental risks for biodiversity with tested methods. *Gaia-Ecological Perspectives for Science and Society, 14*, 69–72.
Shigenobu, S., Watanabe, H., Hattori, M., Sakaki, Y., & Ishikawa, H. (2000). Genome sequence of the endocellular bacterial symbiont of aphids *Buchnera* sp. APS. *Nature, 407*, 81–86.
Slotsbo, S., Heckmann, L. H., Damgaard, C., Roelofs, D., de Boer, T., & Holmstrup, M. (2009). Exposure to mercury reduces heat tolerance and heat hardening ability of the springtail *Folsomia candida*. *Comparative Biochemistry and Physiology Part C: Toxicology and Pharmacology, 150*, 118–123.
Smith, P. H. D., & Jones, T. H. (1998). Effects of elevated CO_2 on the chrysanthemum leafminer, *Chromatomyia syngenesiae*: A greenhouse study. *Global Change Biology, 4*, 287–291.
Stange, E. E., & Ayres, M. P. (2001). *Climate change impacts: insects*. eLS, John Wiley & Sons, Ltd.
Steudel, B., Hector, A., Friedl, T., Löfke, C., Lorenz, M., Wesche, M., & Kessler, M. (2012). Biodiversity effects on ecosystem functioning change along environmental stress gradients. *Ecology Letters, 15*, 1397–1405.
Stiling, P., & Cornelissen, T. (2007). How does elevated carbon dioxide (CO_2) affect plant–herbivore interactions? A field experiment and meta-analysis of CO_2-mediated changes on plant chemistry and herbivore performance. *Global Change Biology, 13*, 1823–1842.
Stireman, J. O., Dyer, L. A., Janzen, D. H., Singer, M. S., Lill, J. T., Marquis, R. J., … Barone, J. A. (2005). Climatic unpredictability and parasitism of caterpillars: Implications of global warming. *Proceedings of the National Academy of Sciences of the United States of America, 102*, 17384–17387.
Strong, D. R., Lawton, J. H., & Southwood, R. (1984). *Insects on plants: Community patterns and mechanisms*. London, UK: Blackwell Scientific.

Sun, Y., Yin, J., Cao, H., Li, C., Kang, L., & Ge, F. (2011). Elevated CO_2 influences nematode-induced defense responses of tomato genotypes differing in the JA pathway. *PLoS One, 6*, e19751.

Sunday, J. M., Bates, A. E., Kearney, M. R., Colwell, R. K., Dulvy, N. K., Longino, J. T., & Huey, R. B. (2014). Thermal safety margins and the necessity of thermoregulatory behavior across latitude and elevation. *Proceedings of the National Academy of Sciences of the United States of America, 111*, 5610–5615.

Tauber, M. J., Tauber, C. A., & Masaki, S. (1986). *Seasonal adaptations of insects*. New York, USA: Oxford University Press.

Terzo, M., & Rasmont, P. (2014). *Atlas of the European bees: genus* Xylocopa (STEP Project, Atlas Hymenoptera, Mons, Gembloux).

Thaler, J. S., Agrawal, A. A., & Halitschke, R. (2010). Salicylate-mediated interactions between pathogens and herbivores. *Ecology, 91*, 1075–1082.

Thaler, J. S., Humphrey, P. T., & Whiteman, N. K. (2012). Evolution of jasmonate and salicylate signal crosstalk. *Trends in Plant Science, 17*, 260–270.

Thies, C., Haenke, S., Scherber, C., Bengtsson, J., Bommarco, R., Clement, L. W., … Hawro, V. (2011). The relationship between agricultural intensification and biological control: Experimental tests across Europe. *Ecological Applications, 21*, 2187–2196.

Thomas, M. B., & Blanford, S. (2003). Thermal biology in insect-parasite interactions. *Trends in Ecology & Evolution, 18*, 344–350.

Tougeron, K., van Baaren, J., Burel, F., & Alford, L. (2016). Comparing thermal tolerance across contrasting landscapes: First steps towards understanding how landscape management could modify ectotherm thermal tolerance. *Insect Conservation and Diversity, 9*, 171–180.

Trotta, V., Calboli, F. C., Ziosi, M., Guerra, D., Pezzoli, M. C., David, J. R., & Cavicchi, S. (2006). Thermal plasticity in *Drosophila Melanogaster*: A comparison of geographic populations. *BMC Evolutionary Biology, 6*, 67.

Tscharntke, T., & Brandl, R. (2004). Plant-insect interactions in fragmented landscapes. *Annual Review of Entomology, 49*, 405–430.

Tsuchida, T., Koga, R., & Fukatsu, T. (2004). Host-plant specialization governed by facultative symbiont. *Science, 303*, 1989.

Turner, M. G. (2005). Landscape ecology: What is the state of the science? *Annual Review of Ecology, Evolution, and Systematics, 36*, 319–344.

Tylianakis, J. M., Didham, R. K., Bascompte, J., & Wardle, D. A. (2008). Global change and species interactions in terrestrial ecosystems. *Ecology Letters, 11*, 1351–1363.

Via, S. (2001). Sympatric speciation in animals: The ugly duckling grows up. *Trends in Ecology & Evolution, 16*, 381–390.

Virilio, P. (2005). *L'accident originel*. Paris, France: Galilée.

Wahid, A., Gelani, S., Ashraf, M., & Foolad, M. R. (2007). Heat tolerance in plants: An overview. *Environmental and Experimental Botany, 61*, 199–223.

Warrington, S. (1989). Ozone enhances the growth rate of cereal aphids. *Agriculture, Ecosystems and Environment, 26*, 65–68.

Welter, S. C. (1989). Arthropod impact on plant gas exchange. *Insect-Plant Interactions, 1*, 135–150.

Whitehorn, P. R., Tinsley, M. C., Brown, M. J. F., Darvill, B., & Goulson, D. (2009). Impacts of inbreeding on bumblebee colony fitness under field conditions. *BMC Evolutionary Biology, 9*, 152.

Whittaker, J. B. (2001). Insects and plants in a changing atmosphere. *Journal of Ecology, 89*, 507–518.

Williams, R. S., Norby, R. J., & Lincoln, D. E. (2000). Effects of elevated CO_2 and temperature-grown red and sugar maple on gypsy moth performance. *Global Change Biology, 6*, 685–695.
With, K. A., & Crist, T. O. (1995). Critical thresholds in species' responses to landscape structure. *Ecology, 76*, 2446–2459.
Woods, H. A. (2013). Ontogenetic changes in the body temperature of an insect herbivore. *Functional Ecology, 27*, 1322–1331.
Woods, H. A., Bonnecaze, R. T., & Zrubek, B. (2005). Oxygen and water flux across eggshells of Manduca sexta. *Journal of Experimental Biology, 208*, 1297–1308.
Wratten, S. D., Gillespie, M., Decourtye, A., Mader, E., & Desneux, N. (2012). Pollinator habitat enhancement: Benefits to other ecosystem services. *Agriculture, Ecosystems and Environment, 159*, 112–122.
Wu, J., & Baldwin, I. T. (2010). New insights into plant responses to the attack from insect herbivores. *Annual Review of Genetics, 44*, 1–24.
Young, A., Boyle, T., & Brown, T. (1996). The population genetic consequences of habitat fragmentation for plants. *Trends in Ecology & Evolution, 11*, 413–418.
Youngsteadt, E., Dale, A. G., Terando, A. J., Dunn, R. R., & Frank, S. D. (2014). Do cities simulate climate change? A comparison of herbivore response to urban and global warming. *Global Change Biology, 21*, 97–105.
Zavala, J. A., Casteel, C. L., DeLucia, E. H., & Berenbaum, M. R. (2008). Anthropogenic increase in carbon dioxide compromises plant defense against invasive insects. *Proceedings of the National Academy of Sciences of the United States of America, 105*, 5129–5133.
Zavala, J. A., Nabity, P. D., & DeLucia, E. H. (2013). An emerging understanding of mechanisms governing insect herbivory under elevated CO_2. *Annual Review of Entomology, 58*, 79–97.
Zvereva, E. L., & Kozlov, M. V. (2006). Consequences of simultaneous elevation of carbon dioxide and temperature for plant-herbivore interactions: A meta-analysis. *Global Change Biology, 12*, 27–41.

CHAPTER TEN

Conservation Biological Control in Agricultural Landscapes

A. Rusch*[,1], R. Bommarco[§], B. Ekbom[§]
*INRA, UMR SAVE 1065, Bordeaux Sciences Agro, Centre INRA de recherches de Bordeaux- Aquitaine, Institut des Sciences de la Vigne et du Vin, Villenave d'Ornon Cedex, France
[§]Swedish University of Agricultural Sciences, Uppsala, Sweden
[1]Corresponding author: E-mail: adrien.rusch@inra.fr

Contents

Abstract

Integrating supporting and regulating ecosystem functions provided by several components of biodiversity into cropping systems has been proposed as a promising way to decrease agrochemical inputs and negative environmental impacts while maximizing crop productivity. In this chapter, we illustrate how agroecological knowledge can be used to revisit crop protection and insect pest management using conservation biological control. We review how key management options, from the plant to the landscape level, affect natural enemy communities, insect pest abundance and the level of biological control. We particularly show that maintaining within-field diversity in space and time, reducing nitrogen fertilization or soil tillage as well as using

Advances in Botanical Research, Volume 81
ISSN 0065-2296
http://dx.doi.org/10.1016/bs.abr.2016.11.001

organic farming practices at the farm scale or maintaining seminatural habitats at the landscape scale generally benefit natural enemies, increase biological control and limit pest abundance. We also summarize the body of knowledge of the relationship between natural enemy community structure and the level of pest control. Future research needs and applied perspectives are highlighted.

1. INTRODUCTION

The key challenges for agriculture have drastically changed these last two decades. Following major ecological and sociological changes, agriculture is now facing issues such as climate change, biodiversity loss, biotic invasions, urbanization, asymmetric demand for food and globalization of food production. These multiple and interacting changes are seriously challenging our current food production model and profound modifications are needed to handle these threats. In Northern industrialized countries, the intensification of agriculture that started in the 1960's is now recognized to have several negative externalities that jeopardize the sustainability of food production systems. This intensification is, for instance, manifested by enlargement of field size, high application levels of synthetic agrochemicals per unit area, intensive soil tillage, fragmentation of seminatural habitats in the landscape, rotation simplification, and decreasing genetic diversity of crops (Tscharntke, Klein, Kruess, Steffan-Dewenter, & Thies, 2005). Although the intensification has been successful in meeting a growing demand for food and fibre, it is now recognized that such changes have severely impacted the environment, human health, and even long-term production goals (Foley et al., 2011; Tilman et al., 2001).

Agroecology offers a promising avenue for handling these challenges and designing innovative productive and environmentally friendly crop production systems. Agroecology has multiple definitions and is considered as a scientific discipline, a social movement, and a management system (Wezel et al., 2009). The term appeared in the scientific literature in the 1930's, and the concept has undergone a significant development during the 1980's (for a complete historic perspective on agroecology see Wezel et al., 2009). From a scientific perspective, agroecology can be seen as a discipline at the interface between agronomy, ecology and social-economical sciences. Two definitions of agroecology highlight these aspects: the first is by Francis et al. (2003) who define agroecology as "the integrative study of ecology of the entire food system, encompassing ecological, economical and social dimensions"; the second is by Gliessman (2007) who defines agroecology as "the science of applying ecological concepts and principles to the design and management of sustainable food systems". In addition to this scientific perspective, the term also recently appeared in several reports from policy-makers at national and international levels and as an orientation plan to reorganize food production systems for several European countries and to limit pesticide use (De Schutter, 2011).

Despite a broad range of terminology, utilization and implications, some key principles provide cornerstones of the scientific vision of agroecology. The most general one is to integrate supporting and regulating ecosystem functions provided by several components of biodiversity into cropping systems, with the aim to decrease agrochemical inputs and negative environmental impacts while maximizing productivity (Bommarco, Kleijn, & Potts, 2013; Rusch, Valantin-Morison, Sarthou, & Roger-Estrade, 2010). This approach is, by definition, particularly relevant to crop protection as ecological knowledge about species assemblages, trophic interactions or population dynamics provides a necessary basis for developing crop protection strategies mainly based on biological pest control services provided by natural enemies instead of chemical pesticides. Biological pest control is a major regulating ecosystem service provided by biodiversity of immense economic value for farmers and society (Costanza et al., 1997; Losey & Vaughan, 2006). Here we review agroecological knowledge about pest management strategies based on Conservation Biological Control (CBC; see Box 3 for definition). We particularly illustrate on-field and off-field key management options to enhance biological pest control services and decrease the use of pesticides in agroecosystems.

2. BASIC PRINCIPLES IN CONSERVATION BIOLOGICAL CONTROL: PROVISIONING OF KEY RESOURCES IN SPACE AND TIME

The idea of biological pest control is based on the recognition that pest populations can be kept in check by the action of other living beneficial organisms (in contrast to harmful organisms that damage crops) that are natural enemies to the pest. The enemies are classified into three types: (1) predators that feed directly on pests; (2) parasitoids that lay their eggs in or on a host, and the host is killed in the process of parasitic development; and (3) pathogens that are microorganisms that cause diseases, which kill or injure their hosts (e.g. see Chapters 3, 4 and 6).

Natural enemies are beneficial organisms that occur naturally in the same environment that also harbours the pests. Natural enemies are found in all agricultural ecosystems and can often be observed consuming prey, for example lady beetles eating aphids. The actions of parasitizing enemies or disease infections can also be seen in aphid colonies in the form of aphid mummies or aphids covered with fungal mycelia. The pest population regulation they thereby exert and the active management to support naturally occurring natural enemies with the purpose to limit the size of pest populations is termed CBC.

Unfortunately, appreciation of the actions of natural enemies in holding pest levels below damaging numbers has decreased after the introduction of synthetic pesticides. The widespread use of chemical control has brought about problems not only of an environmental nature, but also in terms of efficacy (Carson, 1962; Way, 1966). Rapid development of pesticide resistance in the target pest reduces control efficiency worldwide (Denholm, Devine, & Williamson, 2002). Adding to this, a major negative side effect has been resurgence of secondary pests as application of insecticides has killed off natural enemies and degraded natural biological control to the extent that previously harmless herbivores have become pests.

In response, the concept of CBC was put forth emphasizing that beneficial organisms can provide pest control (Barbosa, 1998). In its simplest form, CBC is the avoidance of management practices, for instance the use of broad-spectrum insecticides, which are harmful to natural enemies. But this risk reduction is usually not sufficient because natural enemies must occur in the cropping system in sufficient numbers and at the right time to be able to stop pests from multiplying to harmful amounts.

A pioneer of biological control, Paul DeBach, proposed in 1964 that the environment should be modified to safeguard and increase natural enemy populations; measures that enhance their survival, reproduction and subsequent population size. For successfully reducing the risks and enhancing the natural enemy populations, a thorough understanding of the biology and ecology of the natural enemies, such as their resource requirements for reproduction and survival, is essential. Arthropod natural enemies are mobile, and it is necessary to know how they disperse and are distributed across the season, e.g. how they move in and out of the crop field. It is important to move beyond the boundaries of the arable field, vineyard or orchard and take into consideration the surrounding landscape because the well-being of the natural enemies is often dependent on elements outside the production unit (Ekbom, 2000).

So far the catch-all term for natural enemies has been used. But it is impossible to apply generic solutions to achieve CBC. Knowledge is needed about the community of natural enemies occurring in the agroecosystem and their interactions with the pests. The agroecosystem will have a profound influence on both the species of natural enemies as well as their interaction with crop pests. Crop identity, phenology and biogeography will set the stage for timing of events of importance to the organisms occurring there. The sequence of crop plants, in particular combinations with annual and perennial forms, adds a temporal component that is often not considered in crop protection. The spatial patterns of crops in the landscape might also affect which resources are available to natural enemies (Vasseur et al., 2013). Therefore both local and landscape conditions must be considered in designing CBC efforts.

What are then the key resources for natural enemies? First, they need shelter such as sites for overwintering or refuges from disturbances in cropping systems. Second, they need food, both when pests are not available and for natural enemy life stages that are not directly associated with pests. Providing overwintering and refuge sites for natural enemies is critical to maintain a sufficient level of natural enemy populations in crop fields (Landis, Wratten, & Gurr, 2000; Rusch et al., 2010). Seminatural habitats; such as forests, hedgerows, field margins or fallows; provide shelter to natural enemies as they provide more stable and less disturbed environment than annual crops. It has been demonstrated that these habitats generally house a larger proportion of neutral or beneficial than detrimental arthropods (Denys & Tscharntke, 2002). Indeed, 9 out of 10 beneficial species require a noncrop habitat at one stage of their life cycle whereas only one out of two pest species have such requirements (Keller & Häni, 2000). For instance, woody habitats often provide a more moderate microclimate than the centre of crops, thereby protecting predators or parasitoids against extreme temperature variation during the season (Landis et al., 2000). Moreover, seminatural habitats provide conditions suitable for natural enemies for overwintering, and the spatial distribution of these habitats will thereby affect population dynamics from one year to another (Rusch, Valantin-Morison, Sarthou, & Roger-Estrade, 2013). Several abundant natural enemy species are known to overwinter in woody or grassy habitats within the landscape (Corbett & Rosenheim, 1996; Sarthou, Badoz, Vaissière, Chevallier, & Rusch, 2014). Therefore, the spatial distribution between overwintering habitats and crops often determines the size and net direction of movement and distribution of individuals between habitats in the landscape (Rand, Tylianakis, &

Tscharntke, 2006). In a recent study, Sarthou et al. (2014) examined natural enemy communities (taking into account Araneae, Carabidae, Coccinellidae, Neuroptera, Hemiptera and Staphylinidae) emerging from seven types of seminatural habitats including woody and grassy habitats with different management or plant composition characteristics. They found that managed grass strips were the main source habitat for beneficial organisms in the spring.

Pollen and nectar are vital for many natural enemy species. Hymenopteran parasitoids have, in many cases, been shown to feed on floral nectar and providing them with nectar, both in laboratory and in field conditions, generally increases female longevity and fecundity and therefore the potential for parasitism (Balmer et al., 2014; Wäckers, van Rijn, & Bruin, 2005; Wäckers, Romeis, & van Rijn, 2007). For instance, Winkler, Wäckers, Bukovinszkine-Kiss, and van Lenteren (2006) showed, in field conditions, that nectar feeding determined longevity and fecundity of the parasitoid females, *Diadegma semiclausum* (Hellen, 1949) (Hymenoptera: Ichneumonidae), and that the level of parasitism of the pest, *Plutella xylostella* (Linnaeus, 1758) (Lepidoptera: Plutellidae), was much higher when females were supplied with nectar sources than if they were deprived.

Providing alternative hosts or prey is another option for maintaining natural enemy populations in the crop field and its surroundings. This is particularly efficient during periods in which host and prey densities are low in the crop fields, providing resource continuity for the predators throughout the season (Schellhorn, Gagic, & Bommarco, 2015). For instance, it has been shown that if the aphid infestation of wheat is delayed, populations of the predatory lady beetle *Coccinella septempunctata* Linnaeus, 1758 (Coleoptera: Coccinellidae) became dependent on aphid populations in seminatural habitats (Bianchi & van der Werf, 2004). Hence, lady beetles are more vulnerable to periods of food limitation when prey availability in seminatural habitats is low.

In summary, noncrop habitats will for many beneficial predatory arthropods provide shelter and food. Preservation, restoration and/or establishment of such habitats within a biologically meaningful distance (i.e. within the predators' dispersal range) to the crop fields are a baseline strategy to increase biological control and limit pest populations. Food may be provided by nonpest organisms, for example weeds or companion plants, and may directly or indirectly contain prey that will sustain natural enemies when pests are not present. Thorough knowledge of more specific natural enemy requirements for shelter and food, will allow for the development of targeted management options at local and landscape levels.

3. ON-FIELD MANAGEMENT OPTIONS FOR REDUCING PEST POPULATIONS AND ENHANCING BIOLOGICAL PEST CONTROL

Several on-field management options are known to affect pest populations or natural enemy communities (for more detailed review see Bommarco et al., 2013; Rusch et al., 2010). We highlight here some of these options.

3.1 Diversity in Space and Time at the Field Scale

Plant diversity within a crop field has a major impact on pest and natural enemy communities (Andow, 1991; Letourneau et al., 2011; Tonhasca & Byrne, 1994). Several syntheses of experiments have shown that habitats with higher plant diversity harbour higher natural enemy abundance, smaller herbivore populations and reduced plant damage compared with plant monocultures (Letourneau et al., 2011). This effect is attributed to two nonmutually exclusive hypotheses: the natural enemy hypothesis and the resource concentration hypotheses (Root, 1973). The former states that fewer herbivores and reduced damage occur in more diverse habitats owing to more abundant and/or species rich communities of predators and parasitoids (Langellotto & Denno, 2004). The effect is attributed to a higher attractiveness of diverse habitats for predators as a result of an increased availability of resources. The latter hypothesis states that smaller herbivore populations and reduced plant damage are found in more complex or diverse habitats owing to a lower probability of herbivores finding their host plants and therefore failing to feed and reproduce. The effect is attributed to chemical or physical confusion as well as changes in plant physiological status due to interspecific competition among plant species.

Positive effects of diversification within crop fields has been found for very different diversification options such as intercropping, trap crops, push–pull strategies and flower strips within or around a crop (Cook, Khan, & Pickett, 2006; Letourneau et al., 2011; see also the case study in Box 1). There are multiple examples throughout the world of plant diversification that succeeds in limiting pest abundance. For instance, in eastern Africa a push–pull strategy combines intercropping to repel stem borers from the crop (maize or sorghum) and attracts parasitoids with especially chosen noncrop plants. These plantings result in an aggregation of stem borers on the cultivated noncrop plants that also enhance natural enemies. The approach can successfully control stem borer populations

Box 1 Local Vegetation Management and Landscape Heterogeneity to Limit Insect Pest Density and Enhance Biological Pest Control Services: a Case Study in Vineyards

Increased integration of ecosystem services such as biological control in farming systems to increase the sustainability of cropping systems and limit their negative impacts on the environment is increasingly acknowledged. However, indicative examples of efficient management options to limit pesticide applications remain scarce. Habitat heterogeneity at multiple scales is recognized as a key driver of trophic interactions and pest population dynamics in agricultural landscapes. But management options based on such ecological knowledge remain limited. Grapevines are of major economic importance in around the world and it is also one of the most pesticide-consuming crops in the world. In France, grapevines receive, on average, 13 to 21 pesticide treatments a year per unit area. Wine growers are increasingly challenged by society to limit the amount of pesticide they apply but alternative pest control options remain scarce.

The impact of heterogeneity at the habitat and the landscape scales on biological control and insect pest density was explored in vineyards of south western France. The effect of heterogeneity at these two scales on grape moths, *Lobesia botrana* and *Eupoecilia ambiguella* (Lepidoptera, Tortricidae), was assessed, the two species that are among the most damaging insect pests in European vineyards, and their biological control in 20 vineyards during 3 consecutive years. Local vegetation management (full grass cover vs. partial grass cover) and the proportion of seminatural habitats in the surrounding landscape were used as proxies of heterogeneity at local and landscape scales (Fig. 1). Grape moth density was measured over time, as well as biological control services provided by different groups: birds, invertebrate predators, parasitoids and entomopathogenic fungi.

Figure 1 Two pictures illustrating the two main ways of managing within-field vegetation in southwestern France. The picture on the left shows partial grass cover management whereas the picture on the right shows a full grass cover management. *Credits: INRA, UMR SAVE.*

Box 1 Local Vegetation Management and Landscape Heterogeneity to Limit Insect Pest Density and Enhance Biological Pest Control Services: a Case Study in Vineyards (*cont'd*)

Over the 3 years, lower densities of grape moths were found in vineyards with full grass cover compared to partial grass cover (Fig. 2). Despite these results, biological control of grape moths was not primarily affected by local vegetation management but by habitat heterogeneity in the landscape, and the direction of this effect varied over time. Notably, predation by birds increased with landscape heterogeneity in spring while attacks by entomopathogenic fungi decreased with landscape heterogeneity during winter. The results of this study have important implications for the ecological intensification of vineyard landscapes. Heterogeneity at both the habitat and the landscape scales were found to be good management options to reduce pest pressure and enhance biological pest control. Grape moth attacks seem to be mainly determined by bottom-up processes related to resource localization, and less by top-down processes related to natural enemy action. Maintaining full grass cover within vineyards reduced grape moth density to a level below common economic thresholds used in south western France.

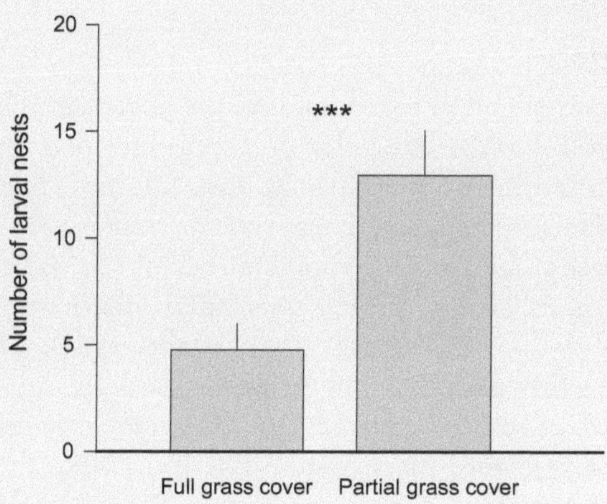

Figure 2 Larval nests per 100 grape clusters (±SD) for the first generation of grape moths (*Lobesia botrana* and *Eupoecilia ambiguella*) over the 3 sampled years was clearly lower in local vegetation management giving full grass cover compared with partial grass cover ***$P < .01$. The economic threshold used in the region is 5 larval nests per 100 grape clusters.

(Khan, Pickett, Berg, Wadhams, & Woodcock, 2000). A similar mechanism has been demonstrated effective in controlling the pollen beetle in oilseed rape using turnip rape as a trap crop (Cook, Watts, Hunter, Smart, & Williams, 2004), and the Colorado potato beetle, *Leptinotarsa decemlineata* (Say 1824) (Coleoptera: Chrysomelidae) with early-planted potato trap crops sprayed with an attractant (Martel, Alford, & Dickens, 2005).

Crop diversity in time, or crop succession at the field scale such as crop rotation or cover crops (i.e., crops that are not harvested but produced to enrich the soil and capture inorganic N), is another key management-option empirically developed by farmers to reduce pests, weeds, and pathogen prevalence, in addition to having several positive effects on soil fertility, nutrient use efficiency and nitrogen retention. The basic principle of managing crop succession for crop protection is to disrupt the temporal cycle of pests or diseases by avoiding the presence of successive crops that serve as hosts to pests and diseases (Ratnadass, Fernandes, Avelino, & Habib, 2012). Moreover, it is known that the abundance, activity, reproductive rates or species richness of beneficial arthropods, such as carabid beetles, increased with the lengthening of rotations combined with lowered inputs of agrochemicals, suggesting that biological pest control services is higher in such cropping systems (Büchs, Harenberg, & Zimmermann, 1997; O'Rourke, Liebman, & Rice, 2008), and more studies are needed to confirm this.

3.2 Fertilization

Nitrogen fertilization and nitrogen status of the host plant is known to play an important role for the dynamics of populations and performance of herbivores by affecting plant resistance, host selection mechanisms, and the ability of plants to compensate from the damage caused by phytophagous insect attacks. Two main hypotheses have been formulated on the impact of host plant quality on pest populations: the *plant stress hypothesis* and the *plant vigor hypothesis* (Price, 1991; White, 1984). The plant stress hypothesis states that physiologically stressed plants are more attacked by pests due to changes in the qualitative and nutritional status of the plant, or a decrease in resistance mechanisms. The plant vigor hypothesis states that phytophagous insects select more vigorous plants because they provide a higher quality source of food. Even if there is empirical support for both the plant stress hypothesis or the plant vigor hypothesis, a majority of cases show herbivorous insects to respond positively to plant vigour and fertilization (Butler, Garratt, & Leather, 2012; De Bruyn, Scheirs, & Verhagen, 2002; Fritz, Crabb, & Hochwender, 2003; Waring & Cobb, 1992).

For instance, Waring and Cobb (1992) reviewed 186 studies investigating the effect of soil or host plant nutrient status on insects and mites and found a majority (about 60%) of responses being positive to nitrogen fertilization. A follow-up synthesis investigated the response to nitrogen, phosphorous and potassium fertilizers in insect populations using both vote-count and *meta*analysis approaches (Butler et al., 2012). This study confirmed an overall positive response of herbivorous insects to nitrogen in particular, with a much stronger response to fertilizers for sucking insects than chewing insects. No significant effect on herbivorous insects was detected for potassium or phosphorous fertilizers.

Fertilization practices can also affect generalist predators through trophic cascades. The addition of crop residues is an important farming practice in organically fertilized agrosystems. Adding crop residues is known to indirectly enhance generalist predator numbers via a positive effect of litter resources on decomposer prey and/or modifications in microhabitat structure and microclimatic conditions (Diehl, Wolters, & Birkhofer, 2012; Halaj & Wise, 2002; Scheu, 2001). However, the consequences of fertilization practices related to residues management on biological control of crop pest are not always known. Nitrogen fertilization can also affect parasitoid performance on its herbivorous host. For instance, a higher proportion of *Plutella xylostella* escaped biological control by *Diadegma insulare* (Cresson 1865) (Hymenoptera: Ichneumonidae) when the pest has developed on plants that received a low level of nitrogen fertilization compared to pests that has developed on plants that received high levels of fertilization (Sarfraz, Dosdall, & Keddie, 2009).

3.3 Soil Tillage

Reduced or absence of tillage are common practices in agroecology that have many virtues such as reducing energy consumption, decreasing soil erosion and compaction, increasing soil microbial activity and carbon sequestration (El Titi, 2003; Holland et al., 2004; Roger-Estrade, Anger, Bertrand, & Richard, 2010). Modifying the soil tillage regime is an on-field management option that can reduce pest pressure and increase biological control. Several aspects of soil tillage regime can affect pest and natural enemy populations: the intensity of soil tillage, the type of tool used, the frequency of operation, or the time period (Rusch et al., 2010). Especially deep tillage induces significant biochemical and biophysical changes that modify habitat quality, remove microhabitats, or decrease prey availability and these changes have indirect effects on soil organisms and their interactions. In addition, soil tillage can have direct lethal effects by injuring, killing, forcing organisms to migrate, or exposing soil organisms to predation (Roger-Estrade et al., 2010). Responses of soil organisms to reduced tillage can be highly variable and involve several parameters of soil tillage, but the overall response pattern is that both abundance and diversity of the soil fauna tend to increase with reduced tillage intensity (El Titi, 2003; Kladivko, 2001; Roger-Estrade et al., 2010).

Depending on the pest species considered, soil tillage can have ambivalent effects. For instance, intensive tillage, involving mouldboard ploughing, is an efficient way to limit slug populations through direct mortality and indirect effects on habitat quality (Roger-Estrade et al., 2010). On the other hand, mulches on the soil surface left in reduced or no-till cropping systems can provide a stable environment for natural enemies of crop pests and thus improve biological control (Kendall, 2003). Reduced tillage and crop residues left on the soil surface have been found to limit insect pest pressure and enhance natural enemy abundance and diversity in small scale plot experiments (Pullaro, Marino, Jackson, Harrison, & Keinath, 2006; Schmidt, Thewes, Thies, & Tscharntke, 2004). In addition to more abundant predator communities, reduced tillage was recently found to support a higher level of cereal aphid predation (biological control was 16% higher in reduced tillage than in conventional tillage) in real field conditions (Tamburini, De Simone, Sigura, Boscutti, & Marini, 2016). Potential mechanisms underpinning these positive effects on pest control are physical barriers disturbing host plant location for the pests, reduced competition between natural enemies, higher availability of alternative prey or hosts, and more favourable microclimate conditions, lower mortality and higher availability of organic matter for natural enemies (Landis et al., 2000; Sirrine et al., 2008; Roger-Estrade et al., 2010).

3.4 Organic Farming

It is well documented that the abundance and species richness of several taxa ranging from plants to mammals and birds are higher in organically than in conventionally managed crop fields (Bengtsson, Ahnström, & Weibull, 2005; Hole et al., 2005; Tuck et al., 2014). In a *meta*analysis, Bengtsson et al. (2005) found that species richness was on average 30% higher in organic fields compared with conventional fields despite variable results among studies. They particularly found that birds, insects, and plants responded positively to organic farming. Organic farming enhances abundance and species richness of taxonomic groups including generalist and specialist natural enemies to crop pests, but it is not know if this generally leads to improved biological pest control (see Section 4).

Contrasting empirical results have been reported about the effect of organic farming on biological control services, with positive, neutral and negative effects of organic farming being found. For instance, Macfadyen et al. (2009) examined the differences in food web structure and biological control services between organic and conventional farms in the United Kingdom. Despite significant differences in the structure of the food webs between organic and conventional farms (with higher parasitoid diversity on organic farms) they found no difference in the level of biological control between the two production systems. Birkhofer et al. (2016) explored the effect of organic and conventional farming on biological control of hemipteran pests in barley along a landscape complexity gradient in southern Sweden. They demonstrated that aphid predation was higher under organic farming and influenced by effects on predator abundance and community composition independent of landscape complexity. Finally, Rusch, Delbac, Muneret, and Thiéry (2015) analysed how landscape composition and organic and conventional farming systems affected abundance of insect pests of grapes and their parasitism rates in southwestern France. They found that farming system and host density were the two main factors determining the level of biological control of tortricid moths by their parasitoids. Surprisingly, their results showed that organic fields had lower parasitism rates compared with conventional ones, and that this rate was negatively correlated to host density in the field.

These contrasting findings can result from the relatively broad definition of organic farming in terms of actual farming practices and their effects on natural enemy communities, or the taxonomic group studied. Hence, there is a need to quantitatively summarize the literature on this topic, and to analyse the context-dependencies determining the emergence of positive, neutral or negative effects of organic farming.

4. PEST ABUNDANCE AND BIOLOGICAL CONTROL AT THE LANDSCAPE SCALE

4.1 Arthropod Dynamics at the Landscape Scale

As illustrated in the previous section, the underlying processes determining population dynamics and trophic interactions involving pests and their natural enemies usually operate at various spatial and temporal scales (Rusch et al., 2010). A landscape-based perspective is therefore needed to understand population dynamics and biological control of pest species in agricultural landscapes (Tscharntke et al., 2007). We highlighted the complementary of crop and seminatural habitats as most natural enemies and insect pests exploit several resources found in both habitat types. This suggests that patterns of the distribution of resources in the landscapes interacting with life-history traits of species (e.g., dispersal abilities) determine population dynamics and trophic interactions.

Metapopulation ecology (i.e., the study of the effects of habitat structure and configuration on metapopulation dynamics) and landscape ecology (i.e., the study of landscape structure and its effect on ecological processes) have both assisted, in complementary ways, to understanding local population dynamics based on processes acting at larger spatial and/or temporal scales (Cronin & Reeve, 2005; Dunning, Danielson, & Pulliam, 1992). Dunning et al. (1992) described four main processes occurring at the landscape scale that affect population dynamics: landscape complementation, landscape supplementation, source/sink dynamics, and neighbourhood effects. Landscape complementation refers to situations where a species needs at least two nonsubstitutable resources located in two different habitat types implying dispersal of the species between these habitat types. In the landscape supplementation process the population of a focal patch may be enhanced if that patch is located close to other patches of similar resource or function. In the source/sink process productive patches or habitats serve as sources of emigrants, which disperse to less productive patches or habitats. In source/sink case, local populations in sink patches cannot be maintained without immigration from more productive patches. Finally, the neighbourhood effect appears when species abundance of a particular habitat patch is more affected by the characteristics of adjacent patches than by those of patches located further away.

Dependence between two adjacent habitat types has been reviewed in the context of biological control with spillover of natural enemies between seminatural and cultivated habitats (Blitzer et al. 2012; Rand et al. 2006). Spillover between adjacent seminatural and cultivated habitats occurs in both directions and the direction and magnitude of the spillover is determined by differences in primary productivity, temporal dynamics, and complementarity in resources between habitats (Blitzer et al. 2012; Rand et al. 2006). But very few empirical studies have measured the actual functional implications of these spillovers by quantifying the level of pest suppression or the amount of crop injury in adjacent habitat types for instance. A recent investigation of spillovers into Australian vineyards of natural enemies and biological control services coming from woody vegetation showed that abundances of coccinellids and parasitoids tend to be higher in vine rows closer to the woody vegetation. Predation and parasitism of sentinel eggs of Light Brown Apple moth [*Epiphyas postvittana* (Walker, 1863) (Lepidoptera: Tortricidae)] follow a similar pattern (Thomson & Hoffmann, 2013). Similarly, higher predation rates of sentinel eggs by ground-dwelling predators were found when coniferous forest was the adjacent habitat compared with having another crop next to the crop field, but only in cool days. In warm days predation rates were high across all interfaces and did not allow for the detection of possible differences between settings (Schneider, Krauss, & Steffan-Dewenter, 2013). These findings suggest that the positive effects of adjacent seminatural habitat, such as forests, can be attributed to predator spillover.

4.2 Effect of Landscape Context on Pest Pressure and Biological Control Services

The need for integrating a landscape perspective to understand population dynamics and trophic interactions has inspired researchers to examine how landscape context affects local populations and communities of both insect pests and their natural enemies. Several *meta*analyses have synthesized current knowledge about the effect of landscape context, especially of landscape composition, on natural enemy communities, pest abundances as well as level of biological control (Bianchi, Booij, & Tscharntke, 2006; Chaplin-Kramer, O'Rourke, Blitzer, & Kremen, 2011; Rusch et al., 2016; Veres, Petit, Conord, & Lavigne, 2013). It is clear that species richness and abundance of natural enemies increases with the amount of seminatural habitats in the landscape, which supports the notion that these habitats provide key resources for natural enemies as explained previously (Bianchi et al., 2006; Chaplin-Kramer et al., 2011). In particular, generalist enemies show consistent positive responses to the proportion of seminatural habitats in the landscape across all spatial extents explored in the study (from a 500 m to a 3 km radius around fields), while specialist natural enemies respond more strongly to the amount of seminatural habitats at smaller extents (Chaplin-Kramer et al., 2011). Adapted management strategies, hence, appear to increase biological control depending on the natural enemy species in the regional pool. Furthermore, it has now been demonstrated that landscape composition (in terms of proportion of crop or noncrop habitat) also drives the level of biological control and pest abundance in crop fields (Rusch et al., 2016; Veres et al., 2013) (see Box 2). More complex landscapes, with higher proportions of seminatural habitats, exhibited a higher

Box 2 Landscape Simplification Reduces the Level of Biological Pest Control: a Quantitative Synthesis

It is well established that landscape simplification, characterized by a reduced proportion land cover of seminatural habitats in the landscape, reduces the abundance and diversity of natural enemies. However, the consequences of this reduction in diversity and abundance of natural enemies on the level of biological pest control remains poorly quantified. Moreover, most research on the effect of landscape simplification on biological control has taken into account only a limited number of taxa and thus have not considered potential interactions among natural enemy species (but see Martin et al., 2013).

A recent quantitative synthesis of the effect of landscape simplification on the level of biological control of aphids used several datasets collected in different cropping systems in Europe and North America (Rusch et al., 2016). All included studies used exclusion cages experiments established along landscape gradients to measure the level of aphid pest control that could be attributed to predators or parasitoids.

A consistent negative effect of landscape simplification was found on the level of biological pest control despite interactions among natural enemies. The authors found biological pest control to be 46% lower in homogeneous landscapes dominated by arable land as compared with more complex landscapes dominated by seminatural habitats (Fig. 1). However, landscape

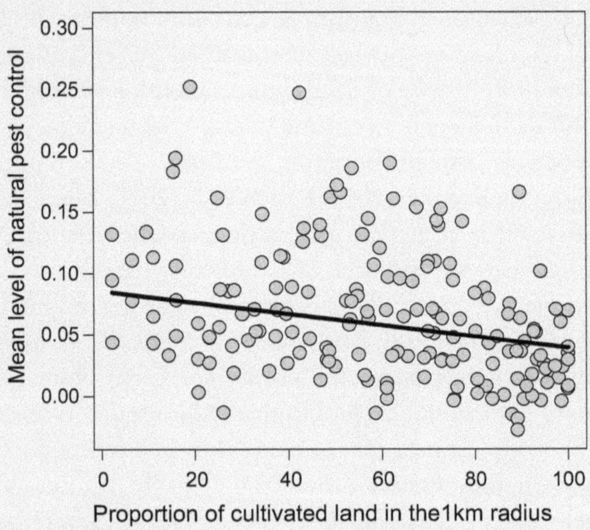

Figure 1 Mean level of biological aphid pest control declined with increasing the proportion of cultivated land in a 1 km radius around fields. The level of pest control was measured by the difference in growth rates of aphids between the total exclusion treatment and the open treatment per day. Each point represents a field site within a study and the line represents the overall regression estimated from the linear mixed effect model. *From Rusch, A., Chaplin-Kramer, R., Gardiner, M. M., Hawro, V., Holland, J., Landis, D., …, Bommarco, R. (2016). Agricultural landscape simplification reduces natural pest control: a quantitative synthesis.* Agriculture, Ecosystems & Environment, 221, *198–204.*

Box 2 Landscape Simplification Reduces the Level of Biological Pest Control: a Quantitative Synthesis (*cont'd*)

simplification did not affect either the interaction strength between ground-dwelling and vegetation-dwelling predators, or the within-field stability of pest control.

The synthesis demonstrates that agricultural intensification through landscape simplification has negative effects on the level of biological pest control with important implications for management to maintain and enhance ecosystem services in agricultural landscapes.

level of biological control and lower pest abundances than more simple landscapes dominated by cultivated land. Biological control by natural enemies was on average 46% lower in extremely simple landscapes (Rusch et al., 2016; see Box 2).

Moreover, in addition to this direct effect of landscape context on trophic interactions, it has recently been hypothesized that landscape context modulates the effects of local management on biodiversity and ecosystem services (Kleijn, Rundlöf, Scheper, Smith, & Tscharntke, 2011). This hypothesis states that the benefits of local management on ecosystem services are smaller in complex landscapes that already support a high level of biodiversity than in simple landscapes. This hypothesis was recently tested in a *meta*analysis of studies examining the effects of agro-environmental schemes on biodiversity that were implemented along landscape gradients. The hypothesis was confirmed in annual cropland but not in perennial grasslands (Batáry, Báldi, Kleijn, & Tscharntke, 2010). Moreover, Tuck et al. (2014) conducted a hierarchical *meta*analysis of studies that compared biodiversity in organic and conventional crop fields. They found that organic farming had a greater effect on biodiversity as the proportion of cultivated land in the landscape increased. These results suggest that the effects of local management (such as those detailed in Section 3) on biodiversity or ecosystem services such as biological control are modulated by the landscape context.

The relationships between arthropod populations and landscape context have generally been explored through a landscape composition approach in the literature as illustrated previously. However, such approaches do not take into account landscape configuration (i.e. the spatial arrangement of habitats) and their specific effects on population dynamics and biological control. It would be interesting to analyse these aspects to provide relevant guidelines for landscape management to increase biological control and reduce pest damage.

5. RELATIONSHIP BETWEEN NATURAL ENEMY COMMUNITY STRUCTURE AND THE LEVEL OF BIOLOGICAL CONTROL

5.1 Emergent Effects of Natural Enemy Species Richness

Several aspects of natural enemy community structure are known to influence the level of suppression of herbivorous species by predation. It has been reported that the abundance of beneficial organisms is more important than species richness for short-term effects on pest control (Duelli & Obrist, 2003). However, when longer temporal scales are considered, the diversity of natural enemy communities might be more important. Several possible relationships, involving different ecological processes, between the number of natural enemy species and the level of suppression of herbivorous pests have been reported (Letourneau, Jedlicka, Bothwell, & Moreno, 2009).

First, a positive relationship between natural enemy species richness and pest mortality may emerge due to a complementarity effect, a sampling effect or a combination of these (Straub, Finke, & Snyder, 2008). The complementarity model predicts that pest mortality resulting from different predator and parasitoid species is equal or greater than the sum of mortalities caused by each natural enemy species considered individually. This relationship between species richness and the level of pest control can be explained by resource complementarity between natural enemy species due to facilitation or niche partitioning (Finke & Snyder, 2010; Otto, Berlow, Rank, Smiley, & Brose, 2008; Straub et al., 2008). Examples of niche partitioning include predation of different prey life stages, at different periods during the season or at different locations, or foraging behaviour that facilitates predation from one species by another. The sampling effect is explained by a larger number of species included an assemblage increasing the probability of including a species that contributes more than the other species to a given ecosystem function (Loreau & Hector, 2001; Straub et al., 2008). The observation of a particular natural enemy species causing the greater part of the mortality of a herbivore species has been reported in the literature suggesting that sampling effect can emerge in natural enemy assemblages (Chang, 1996; Myers, Higgins, & Kovacs, 1989).

Second, a negative relationship between natural enemy species richness and pest mortality can emerge instead. The sampling effect mentioned above, can also lead to a negative relationship between species richness and the level of pest control. An assemblage with a large number of predators species can increase the probability of including a species that negatively interacts with other predators thereby decreasing herbivore mortality, for example due to intraguild predation (i.e. the killing and eating of potential competitors), hyperparasitism (i.e. parasitoid attacking another parasitoid), or behavioral interference (see Box 3 for definitions) (Letourneau et al., 2009; Martin, Reineking, Seo, & Steffan-Dewenter, 2013). There is evidence for each these processes occurring in nature (Finke & Denno, 2005; Perez-Lachaud et al., 2004; Schmitz, 2007; Snyder & Ives, 2001).

Third, no relationship between natural enemy species richness and pest suppression has been observed because of functional redundancy between species and minimal interactions among them or by a balance between positive and negative effects (Straub et al., 2008; Wilby & Thomas, 2002). Increasing the number of predator species has been observed not to change prey suppression as compared with lower numbers of predator species (Evans, 1991; Straub & Snyder, 2006).

As mentioned above ample evidence for each of these potential relationships and ecological processes have been found (Cardinale et al., 2006; Schmitz, 2009; Snyder, Snyder, Finke, & Straub, 2006; Straub & Snyder, 2006; Wilby, Villareal, Lan, Heong, & Thomas, 2005). A recent synthesis, using a *meta*analysis approach, including numerous examinations

Box 3 Definitions and Terminology Used in Section 4

Conservation biological control: According to Landis et al. (2000), conservation biological control "involves manipulation of the environment to enhance the survival, fecundity, longevity, and behaviour of natural enemies to increase their effectiveness".

Evenness: An index that quantifies the relative abundance (or the distribution) of each species within a community. See Hillebrand et al. (2008) for a review on the effect of evenness on ecosystem processes.

Functional traits: Observable and/or operationally defined phenotypic characteristics that influence species performance and/or ecosystem processes. See Rusch, Birkhofer, et al. (2015) for illustration of how trait composition can affect the level of biological pest control.

Intraguild predation: A special case of omnivory in which a given predator species kills and eats another predator that shares the same prey species. See Finke and Denno (2005).

Hyperparasitism: The development of a secondary parasitoid at the expense of a primary parasitoid, which already parasitized a host.

of the consequences of natural enemy species richness on prey mortality demonstrated that a large majority of studies (almost 70% of 266 case studies) reported a higher level of prey suppression by richer natural enemy assemblages (Letourneau et al., 2009). This suggests that positive, complementary interactions between natural enemy species dominate over negative or neutral interactions in trophic interactions in terrestrial ecosystems. The results demonstrate that conservation of natural enemy biodiversity and pest control are compatible goals, but the fact that neutral or negative relationships can emerge in a non negligible amount of cases highlights the importance of understanding the context-dependency of such relationships if we are to optimize the level of pest control based on CBC.

5.2 Effects of Community Evenness and Functional Diversity

In addition to this large body of evidence about the effect of species richness, several recent studies have demonstrated that key facets of natural enemy communities, other than the number of species they hold, are important drivers of biological pest control services. Not all species contribute equally to ecosystem functions and an increasing amount of evidence indicates that by taking community composition into account, in terms of relative abundance and functional traits of species, major insights on processes shaping emergent functions of assemblages can be understood (Cadotte, Carscadden, & Mirotchnick, 2011; Gagic et al., 2015; Hillebrand, Bennett, & Cadotte, 2008). Crowder, Northfield, Strand, and Snyder (2010) demonstrated that a skewed relative abundance distribution of species in a natural enemy community generally weakens biological control compared with more even abundance distribution of species in natural enemy assemblages. Relatively small increases in evenness of predator and pathogen communities significantly improved control of the Colorado potato beetle generating a strong trophic cascade and larger potato plants. Intraspecific competition was reduced in more even communities, leading to increased natural enemy survival in even communities and greater pest control (Crowder et al., 2010). Moreover, an increasing body of evidence indicates that functional diversity and specific traits such as habitat domain, hunting mode, or body size, in natural enemy communities can help in understanding and predicting trophic interactions (Griffin, Byrnes, & Cardinale, 2013; Rusch, Birkhofer, Bommarco, Smith, & Ekbom, 2015; Schmitz, 2007; Schneider, Scheu, & Brose, 2012). For example, Schmitz (2007) performed a synthesis of experiments and revealed that the strength and direction of multiple predator effects on prey can be predicted based on the habitat domain of both predator and prey. The habitat domain of a given species considers the choice of microhabitat and the extent of movement in space and time and can be inferred based on specific traits such as hunting mode.

Three kinds of multiple predator effects on prey can be expected: substitutive, risk-enhancing or risk-reducing effects (Schmitz, 2007). Predators will have substitutive effects on prey when predators have spatially or temporally complementary habitat domains and prey species have broad habitat domains. In this case, there is a low probability that predators will engage in interspecific interactions such as intraguild predation. Increased risk enhancement for the prey will occur when prey species have narrow habitat domains and predatory species have broad and overlapping habitat domain. In this case, prey species have a low probability of escaping their predators and predators roam more broadly than their prey and can exploit alternative hosts to subsidize their needs and limit interspecific interaction such as intraguild predation. Risk for the prey is reduced by negative interactions among predators. This appears when: (1) prey and predator have either broad or narrow habitat domains but with complete overlap in space, or (2) when predators have overlapping narrow habitat domains with a shared prey species with a broad habitat domain. In the latter case, the prey can therefore escape predators favoring intraguild predation (Schmitz, 2007).

Using empirical data replicated at the landscape scale, Rusch, Birkhofer, et al. (2015) examined the explanatory power of several aspects of ground-dwelling predator communities (e.g., activity-density, species richness, evenness, community weighted mean of various traits, functional diversity) on predation rates of aphids in cereal fields. They demonstrated that functional diversity explained a greater part of variation in predation rates than any other taxonomic or activity-density metrics. In particular, the community-average value of body-size of ground-dwelling predators was negatively related to predation rates of aphids, whereas the proportion of spiders with a preference for arable land was positively related to predation rates. Additional analyses of body-size distributions of ground-dwelling predators suggested that intraguild predation was a key process shaping the relationship between predator community composition and the level of aphid pest control. This example illustrates functional trait aspects of predator communities as a critical characteristic for mechanistically understanding top-down control of pest populations by their natural enemies.

We have shown that several non independent features of natural enemy communities affect the level of top-down control of pests and we have summarized ecological evidence indicating that maintaining species rich, evenly distributed or functionally complementary assemblages may lead to biological pest control enhancement. However, the context-dependency of these relationships indicates that more research is needed to identify primary objectives in terms of natural enemy community structure leading to the most effective strategies to reduce pest damage and limit pesticide use in crops.

6. CONCLUSIONS AND FUTURE CHALLENGES

This chapter reviews findings about CBC and examines management options to enhance biological pest control in agricultural landscapes. The last decades of CBC research has moved beyond the plant and field scales to the landscape scale and provided valuable knowledge about variables that shape trophic interactions in cropping systems. This is an important step forward, and not the least, because we in a near future will have to rely even more heavily on CBC for plant protection; resistance against chemical pesticides is ubiquitous, few new active ingredients are developed and consumers are asking for food produced without the help of pesticides. Our review particularly showed management options that generally benefit natural enemies, increase biological control and limit pest abundances include: maintaining within-field diversity in space and time, reducing nitrogen fertilization and soil tillage, decreasing risk from insecticide use, employing organic farming practices at the farm scale and maintaining seminatural habitats in the landscape. However, our review also pinpoints the strong context-dependency of CBC outcomes in agricultural landscapes. There are clear knowledge gaps and research on how context-dependency affects natural enemies and their prey is needed to provide concrete advice for ecological intensification of modern farming systems. The consequences of management options on crop damage, yield, or income have received comparably little attention. A key challenge is to demonstrate the impact of natural enemy activity on crop damage, crop yield, and farmer income. Moreover, pest control services have generally been examined at a single time during the season and with relatively low replication over time. Exploring how biological control services and pest abundances vary within a year and between years is key. We therefore advocate a more explicit consideration of pest and natural enemy population dynamics and biological control processes occurring over time. Similarly, the role of natural enemy species in terms of their impact on pest populations as well as their functional traits or their ecology is not known for a majority of species. Such knowledge is, however, fundamental information to understand community and population dynamics, and to successfully implement management options that will benefit natural enemies and biological pest control. Finally, exploration of multiscale effects (and multiscale interactions) among farming practices and landscape management on biological pest control in agricultural landscapes is needed if we are to extend our knowledge about how ecological functions, such as predation, respond to large scale land use changes.

ACKNOWLEDGEMENTS

AR acknowledges support from the FRB (the French Foundation for Research on Biodiversity) and AR and RB from ERA-NET to the project "Rural development through governance of multifunctional agricultural land use - MULTAGRI".

REFERENCES

Andow, D. A. (1991). Vegetational diversity and arthropod population response. *Annual Review of Entomology, 36*, 561–586.

Balmer, O., Géneau, C. E., Belz, E., Weishaupt, B., Förderer, G., Moos, S., ... Luka, H. (2014). Wildflower companion plants increase pest parasitation and yield in cabbage fields: experimental demonstration and call for caution. *Biological Control, 76*, 19–27.

Barbosa, P. (1998). *Conservation biological control.* San Diego: Academic Press.

Batáry, P., Báldi, A., Kleijn, D., & Tscharntke, T. (2010). Landscape-moderated biodiversity effects of agri-environmental management: a meta-analysis. *Proceedings of the Royal Society of London B: Biological Sciences.* http://dx.doi.org/10.1098/rspb.2010.1923.

Bengtsson, J., Ahnström, J., & Weibull, A. C. (2005). The effects of organic agriculture on biodiversity and abundance: a meta-analysis. *Journal of Applied Ecology, 42*, 261–269.

Bianchi, F. J., Booij, C. J., & Tscharntke, T. (2006). Sustainable pest regulation in agricultural landscapes: a review on landscape composition, biodiversity and natural pest control. *Proceedings of the Royal Society of London B: Biological Sciences, 273*, 1715–1727.

Bianchi, F. J., & van der Werf, W. (2004). Model evaluation of the function of prey in non-crop habitats for biological control by ladybeetles in agricultural landscapes. *Ecological Modelling, 171*, 177–193.

Birkhofer, K., Arvidsson, F., Ehlers, D., Mader, V. L., Bengtsson, J., & Smith, H. G. (2016). Organic farming affects the biological control of hemipteran pests and yields in spring barley independent of landscape complexity. *Landscape Ecology, 31*, 567–579.

Blitzer, E. J., Dormann, C. F., Holzschuh, A., Klein, A. M., Rand, T. A., & Tscharntke, T. (2012). Spillover of functionally important organisms between managed and natural habitats. *Agriculture, Ecosystems & Environment, 146*, 34–43.

Bommarco, R., Kleijn, D., & Potts, S. G. (2013). Ecological intensification: harnessing ecosystem services for food security. *Trends in Ecology & Evolution, 28*, 230–238.

Büchs, W., Harenberg, A., & Zimmermann, J. (1997). The invertebrate ecology of farmland as a mirror of the intensity of the impact of man?—An approach to interpreting results of field experiments carried out in different crop management intensities of a sugar beet and an oil seed rape rotation including set-aside. *Biological Agriculture & Horticulture, 15*, 83–107.

Butler, J., Garratt, M. P. D., & Leather, S. R. (2012). Fertilisers and insect herbivores: a meta-analysis. *Annals of Applied Biology, 161*, 223–233.

Cadotte, M. W., Carscadden, K., & Mirotchnick, N. (2011). Beyond species: functional diversity and the maintenance of ecological processes and services. *Journal of Applied Ecology, 48*, 1079–1087.

Cardinale, B. J., Srivastava, D. S., Duffy, J. E., Wright, J. P., Downing, A. L., Sankaran, M., & Jouseau, C. (2006). Effects of biodiversity on the functioning of trophic groups and ecosystems. *Nature, 443*, 989–992.

Carson, R. (1962). *Silent spring*. Boston. MA: Houghton Mifflin.

Chang, G. C. (1996). Comparison of single versus multiple species of generalist predators for biological control. *Environmental Entomology, 25*, 207–212.

Chaplin-Kramer, R., O'Rourke, M. E., Blitzer, E. J., & Kremen, C. (2011). A meta-analysis of crop pest and natural enemy response to landscape complexity. *Ecology Letters, 14*, 922–932.

Cook, S. M., Khan, Z. R., & Pickett, J. A. (2006). The use of push-pull strategies in integrated pest management. *Annual Review of Entomology, 52*, 375.

Cook, S. M., Watts, N. P., Hunter, F., Smart, L. E., & Williams, I. H. (2004). Effects of a turnip rape trap crop on the spatial distribution of *Meligethes aeneus* and *Ceutorhynchus assimilis* in oilseed rape. *IOBC/WPRS Bulletin, 27*, 199–206.

Corbett, A., & Rosenheim, J. A. (1996). Impact of a natural enemy overwintering refuge and its interaction with the surrounding landscape. *Ecological Entomology, 21*, 155–164.

Costanza, R., d' Arge, R., De Groot, R., Faber, S., Grasso, M., Hannon, B., ... Paruelo, J. (1997). The value of the world's ecosystem services and natural capital. *Nature, 387*, 253–260.

Cronin, J. T., & Reeve, J. D. (2005). Host–parasitoid spatial ecology: a plea for a landscape-level synthesis. *Proceedings of the Royal Society of London B: Biological Sciences, 272*, 2225–2235.

Crowder, D. W., Northfield, T. D., Strand, M. R., & Snyder, W. E. (2010). Organic agriculture promotes evenness and natural pest control. *Nature, 466*, 109–112.

De Bruyn, L., Scheirs, J., & Verhagen, R. (2002). Nutrient stress, host plant quality and herbivore performance of a leaf-mining fly on grass. *Oecologia, 130*, 594–599.

De Schutter, O. (2011). Agroecology and the right to food. In *Report presented at the 16th session of the United Nations Human Rights Council*, 23 p.

DeBach, P. (1964). *Biological control of insect pests and weeds*. London: Chapman & Hall.

Denholm, I., Devine, G. J., & Williamson, M. S. (2002). Insecticide resistance on the move. *Science, 297*(5590), 2222–2223.

Denys, C., & Tscharntke, T. (2002). Plant-insect communities and predator-prey ratios in field margin strips, adjacent crop fields, and fallows. *Oecologia, 130*, 315–324.

Diehl, E., Wolters, V., & Birkhofer, K. (2012). Arable weeds in organically managed wheat fields foster carabid beetles by resource-and structure-mediated effects. *Arthropod-Plant Interactions, 6*, 75–82.

Duelli, P., & Obrist, M. K. (2003). Biodiversity indicators: the choice of values and measures. *Agriculture, Ecosystems & Environment, 98*, 87–98.

Dunning, J. B., Danielson, B. J., & Pulliam, H. R. (1992). Ecological processes that affect populations in complex landscapes. *Oikos*, 169–175.

Ekbom, B. (2000). Interchanges of insects between agricultural and surrounding landscapes. In B. Ekbom, M. E. Irwin, & Y. Robert (Eds.), *Interchanges of insects between agricultural and surrounding landscapes* (pp. 1–3). Dordrecht: Kluwer.

El Titi, A. (2003). Implications of soil tillage for weed communities. In A. El Titi (Ed.), *Soil tillage in agroecosystems* (pp. 147–185). Boca Raton, Florida, USA: CRC Press.

Evans, E. W. (1991). Intra versus interspecific interactions of ladybeetles (Coleoptera: Coccinellidae) attacking aphids. *Oecologia, 87*, 401–408.

Finke, D. L., & Denno, R. F. (2005). Predator diversity and the functioning of ecosystems: the role of intraguild predation in dampening trophic cascades. *Ecology Letters, 8*, 1299–1306.

Finke, D. L., & Snyder, W. E. (2010). Conserving the benefits of predator biodiversity. *Biological Conservation, 143*, 2260–2269.

Foley, J. A., Ramankutty, N., Brauman, K. A., Cassidy, E. S., Gerber, J. S., Johnston, M., … Zaks, D. P. M. (2011). Solutions for a cultivated planet. *Nature, 478*, 337–342.

Francis, C., Lieblein, G., Gliessman, S., Breland, T. A., Creamer, N., Harwood, R., … Poincelot, R. (2003). Agroecology: the ecology of food systems. *Journal of Sustainable Agriculture, 22*, 99–118.

Fritz, R. S., Crabb, B. A., & Hochwender, C. G. (2003). Preference and performance of a gall-inducing sawfly: plant vigor, sex, gall traits and phenology. *Oikos, 102*, 601–613.

Gagic, V., Bartomeus, I., Jonsson, T., Taylor, A., Winqvist, C., Fischer, C., … Bommarco, R. (2015). Functional identity and diversity of animals predict ecosystem functioning better than species-based indices. *Proceedings of the Royal Society of London B: Biological Sciences, 282*, 20142620. http://dx.doi.org/10.1098/rspb.2014.2620.

Gliessman, S. R. (2007). *Agroecology: the ecology of sustainable food systems*. New York, USA: CRC Press, Taylor & Francis, 384 p.

Griffin, J. N., Byrnes, J. E., & Cardinale, B. J. (2013). Effects of predator richness on prey suppression: a meta-analysis. *Ecology, 94*, 2180–2187.

Halaj, J., & Wise, D. H. (2002). Impact of a detrital subsidy on trophic cascades in a terrestrial grazing food web. *Ecology, 83*, 3141–3151.

Hillebrand, H., Bennett, D. M., & Cadotte, M. W. (2008). Consequences of dominance: a review of evenness effects on local and regional ecosystem processes. *Ecology, 89*, 1510–1520.

Hole, D. G., Perkins, A. J., Wilson, J. D., Alexander, I. H., Grice, P. V., & Evans, A. D. (2005). Does organic farming benefit biodiversity? *Biological Conservation, 122*, 113–130.

Holland, J. M. (2004). The environmental consequences of adopting conservation tillage in Europe: reviewing the evidence. *Agriculture, Ecosystems & Environment, 103*, 1–25.

Keller, S., & Häni, F. (2000). Ansprüche von Nützlingen und Schädlingen an den Lebensraum. In W. Netwig (Ed.), *Streifenförmige ökologische Ausgleichsflächen in der Kulturlandschaft: Ackerkrautstreifen, Buntbrache, Feldränder* (pp. 199–217). Bern, Hannover: Verlag Agrarökologie.

Kendall, D. A. (2003). Soil tillage and epigeal predatory arthropods. In A. El Titi (Ed.), *Soil tillage in agroecosystems* (pp. 297–342). Boca Raton, Florida, USA: CRC Press.

Khan, Z. R., Pickett, J. A., Berg, J. V. D., Wadhams, L. J., & Woodcock, C. M. (2000). Exploiting chemical ecology and species diversity: stem borer and striga control for maize and sorghum in Africa. *Pest Management Science, 56*, 957–962.

Kladivko, E. J. (2001). Tillage systems and soil ecology. *Soil and Tillage Research, 61*, 61–76.

Kleijn, D., Rundlöf, M., Scheper, J., Smith, H. G., & Tscharntke, T. (2011). Does conservation on farmland contribute to halting the biodiversity decline? *Trends in Ecology & Evolution, 26*, 474–481.

Landis, D. A., Wratten, S. D., & Gurr, G. M. (2000). Habitat management to conserve natural enemies of arthropod pests in agriculture. *Annual Review of Entomology, 45*, 175–201.

Langellotto, G. A., & Denno, R. F. (2004). Responses of invertebrate natural enemies to complex-structured habitats: a meta-analytical synthesis. *Oecologia, 139*, 1–10.

Letourneau, D. K., Armbrecht, I., Rivera, B. S., Lerma, J. M., Carmona, E. J., Daza, M. C., … Mejía, J. L. (2011). Does plant diversity benefit agroecosystems? A synthetic review. *Ecological Applications, 21*, 9–21.

Letourneau, D. K., Jedlicka, J. A., Bothwell, S. G., & Moreno, C. R. (2009). Effects of natural enemy biodiversity on the suppression of arthropod herbivores in terrestrial ecosystems. *Annual Review of Ecology, Evolution, and Systematics, 40*, 573–592.

Loreau, M., & Hector, A. (2001). Partitioning selection and complementarity in biodiversity experiments. *Nature, 412*, 72–76.
Losey, J. E., & Vaughan, M. (2006). The economic value of ecological services provided by insects. *Bioscience, 56*, 311–323.
Macfadyen, S., Gibson, R., Polaszek, A., Morris, R. J., Craze, P. G., Planque, R., … Memmott, J. (2009). Do differences in food web structure between organic and conventional farms affect the ecosystem service of pest control? *Ecology Letters, 12*, 229–238.
Martel, J. W., Alford, A. R., & Dickens, J. C. (2005). Synthetic host volatiles increase efficacy of trap cropping for management of Colorado potato beetle, *Leptinotarsa decemlineata* (Say). *Agricultural and Forest Entomology, 7*, 79–86.
Martin, E. A., Reineking, B., Seo, B., & Steffan-Dewenter, I. (2013). Natural enemy interactions constrain pest control in complex agricultural landscapes. *Proceedings of the National Academy of Sciences of the United States of America, 110*, 5534–5539.
Myers, J. H., Higgins, C., & Kovacs, E. (1989). How many insect species are necessary for the biological control of insects? *Environmental Entomology, 18*, 541–547.
Otto, S. B., Berlow, E. L., Rank, N. E., Smiley, J., & Brose, U. (2008). Predator diversity and identity drive interaction strength and trophic cascades in a food web. *Ecology, 89*, 134–144.
O'Rourke, M. E., Liebman, M., & Rice, M. E. (2008). Ground beetle (Coleoptera: Carabidae) assemblages in conventional and diversified crop rotation systems. *Environmental Entomology, 37*, 121–130.
Pérez-Lachaud, G., Batchelor, T. P., & Hardy, I. C. (2004). Wasp eat wasp: facultative hyperparasitism and intra-guild predation by bethylid wasps. *Biological Control, 30*, 149–155.
Price, P. W. (1991). The plant vigor hypothesis and herbivore attack. *Oikos*, 244–251.
Pullaro, T. C., Marino, P. C., Jackson, D. M., Harrison, H. F., & Keinath, A. P. (2006). Effects of killed cover crop mulch on weeds, weed seeds, and herbivores. *Agriculture, Ecosystems & Environment, 115*, 97–104.
Rand, T. A., Tylianakis, J. M., & Tscharntke, T. (2006). Spillover edge effects: the dispersal of agriculturally subsidized insect natural enemies into adjacent natural habitats. *Ecology Letters, 9*, 603–614.
Ratnadass, A., Fernandes, P., Avelino, J., & Habib, R. (2012). Plant species diversity for sustainable management of crop pests and diseases in agroecosystems: a review. *Agronomy for Sustainable Development, 32*, 273–303.
Roger-Estrade, J., Anger, C., Bertrand, M., & Richard, G. (2010). Tillage and soil ecology: partners for sustainable agriculture. *Soil and Tillage Research, 111*, 33–40.
Root, R. B. (1973). Organization of a plant-arthropod association in simple and diverse habitats: the fauna of collards (*Brassica oleracea*). *Ecological Monographs, 43*, 95–124.
Rusch, A., Birkhofer, K., Bommarco, R., Smith, H. G., & Ekbom, B. (2015). Predator body sizes and habitat preferences predict predation rates in an agroecosystem. *Basic and Applied Ecology, 16*, 250–259.
Rusch, A., Chaplin-Kramer, R., Gardiner, M. M., Hawro, V., Holland, J., Landis, D., … Bommarco, R. (2016). Agricultural landscape simplification reduces natural pest control: a quantitative synthesis. *Agriculture, Ecosystems & Environment, 221*, 198–204.
Rusch, A., Delbac, L., Muneret, L., & Thiéry, D. (2015). Organic farming and host density affect parasitism rates of tortricid moths in vineyards. *Agriculture, Ecosystems & Environment, 214*, 46–53.
Rusch, A., Valantin-Morison, M., Sarthou, J. P., & Roger-Estrade, J. (2010). Biological control of insect pests in agroecosystems: effects of crop management, farming systems,

and semi-natural habitats at the landscape scale: a review. *Advances in Agronomy, 109*, 219–258.
Rusch, A., Valantin-Morison, M., Sarthou, J. P., & Roger-Estrade, J. (2013). Effect of crop management and landscape context on insect pest populations and crop damage. *Agriculture, Ecosystems & Environment, 166*, 118–125.
Sarfraz, R. M., Dosdall, L. M., & Keddie, A. B. (2009). Bottom-up effects of host plant nutritional quality on *Plutella xylostella* (Lepidoptera: Plutellidae) and top-down effects of herbivore attack on plant compensatory ability. *European Journal of Entomology, 106*, 583.
Sarthou, J. P., Badoz, A., Vaissière, B., Chevallier, A., & Rusch, A. (2014). Local more than landscape parameters structure natural enemy communities during their overwintering in semi-natural habitats. *Agriculture, Ecosystems & Environment, 194*, 17–28.
Schellhorn, N. A., Gagic, V., & Bommarco, R. (2015). Time will tell: resource continuity bolsters ecosystem services. *Trends in Ecology & Evolution, 30*, 524–530.
Scheu, S. (2001). Plants and generalist predators as links between the below-ground and above-ground system. *Basic and Applied Ecology, 2*, 3–13.
Schmidt, M. H., Thewes, U., Thies, C., & Tscharntke, T. (2004). Aphid suppression by natural enemies in mulched cereals. *Entomologia Experimentalis et Applicata, 113*, 87–93.
Schmitz, O. J. (2007). Predator diversity and trophic interactions. *Ecology, 88*, 2415–2426.
Schmitz, O. J. (2009). Effects of predator functional diversity on grassland ecosystem function. *Ecology, 90*, 2339–2345.
Schneider, G., Krauss, J., & Steffan-Dewenter, I. (2013). Predation rates on semi-natural grasslands depend on adjacent habitat type. *Basic and Applied Ecology, 14*, 614–621.
Schneider, F. D., Scheu, S., & Brose, U. (2012). Body mass constraints on feeding rates determine the consequences of predator loss. *Ecology Letters, 15*, 436–443.
Sirrine, J. R., Letourneau, D. K., Shennan, C., Sirrine, D., Fouch, R., Jackson, L., & Mages, A. (2008). Impacts of groundcover management systems on yield, leaf nutrients, weeds, and arthropods of tart cherry in Michigan, USA. *Agriculture, Ecosystems & Environment, 125*, 239–245.
Snyder, W. E., & Ives, A. R. (2001). Generalist predators disrupt biological control by a specialist parasitoid. *Ecology, 82*, 705–716.
Snyder, W. E., Snyder, G. B., Finke, D. L., & Straub, C. S. (2006). Predator biodiversity strengthens herbivore suppression. *Ecology Letters, 9*, 789–796.
Straub, C. S., Finke, D. L., & Snyder, W. E. (2008). Are the conservation of natural enemy biodiversity and biological control compatible goals? *Biological Control, 45*, 225–237.
Straub, C. S., & Snyder, W. E. (2006). Species identity dominates the relationship between predator biodiversity and herbivore suppression. *Ecology, 87*, 277–282.
Tamburini, G., De Simone, S., Sigura, M., Boscutti, F., & Marini, L. (2016). Conservation tillage mitigates the negative effect of landscape simplification on biological control. *Journal of Applied Ecology, 53*, 233–241.
Thomson, L. J., & Hoffmann, A. A. (2013). Spatial scale of benefits from adjacent woody vegetation on natural enemies within vineyards. *Biological Control, 64*, 57–65.
Tilman, D., Fargione, J., Wolff, B., D'Antonio, C., Dobson, A., Howarth, R., … Swackhamer, D. (2001). Forecasting agriculturally driven global environmental change. *Science, 292*, 281–284.
Tonhasca, A., & Byrne, D. N. (1994). The effects of crop diversification on herbivorous insects: a meta-analysis approach. *Ecological Entomology, 19*, 239–244.
Tscharntke, T., Bommarco, R., Clough, Y., Crist, T. O., Kleijn, D., Rand, T. A., … van & Vidal, S. (2007). Conservation biological control and enemy diversity on a landscape scale. *Biological Control, 43*, 294–309.

Tscharntke, T., Klein, A. M., Kruess, A., Steffan-Dewenter, I., & Thies, C. (2005). Landscape perspectives on agricultural intensification and biodiversity – ecosystem service management. *Ecology Letters, 8*, 857–874.

Tuck, S. L., Winqvist, C., Mota, F., Ahnström, J., Turnbull, L. A., & Bengtsson, J. (2014). Land-use intensity and the effects of organic farming on biodiversity: a hierarchical meta-analysis. *Journal of Applied Ecology, 51*, 746–755.

Vasseur, C., Joannon, A., Aviron, S., Burel, F., Meynard, J. M., & Baudry, J. (2013). The cropping systems mosaic: how does the hidden heterogeneity of agricultural landscapes drive arthropod populations? *Agriculture, Ecosystems & Environment, 166*, 3–14.

Veres, A., Petit, S., Conord, C., & Lavigne, C. (2013). Does landscape composition affect pest abundance and their control by natural enemies? A review. *Agriculture, Ecosystems & Environment, 166*, 110–117.

Wäckers, F. L., Romeis, J., & van Rijn, P. (2007). Nectar and pollen feeding by insect herbivores and implications for multitrophic interactions. *Annual Review of Entomology, 52*, 301–323.

Wäckers, F. L., van Rijn, P. C. J., & Bruin, J. (2005). *Plant-provided food for carnivorous insects: A protective mutualism and its applications*. Cambridge University Press.

Waring, G. L., & Cobb, N. S. (1992). The impact of plant stress on herbivore population dynamics. *Insect-Plant Interactions, 4*, 167–226.

Way, M. J. (1966). The natural environment and integrated methods of pest control. *Journal of Applied Ecology, 3*, 29–32.

Wezel, A., Bellon, S., Doré, T., Francis, C., Vallod, D., & David, C. (2009). Agroecology as a science, a movement and a practice. A review. *Agronomy for Sustainable Development, Springer Verlag/EDP Sciences/INRA, 29*, 503–515. hal.archives-ouvertes.fr/hal-00886499.

White, T. C. R. (1984). The abundance of invertebrate herbivores in relation to the availability of nitrogen in stressed food plants. *Oecologia, 63*, 90–105.

Wilby, A., & Thomas, M. B. (2002). Natural enemy diversity and pest control: patterns of pest emergence with agricultural intensification. *Ecology Letters, 5*, 353–360.

Wilby, A., Villareal, S. C., Lan, L. P., Heong, K. L., & Thomas, M. B. (2005). Functional benefits of predator species diversity depend on prey identity. *Ecological Entomology, 30*, 497–501.

Winkler, K., Wäckers, F., Bukovinszkine-Kiss, G., & van Lenteren, J. (2006). Sugar resources are vital for *Diadegma semiclausum* fecundity under field conditions. *Basic and Applied Ecology, 7*, 133–140.

SUBJECT INDEX

'*Note*: Page numbers followed by "f" indicate figures, "t" indicate tables and "b" indicate boxes.'

R

S

AUTHOR INDEX

'*Note*: Page numbers followed by "f" indicate figures "t" indicate tables, and "b" indicate boxes.'

C

D

E

G

I

K

L

M

N

O

P

Q

R

S

T

W

X

Y

Z

FSC
www.fsc.org
MIX
Papier | Fördert gute Waldnutzung
FSC® C083411

Kolibri 360 GmbH
Ferdinand-Jühlke-Straße 7
99095 Erfurt, Deutschland
produktsicherheit@kolibri360.de